T0205486

Nečas Center Series

 Birkhäuser

The Nečas Center Series aims to publish high-quality monographs, textbooks, lecture notes, habilitation and Ph.D. theses in the field of mathematics and related areas in the natural and social sciences and engineering. There is no restriction regarding the topic, although we expect that the main fields will include continuum thermodynamics, solid and fluid mechanics, mixture theory, partial differential equations, numerical mathematics, matrix computations, scientific computing and applications. Emphasis will be placed on viewpoints that bridge disciplines and on connections between apparently different fields. Potential contributors to the series are encouraged to contact the editor-in-chief and the manager of the series.

Vít Dolejší • Georg May

Anisotropic *hp*-Mesh Adaptation Methods

Theory, implementation and applications

 Birkhäuser

Vít Dolejší
Faculty of Mathematics & Physics
Charles University
Praha 8, Czech Republic

Georg May
Aeronautics and Aerospace
Von Karman Institute for Fluid Dynamics
Sint-Genesius-Rode, Belgium

ISSN 2523-3343   ISSN 2523-3351   (electronic)
Nečas Center Series
ISBN 978-3-031-04278-2   ISBN 978-3-031-04279-9   (eBook)
https://doi.org/10.1007/978-3-031-04279-9

Mathematics Subject Classification: 65N50, 65M50, 65N30

This book is published under the imprint Birkhäuser, www.birkhauser-science.com by the registered company Springer Nature Switzerland AG
The registered company address is: Gewerbestrasse 11, 6330 Cham, Switzerland

Preface

Partial differential equations (PDE) govern many problems of scientific and engineering interest. Consequently, numerical solution of PDE has become a key technology in the development of new products in such fields as aerospace engineering, turbomachinery, aeroacoustics, automotive engineering, biomechanics, chemical engineering, environmental protection, and many others. However, the discretization of the governing equations by suitable numerical methods leads to potentially very large systems of algebraic equations which are expensive to solve even on supercomputers with thousands of processors.

One effective way to alleviate the computational costs is the reduction of the *number of degrees of freedom* by the use of *higher-order methods* in combination with *mesh adaptation*. Moreover, for schemes using piecewise polynomial approximation, additional flexibility may be gained by using *hp-adaptation*. The resulting algebraic systems are smaller and easier to solve, whereas the accuracy of the approximation is retained. Over the last 10 years, we have developed *anisotropic hp-mesh adaptation* techniques, which allow considerable freedom in the construction of finite dimensional approximation spaces.

In this book we summarize, unify, and extend our results that were published over the years in several journal papers. First, we derive computable estimates of interpolation errors which take into account the geometry of mesh elements as well as the anisotropic features of the interpolated function. These estimates are then used for the optimization of corresponding finite element spaces. We also focus on practical and implementation aspects of anisotropic *hp*-mesh adaptation, including several algorithms. The performance of these adaptive techniques is demonstrated with numerical experiments, which are obtained using the *discontinuous Galerkin method*. While discontinuous Galerkin methods are perfectly suitable for anisotropic meshes with varying polynomial approximation degrees, the theoretical and practical results can be extended to another discretization scheme using piecewise polynomial approximation.

This book is intended to serve as an introduction to the subject for scientists and researchers, including doctoral and master-level students. Parts of the text can also be used as study material for advanced university lectures concerning a posteriori error analysis and mesh adaptation. We expect that the readers of this book are familiar with basic calculus, functional analysis, the Lebesgue and Sobolev spaces, and finite element methods.

We are grateful to our former and current Ph.D. students Ajay Rangarajan, Aravind Balan, Filip Roskovec, and Ondřej Bartoš for the fruitful collaboration and implementation of the methods and providing some numerical experiments. Moreover, we are thankful to editors-in-chief of Nečas Center Series Prof. Josef Málek and Prof. Endre Süli and the executive editor Mrs. Beata Kubiś. Last but not least, we thank Christopher Tominich, the editor of Birkhäuser.

We also acknowledge the activities of Nečas Center for Mathematical Modeling http://ncmm.karlin.mff.cuni.cz which gave rise to this book. Our families gave us considerable support during work on this book. We wish to express our gratitude for their patience and understanding.

The work on this book was financially supported by the grant of the Czech Science Foundation No. 20-01074S.

Prague, Czech Republic   Vít Dolejší
Sint-Genesius-Rode, Belgium   Georg May
December 2021

Contents

1 Introduction 1
  1.1   Numerical Solution of Partial Differential Equations 1
  1.2   Basic Concepts of Finite Element Discretization 3
       1.2.1 Variational Formulations 4
       1.2.2 Galerkin Approximations 6
       1.2.3 Abstract Error Estimates 7
  1.3   Domain Partition and Finite Element Spaces 8
       1.3.1 Basic Terms and Notations 8
       1.3.2 Approximation Properties of Finite Element Spaces 10
  1.4   Illustrative Examples 12
       1.4.1 Poisson Problem 13
       1.4.2 Linear Advection-Diffusion Equation: Triple
              Layer Problem 15
       1.4.3 High-Speed Flow Through a Scramjet 16

2 Metric Based Mesh Representation 19
  2.1   Uniform Meshes 19
  2.2   Uniform Meshes Under a Riemannian Metric 21
  2.3   Space Anisotropy in Two Dimensions 23
       2.3.1 Space Anisotropy Described by an Ellipse 23
       2.3.2 Geometry of a Triangle 25
       2.3.3 Equilateral Triangle with Respect to Matrix $M \in Sym$ ... 28
       2.3.4 Steiner Ellipse 30
       2.3.5 Summary of the Description of Space Anisotropy 31
  2.4   Space Anisotropy in Three Dimensions 31
       2.4.1 Space Anisotropy Described by an Ellipsoid 31
       2.4.2 Geometry of a Tetrahedron 33
       2.4.3 Equilateral Tetrahedra with Respect to Matrix
              $M \in Sym$ 35
  2.5   Mesh Generation by a Metric Field 36
       2.5.1 Setting of the Nodal Metric 36

| | 2.5.2 | Interpolation of the Metric .... | 37 |
| | 2.5.3 | Mesh Optimization .... | 38 |

3   Interpolation Error Estimates for Two Dimensions 43
  3.1   Geometry of a Triangle 43
  3.2   Interpolation Error Function and Its Anisotropic Bound 44
        3.2.1   Interpolation Operator 45
        3.2.2   Interpolation Error Function 45
        3.2.3 s-Homogeneous Functions 46
        3.2.4   Anisotropic Bound of Interpolation Error
                Function for $p = 1$ 50
        3.2.5   Anisotropic Bound of Interpolation Error
                Function for $p > 1$ 52
  3.3   Interpolation Error Estimates Including the Triangle Geometry ... 62
        3.3.1   Estimate in the $L^q(K)$-Norm, $1 \leq q < \infty$ 64
        3.3.2   Estimate in the $L^\infty(K)$-Norm 66
        3.3.3   Estimate in the $H^1(K)$-Seminorm 67
        3.3.4   Estimate in the $L^2(\partial K)$-Norm 70
        3.3.5   Estimate in the $H^1(\partial K)$-Seminorm 72
  3.4   Numerical Verification 73

4   Interpolation Error Estimates for Three Dimensions 77
  4.1   Geometry of a Tetrahedron 77
  4.2   Interpolation Error Function and Its Anisotropic Bound 78
        4.2.1   Interpolation Error Function 78
        4.2.2   Anisotropic Bound of Interpolation Error Function 79
  4.3   Interpolation Error Estimates Including the Tetrahedra
        Geometry 83
        4.3.1   Estimate in the $L^q(K)$-Norm, $1 \leq q < \infty$ 85
        4.3.2   Estimate in the $L^2(\partial K)$-Norm 86
        4.3.3   Estimate in the $H^1(\partial K)$-Seminorm 87

5   Anisotropic Mesh Adaptation Method, h-Variant 89
  5.1   Optimization of the Mesh Element Anisotropy (2D) 90
  5.2   Optimization of the Mesh Element Anisotropy (3D) 98
  5.3   Mesh Adaptation Based on the Equidistribution Principle 103
  5.4   Continuous Mesh Model 105
  5.5   Continuous Mesh Optimization 107
        5.5.1   Solution of Problem 5.24 109
        5.5.2   Solution of Problem 5.25 110
  5.6   Adaptive Solution of Partial Differential Equations 113
        5.6.1   Anisotropic Mesh Adaptation Algorithms for PDEs 113
        5.6.2   Setting of Optimal Size of Mesh Elements 115
  5.7   Numerical Experiments 121
        5.7.1   Boundary Layer Problem 121
        5.7.2   Multiple Difficulties Problem 129
        5.7.3   Summary of the Numerical Examples 131

6   Anisotropic Mesh Adaptation Method, hp-Variant 133
  6.1   Continuous Mesh Model 133
  6.2   Semi-analytical Optimization 136
  6.3   Anisotropic hp-Mesh Adaptation Algorithm 139
  6.4   Numerical Experiments 142
     6.4.1 L^2-Projection of Piecewise Polynomial Function 142
     6.4.2   Boundary Layers Problem 142
     6.4.3   Multiple Difficulties Problem 146
     6.4.4   Mixed Hyperbolic-Elliptic Problem 148
     6.4.5   Convection-Dominated Problem 151

7   Framework of the Goal-Oriented Error Estimates 155
  7.1   Goal-Oriented Error Estimates for Linear PDEs 155
     7.1.1   Primal Problem 156
     7.1.2   Quantity of Interest and the Adjoint Problem 157
     7.1.3   Abstract Goal-Oriented Error Estimates 158
     7.1.4   Computable Goal-Oriented Error Estimates 159
  7.2   Goal-Oriented Error Estimates for Nonlinear PDEs 160
     7.2.1   Primal Problem 161
     7.2.2   Quantity of Interest and Adjoint Problem Based
         on Differentiation 161
     7.2.3   Quantity of Interest and Adjoint Problem Based
         on Linearization 163
     7.2.4   Goal-Oriented Error Estimates 164
  7.3   Error Estimates for Linear Convection–Diffusion Equation 165
     7.3.1   Problem Formulation 165
     7.3.2   Triangulation and Finite Element Spaces 166
     7.3.3   Discretization of the Primal Problem 168
     7.3.4   Consistency 169
     7.3.5   Quantity of Interest and Adjoint Problem 171
     7.3.6   Adjoint Consistency 172
     7.3.7   Goal-Oriented Error Estimates 174
  7.4   Error Estimates for Nonlinear Convection–Diffusion
     Equations 177
     7.4.1   Problem Formulation 177
     7.4.2   Target Functional and Adjoint Problem 180
     7.4.3   Goal-Oriented Error Estimates 182
  7.5   Compressible Euler Equations 183

8   Goal-Oriented Anisotropic Mesh Adaptation 185
  8.1   Goal-Oriented Estimates Including the Geometry of Elements 185
     8.1.1   Estimates Including the Geometry of Elements
         for 2D 186
     8.1.2   Goal-Oriented Optimization of the Anisotropy
         of Triangles 188

        8.1.3   Estimates Including the Geometry of Elements
                for 3D 190
        8.1.4   Goal-Oriented Optimization of the Anisotropy
                of Tetrahedra 192
    8.2 Goal-Oriented Anisotropic hp-Mesh Adaptive Algorithm 193
        8.2.1   Setting of Element Size 194
        8.2.2   Setting of Polynomial Approximation Degree
                and Element Shape 198
    8.3 Numerical Experiments 201
        8.3.1   Elliptic Problem on a "Cross" Domain 202
        8.3.2   Mixed Hyperbolic-Elliptic Problem 204
        8.3.3   Convection-Dominated Problem 209

9   Implementation Aspects 217
    9.1 Higher-Order Reconstruction Techniques 217
        9.1.1   Weighted Least-Square Reconstruction 217
        9.1.2   Reconstruction Based on the Solution
                of Local Problems 220
    9.2 Anisotropic Mesh Adaptation for Time-Dependent Problems 223
        9.2.1   Space-Time Discontinuous Galerkin Method 223
        9.2.2   Interpolation Error Estimates for
                Time-Dependent Problems 225
        9.2.3   Setting of the Time Step 226
        9.2.4   Adaptive Solution of Time-Dependent Problem 227

10 Applications 229
    10.1 Steady-State Inviscid Compressible Flow 229
        10.1.1 Subsonic Flow 231
        10.1.2 Transonic Flow 232
    10.2 Viscous Shock-Vortex Interaction 234
    10.3 Transient Flow Through a Nonhomogeneous Landfill Dam 236

Conclusion 243

References 245

Index 251

Chapter 1
Introduction

The goal of the first chapter is to present a brief motivation for the anisotropic mesh adaptation method and to illustrate its potential. We start the exposition by recalling several well-known facts concerning the numerical analysis of Galerkin (or finite element) methods, including approximation properties of finite element spaces. Finally, we show several motivational examples.

1.1 Numerical Solution of Partial Differential Equations

Many problems of practical interest are described by (a system of) partial differential equations (PDEs), defined on a computational domain Ω together with suitable conditions on the boundary of Ω. These equations are often complicated enough to defy analytical solution, so that a numerical solution is attempted instead. This means that the unknown solution is approximated by a function from a finite-dimensional space, denoted W_h. Its dimension is called the number of *degrees of freedom* (DoF). The finite-dimensional spaces W_h consist typically of piecewise polynomial functions, defined on a suitable partition $\mathscr{T}_h$ (= mesh) of Ω. This approximation (= discretization of the original problem) leads to a large system of algebraic equations which has to be solved numerically. The size of the algebraic system is equal to DoF.

In general, a finer mesh (larger DoF) results in a better approximation (smaller computational error) but leads to increased computational cost. It is necessary to balance these aspects, which is a challenging task.

The ultimate goal of the numerical solution of PDEs is to achieve a given error tolerance using the shortest possible computational time.

© The Author(s), under exclusive license to Springer Nature Switzerland AG 2022
V. Dolejší, G. May, *Anisotropic* hp-*Mesh Adaptation Methods*, Nečas Center Series,
https://doi.org/10.1007/978-3-031-04279-9_1

One way to reduce the computational cost is a suitable (adaptive) setting of the space W_h, usually through control of its dimension DoF. We note that the demand for the "shortest possible computational time" and the "minimal number of DoF" are not equivalent since algebraic systems arising from strongly adapted finite element space are usually more difficult to solve. Nevertheless, the reduction of DoF tends to give a large benefit in terms of the computational cost.

The core idea of the adaptive solution of PDEs is to perform the computation on an initial mesh (and corresponding space W_h), estimate the error of interest, adapt (refine) the mesh $\mathscr{T}_h$ and space W_h using available information, and repeat the computation until the prescribed error tolerance is achieved. An abstract *mesh adaptive algorithm* is given in Algorithm 1.1.

Algorithm 1.1: Abstract mesh adaptive algorithm

1: set tolerance → TOL > 0
2: generate initial mesh $\mathscr{T}_h$ with associated approximation space W_h
3: solve problem on W_h
4: estimate the error of interest → EST ≥ 0
5: **if** EST ≤ TOL **then**
6:     stop the computational process
7: **else**
8:     adapt (refine) mesh $\mathscr{T}_h$ and space W_h
9:     go to step 3
10: **end if**

There exist several basic mesh adaptation strategies (step 8 of Algorithm 1.1) which can be employed:

- *h-adaptation*, which allows the local refinement (or coarsening) of the mesh elements,
- *p-adaptation*, which allows the local increase (or decrease) of the polynomial approximation degrees on mesh elements,
- *r-adaptation*, which only adapts the position of the mesh nodes, keeping the number of DoF,
- *anisotropic mesh adaptation*, which generates (through a combination of *h*- and *r*-adaptation) meshes which may include long and thin mesh elements to better align with anisotropic solution features.

Obviously, it is possible to combine some of these strategies. For instance, *hp-adaptive methods* are very efficient techniques combining *h*- and *p*-adaptation, thus allowing adaptation of the element size *h*, as well as the polynomial degree of approximation *p*. Theoretical results, as well as numerical studies, show that the computational error of an *hp*-method may converge at an exponential rate with respect to DoF [11, 35, 94, 106, 108].

On the other hand, the benefit of *anisotropic mesh adaptation* for problems whose solutions contain discontinuities and/or interior and boundary layers was

clearly demonstrated in many works, e.g., [4, 8, 39, 46, 55, 64, 73, 90, 107, 117].
For the analysis of anisotropic elements in the context of the finite element method,
we refer to [2, 5, 12, 30, 61, 62].

However, for the most part, the anisotropic mesh adaptation methods developed
in the aforementioned papers deal with at most piecewise linear approximation,
using finite volume or finite element methods. For higher-order approximation, it
seems promising to combine anisotropic mesh adaptation with hp-adaptation. In
the present book, we develop this concept, to which we refer as *anisotropic hp-
mesh adaptation* methods.

1.2 Basic Concepts of Finite Element Discretization

For the numerical solution of partial differential equations (PDE), we consider
(finite element) Galerkin methods, where the solution is approximated by a function
from a finite-dimensional solution space. Hereafter, we present only the main results
that are needed to underpin the anisotropic mesh adaptation technique. A more
detailed coverage of numerical solution methods for PDE can be found in many
monographs, e.g., [21, 31, 56, 99]. The majority of the numerical experiments
presented in this book are carried out using the discontinuous Galerkin method,
which is suitable for hp-adaptation   [37, 45, 104]. These are generally nonconforming methods, in which the space of trial functions, where the approximate
solution is sought, is not a subspace of the space in which the exact (weak) solution
of the PDE is found.

In the following, we use standard notation and known facts from functional
analysis in Banach and Hilbert spaces, as well as some properties concerning
Sobolev spaces, see [1] or [97], for example. In particular, let $M \subset \mathbb{R}^d, d = 2, 3$ be
a domain. Then

$$L^q(M) = \{v : M \to \mathbb{R}; \ \|v\|_{L^q(M)} < \infty\}, \qquad 1 \le q \le \infty, \tag{1.1}$$

denotes the Lebesgue spaces with the norms

$$\|v\|_{L^q(M)} = \left(\int_M |v|^q \, dx \right)^{1/q}, \quad 1 \le q < \infty, \qquad \|v\|_{L^\infty(M)} = \operatorname*{ess\,sup}_M |v|.$$

By $(\cdot, \cdot)_M$ we denote the L^2-scalar product over M, i.e., $(u, v)_M = \int_M uv \, dx$, and
sometimes we write $\| \cdot \|_M := \| \cdot \|_{L^2(M)}$ for simplicity. Further,

$$H^k(M) = W^{k,2}(M) = \{v : M \to \mathbb{R}; \ \|v\|_{H^k(M)} < \infty\}, \quad k = 1, 2 \ldots \tag{1.2}$$

is the Sobolev space based on L^2 with the norm and seminorm

$$\|v\|_{H^k(M)} = \left(\sum_{|\alpha| \leq k} \|D^\alpha v\|^2_{L^2(M)} \right)^{1/2} \text{ and } |v|_{H^k(M)} = \left(\sum_{|\alpha| = k} \|D^\alpha v\|^2_{L^2(M)} \right)^{1/2},$$

respectively, where $\alpha = (\alpha_1, \ldots, \alpha_d)$, $\alpha_i \geq 0$, $i = 1, \ldots, d$ is a multi-index with length $|\alpha| = \alpha_1 + \cdots + \alpha_d$ and D^α denotes the multidimensional derivative of order $|\alpha|$. Further, the symbols ∇ and $\nabla\cdot$ are used for the gradient and divergence operators, respectively, and $\Delta = \nabla \cdot \nabla$ is the Laplace operator.

1.2.1 Variational Formulations

For a linear PDE of interest, we consider the following abstract variational (weak) formulation: find $u \in W$ such that

$$a(u, v) = \ell(v) \qquad \forall v \in V, \tag{1.3}$$

where

(i) W and V are Banach spaces equipped with norms $\|\cdot\|_W$ and $\|\cdot\|_V$, respectively.
(ii) $a : W \times V \to \mathbb{R}$ is a continuous bilinear form.
(iii) $\ell : V \to \mathbb{R}$ is a continuous linear functional, i.e., $\ell \in V'$, where V' denotes the dual space to V.

From a theoretical as well as a practical point of view, it is reasonable to require that problem (1.3) be well-posed, as introduced by Hadamard [74].

Definition 1.1 (Well-Posedness by Hadamard) Problem (1.3) is said to be *well-posed* if, for any $\ell \in V'$, there exists a unique solution $u \in W$ and the following bound holds:

$$\exists c > 0 : \forall \ell \in V', \quad \|u\|_W \leq c\|\ell\|_{V'}. \tag{1.4}$$

The well-posedness of problem (1.3) is guaranteed by the following theorem which is often called the Banach–Nečas–Babuška (BNB) theorem.

Theorem 1.2 *Let W be a Banach space and V be a reflexive Banach space. Let $a : W \times V \to \mathbb{R}$ be a continuous bilinear form and $\ell \in V'$. Then problem (1.3) is well-posed in the sense of Definition 1.1 if and only if the following two conditions hold:*

$$\exists \alpha > 0 : \quad \inf_{w \in W} \sup_{v \in V} \frac{a(w, v)}{\|w\|_W \|v\|_V} \geq \alpha, \tag{1.5a}$$

$$(v \in V, \quad a(w, v) = 0 \; \forall w \in W) \implies v = 0. \tag{1.5b}$$

Moreover, bound (1.4) *is valid with* $c := 1/\alpha$.

Proof We refer to [56, Theorem 2.6] □

Example 1.3 Let $\Omega \subset \mathbb{R}^d$, $d = 2, 3$ be a bounded domain with Lipschitz boundary $\Gamma := \partial\Omega$ and let $g \in L^2(\Omega)$ be given. The Poisson problem with homogeneous Dirichlet boundary conditions reads: Find $u : \Omega \to \mathbb{R}$ such that

$$-\Delta u = g \quad \text{in } \Omega,$$
$$u = 0 \quad \text{on } \Gamma. \tag{1.6}$$

In order to formulate (1.6) in the variational form (1.3), we set $W = V = H_0^1(\Omega)$ where $H_0^1(\Omega)$ is the Sobolev space of functions with square integrable first order weak derivatives and zero traces on Γ. The seminorm $|v|_1 := \left(\int_\Omega |\nabla u|^2 \, dx \right)^{1/2}$ defines a proper norm on $W = V$. Moreover, we set

$$a(u, v) := \int_\Omega \nabla u \cdot \nabla v \, dx, \qquad \ell(v) := \int_\Omega g \, v \, dx. \tag{1.7}$$

It is imminent that the bilinear form in (1.7) is *coercive*, i.e.,

$$a(v, v) \geq c|v|_1^2 \quad \forall v \in H_0^1(\Omega), \tag{1.8}$$

which implies the inf-sup condition (1.5a). In fact, (1.8) is satisfied with equality and $c = 1$. Similarly, (1.5b) is trivially valid and hence the variational problem (1.3) corresponding to (1.6) is well-posed.

Example 1.4 Let $\Omega \subset \mathbb{R}^d$, $d = 2, 3$ be a bounded domain with Lipschitz boundary $\Gamma := \partial\Omega$. Let $\boldsymbol{n}$ denote the unit outer normal to Γ, and $\boldsymbol{\beta} \in [C^1(\overline{\Omega})]^d$ be a given vector field. We denote by $\Gamma^- := \{x \in \Gamma : (\boldsymbol{\beta} \cdot \boldsymbol{n})(x) < 0\}$ the so-called inflow boundary. For given $g \in L^2(\Omega)$, we consider the following advection problem: Find $u : \Omega \to \mathbb{R}$ such that

$$\boldsymbol{\beta} \cdot \nabla u = g \quad \text{in } \Omega,$$
$$u = 0 \quad \text{on } \Gamma^-. \tag{1.9}$$

Problem (1.9) can be formulated in the form (1.3) by setting

$$W := \{u \in L^2(\Omega); \; \boldsymbol{\beta} \cdot \nabla u \in L^2(\Omega), \; u = 0 \text{ on } \Gamma^-\}, \quad V = L^2(\Omega), \tag{1.10}$$

$$a(u, v) = \int_\Omega v(\boldsymbol{\beta} \cdot \nabla u) \, dx, \qquad \ell(v) = \int_\Omega g \, v \, dx.$$

Note that the spaces W and V differ. For the well-posedness of this example, we refer to [56, Section 5.2.3].

1.2.2   Galerkin Approximations

The main idea of the approximation of the solution of (1.3) by a Galerkin method is to replace spaces W and V by finite-dimensional spaces W_h and V_h, respectively. The subscript h formally represents the discretization parameters. Spaces W_h and V_h are equipped with norms denoted by $\|\cdot\|_{W_h}$ and $\|\cdot\|_{V_h}$, respectively.

Since $W_h \not\subset W$ for the discontinuous Galerkin method (DGM), which is frequently used in this book, we define the space

$$W(h) = W + W_h \tag{1.11}$$

equipped with the norm $\|\cdot\|_{W(h)}$ satisfying:

(i) $\|w_h\|_{W(h)} = \|w_h\|_{W_h}$ for all $w_h \in W_h$.
(ii) $\|w\|_{W(h)} \le c\|w\|_W$ for all $w \in W$, i.e., W is continuously embedded in $W(h)$.

In order to define an approximate solution of (1.3), we introduce a bilinear form $a_h : W_h \times V_h \to \mathbb{R}$, which serves an approximation of a, and a functional $\ell_h : V_h \to \mathbb{R}$, which is an approximation of ℓ. We assume that a_h can be extended to $W(h) \times V_h$.

Definition 1.5 A function $u_h \in W_h$ is called the *approximate solution* of (1.3) if

$$a_h(u_h, v_h) = \ell_h(v_h) \qquad \forall v_h \in V_h. \tag{1.12}$$

Remark 1.6 If $W_h = V_h$, then the corresponding method is usually called the *standard Galerkin method*, otherwise we refer to the *Petrov–Galerkin* or *non-standard Galerkin method*. However, we use only the term Galerkin method, for simplicity.

Moreover, we assume that the discrete formulation (1.12) is *consistent*, which means that (1.12) is fulfilled for the exact solution, i.e., the solution $u \in W$ of (1.3):

$$a_h(u, v_h) = \ell_h(v_h) \qquad \forall v_h \in V_h. \tag{1.13}$$

It would be possible to relax this assumption and replace the consistency property (1.13) by the *asymptotic consistency*. Then the forthcoming estimate (1.15) will contain an additional term, called the consistency error. However, the DGM used in this book satisfies (1.13). Consequently, we restrict ourselves to this case only.

The existence of the approximate solution (1.12) is guaranteed by an analogue of Theorem 1.2, namely conditions (1.5) are replaced by

$$\exists \alpha_h > 0 : \quad \inf_{w_h \in W_h} \sup_{v_h \in V_h} \frac{a_h(w_h, v_h)}{\|w_h\|_{W_h} \|v_h\|_{V_h}} \ge \alpha_h, \tag{1.14a}$$

$$(v_h \in V_h, \quad a_h(w_h, v_h) = 0 \ \forall w_h \in W_h) \implies v_h = 0. \tag{1.14b}$$

Moreover, it can be proved that if dim $W_h = $ dim V_h, then (1.14a) is equivalent with (1.14b), [56, Proposition 2.21].

Then the well-posedness of the Galerkin approximation problem (1.12) is established by the following theorem.

Theorem 1.7 *Let W_h and V_h be finite-dimensional spaces equipped with the norms $\|\cdot\|_{W_h}$ and $\|\cdot\|_{V_h}$, respectively. We assume that*

(i) a_h is bounded on $W_h \times V_h$.
(ii) ℓ_h is continuous on V_h.
(iii) the discrete inf–sup condition (1.14a) is valid.
(iv) dim $W_h = $ dim V_h.

Then the approximate problem (1.12) is well-posed, i.e., there exists a unique $u_h \in W_h$ solving (1.12) and $\|u_h\|_{W_h} \leq \frac{1}{\alpha}\|\ell_h\|_{V_h'}$, where V_h' is the dual space to V_h.

Proof See [56, Theorem 2.22]. □

1.2.3 Abstract Error Estimates

Finally, we present the abstract a priori error estimate which bounds the difference between the exact and approximate solutions in the $W(h)$-norm. It is necessary to assume that condition (1.14a) is valid uniformly in h, i.e., there exists $\alpha > 0$ such that $\alpha_h \geq \alpha$ for all $h > 0$, where α_h is the constant from (1.14a).

Theorem 1.8 *We assume that*

(i) a_h is bounded uniformly on $W_h \times V_h$,
(ii) condition (1.14a) is valid uniformly in h,
(iii) dim $W_h = $ dim V_h
(iv) the approximate problem (1.12) is consistent (see (1.13)).

Then there exists $C > 0$ independent of h such that

$$\|u - u_h\|_{W(h)} \leq C \inf_{w_h \in W_h} \|u - w_h\|_{W(h)}. \tag{1.15}$$

Proof We refer to [56, Theorem 2.24] where a more general case is considered. □

Remark 1.9 Obviously, constant C from Theorem 1.8 is bounded from below ($C \geq 1$) since $u_h \in W_h$.

Definition 1.10 Let $u \in W(h)$ be given. The quantity $\inf_{w_h \in W_h} \|u - w_h\|_{W(h)}$ is called the *best-approximation error* of u in W_h and the norm $\|\cdot\|_{W(h)}$.

Corollary 1.11 *Theorem 1.8 implies that a sufficient condition for the convergence of u_h to u as $h \to 0$ is given by the following approximation property:*

$$\lim_{h \to 0} \inf_{w_h \in W_h} \|w - w_h\|_{W(h)} = 0 \qquad \forall w \in W. \tag{1.16}$$

The condition (1.16) guarantees the convergence of the Galerkin approximations for any data satisfying the assumptions of previous theorems and it appears in many monographs as the *approximation* property or *approximability*. However, there are many ways to satisfy (1.16), including the global refinement of grids on which the spaces $\{W_h,\ h > 0\}$ are defined. Therefore, the approximability condition does not give any guidelines regarding the efficient construction of trial spaces in practice. Nevertheless, for a given problem, in virtue of (1.15), it is sufficient to construct spaces $\{W_h,\ h > 0\}$ such that

$$\lim_{h \to 0} \inf_{w_h \in W_h} \|u - w_h\|_{W(h)} = 0 \qquad (1.17)$$

for each particular (unknown) solution u. Moreover, we are interested not only in satisfying (1.17) but also in the rate of convergence.

Estimate (1.15) is the core of many anisotropic mesh adaptation methods. The idea is to define the finite element space W_h (cf. Sect. 1.3) such that

$$\inf_{w_h \in W_h} \|u - w_h\|_{W(h)} \le \omega, \qquad (1.18)$$

where $u \in W$ is the (unknown) exact solution, and $\omega > 0$ is a given tolerance. In virtue of Remark 1.9, condition (1.18) is clearly a *necessary condition* for the common requirement that the error of the numerical scheme satisfies

$$\|u - u_h\|_{W(h)} \le \omega. \qquad (1.19)$$

Therefore, if condition (1.18) (or its approximation) is valid, then the numerical method seeking the approximate solution in W_h has a chance to satisfy (1.19).

In the rest of this chapter, we give examples of finite-dimensional spaces W_h and show by numerical experiments that their construction with respect to (1.18) can significantly reduce the computational cost.

1.3 Domain Partition and Finite Element Spaces

1.3.1 Basic Terms and Notations

In this section, we specify the finite element spaces in which we seek approximate solutions to variational problems. These spaces consist of piecewise polynomial functions defined on partitions of the computational domains into simplicial meshes. First, we introduce several basic terms using nomenclature very similar to that of standard finite element literature [21, 31, 56, 99].

Let $\Omega \subset \mathbb{R}^d$, $d = 2, 3$ be a bounded polygonal (for $d = 2$) or polyhedral (for $d = 3$) domain. Needless to say, in practical applications, the domain Ω is often not polygonal (polyhedral) but instead has curved boundaries. We touch this aspect briefly in Sect. 2.5.3 but otherwise we restrict ourselves to polygonal (polyhedral) domains for simplicity. We consider problem (1.3) whose solution is an unknown function $u \in W$, $u : \Omega \to \mathbb{R}$. In order to produce a finite-dimensional approximation of u, we introduce a partition of the computational domain Ω by simplexes (triangles for $d = 2$ and tetrahedra for $d = 3$), see Fig. 1.1. Further, we restrict ourselves to *conforming* grids.

Definition 1.12 Let $\mathscr{T}_h$ ($h > 0$) denote a partition of the closure $\overline{\Omega}$ of the domain $\Omega \subset \mathbb{R}^d$, $d = 2, 3$, into a finite number of closed d-dimensional simplexes K with mutually disjoint interiors, i.e., $\overline{\Omega} = \cup_{K \in \mathscr{T}_h} K$. The set $\mathscr{T}_h = \{K\}_{K \in \mathscr{T}_h}$ is called a *conforming grid (mesh)* if an intersection of any K, $K' \in \mathscr{T}_h$, $K \neq K'$, is either the empty set, or a vertex, or an entire common edge, or an entire common face (only for $d = 3$). For $d = 2$, the set of elements $\mathscr{T}_h$ is called *triangulation* and for $d = 3$, we call $\mathscr{T}_h$ a *tetrahedrization*.

Therefore, we do not admit hanging nodes and hanging faces (for $d = 3$), see Fig. 1.2. Throughout the book, we employ the following notation:

- ∂K denotes the boundary of the triangle/tetrahedron $K \in \mathscr{T}_h$.
- $|K|$ denotes the d-dimensional Lebesgue measure of K, i.e., the area for $d = 2$ and volume for $d = 3$.
- h_K denotes the diameter of K defined by $\mathrm{diam}(K) := \max\{|x - y|, \ x, y \in K\}$, where $|\cdot|$ denotes the Euclidean length (distance).

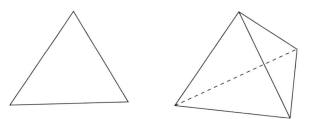

Fig. 1.1 Simplex for $d = 2$ (left) and $d = 3$ (right)

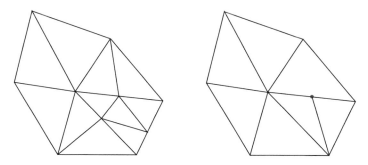

Fig. 1.2 Admissible triangulation (left) and non-admissible triangulation (right) due to the hanging node (blue node)

- $h := \max_{K \in \mathscr{T}_h} h_K$ denotes the *step size* of mesh $\mathscr{T}_h$.
- $\#\mathscr{T}_h$ denotes the number of elements (triangles/tetrahedra) of $\mathscr{T}_h$.
- $\mathscr{F}_h$ denotes the set of all edges of the grid $\mathscr{T}_h$, $\mathbf{e} \in \mathscr{F}_h$ is considered as a vector in $\mathbb{R}^d$ defined by its initial and final vertices (nodes) of $\mathscr{T}_h$. The orientation is arbitrary.

Let $M \subset \Omega$ be a domain. By $P^p(M)$ we denote the space of polynomial functions of total degree at most p on the domain M, i.e.,

$$P^p(M) := \left\{ v_h : M \to \mathbb{R}; \; v_h(x) = \sum_{i_1,\ldots,i_d=0}^{i_1+\cdots+i_d \leq p} a_{1,\ldots,d} \, x_1^{i_1} \ldots x_d^{i_d}, \; a_{1,\ldots,d} \in \mathbb{R} \right\},$$

(1.20)

where $x = (x_1, \ldots, x_d) \in M$. Obviously, any polynomial function from $P^p(M)$ can be smoothly extended to $\mathbb{R}^d$.

We introduce the space of *continuous piecewise polynomial* functions of degree $p \geq 1$ on mesh $\mathscr{T}_h$ by

$$Z_{hp} := \{ v_h \in C(\overline{\Omega}); \; v_h|_K \in P^p(K) \; \forall K \in \mathscr{T}_h \}, \tag{1.21}$$

where $v_h|_K$ denotes the restriction of v_h on K. Moreover, for given $p \in \mathbb{N}$, we consider the space of *discontinuous piecewise polynomial* functions on $\mathscr{T}_h$ given by

$$S_{hp} := \{ v_h \in L^2(\Omega); \; v_h|_K \in P^p(K) \; \forall K \in \mathscr{T}_h \}. \tag{1.22}$$

Furthermore, for discontinuous piecewise polynomial approximation, it is easy to consider a space with locally varying polynomial degree. Let $\boldsymbol{p} = \{ p_K, \; K \in \mathscr{T}_h \}$ be a set of integers where $p_K \geq 0$ is the polynomial approximation degree assigned to simplex $K \in \mathscr{T}_h$. Then we set

$$S_{h\boldsymbol{p}} := \{ v_h \in L^2(\Omega); \; v_h|_K \in P^{p_K}(K) \; \forall K \in \mathscr{T}_h \}. \tag{1.23}$$

The spaces Z_{hp} are used for conforming finite element methods, whereas spaces S_{hp} and $S_{h\boldsymbol{p}}$ are typical spaces for discontinuous Galerkin methods (DGM). We note that for DGM the restriction of the conformity of meshes (cf. Definition 1.12) can be easily relaxed.

1.3.2   Approximation Properties of Finite Element Spaces

There exist many results concerning estimates of the interpolation error, i.e., the difference between a given function and its projection onto the finite element space (see Definition 1.10). As an example, consider the $L^2(K)$ projection: For $K \in \mathscr{T}_h$

and $v \in L^2(K)$, we define the function $\pi_K v \in P^p(K)$ by

$$\int_K (\pi_K v)\, \phi \, \mathrm{d}x = \int_K v\, \phi \, \mathrm{d}x \qquad \forall \phi \in P^p(K). \tag{1.24}$$

Moreover, we assume that the mesh $\mathscr{T}_h$ is *shape regular*, i.e., there exists $c > 0$, such that

$$h_K / r_K \le c \qquad \forall K \in \mathscr{T}_h, \, h > 0, \tag{1.25}$$

where r_K denotes the radius of the maximal ball inscribed into K. Then the following interpolation error estimates are valid:

Lemma 1.13 *Let $K \in \mathscr{T}_h$ be shape regular, $v \in H^s(K)$, $s \ge 1$ and $\pi_K v \in P^p(K)$ be the interpolation given by* (1.24). *Then there exists a constant $c > 0$, independent of v, s, and p, such that*

$$|v - \pi_K v|_{H^m(K)} \le c h_K^{\mu - m} |u|_{H^\mu(K)}, \tag{1.26}$$

where $\mu = \min\{p+1, s\}$ and $m = 0, 1, \ldots, \mu$.

Proof See [31, Theorem 3.1.4]. □

Using Lemma 1.13 and an elementwise global projection of functions defined on Ω into spaces Z_{hp} or S_{hp}, we obtain the global interpolation estimate in a straightforward way. Consequently, combining this result with Theorem 1.8 yields the rate of convergence of the numerical solution with respect to the exact solution.

However, estimate (1.26) does not take into account either the shape of element K nor the possible anisotropy of the function u, since terms contained in $|u|_{H^\mu(K)}$ contain all possible weak derivatives of u, up to order μ, with the same multiplicative factor—a power of the element diameter h_K.

In [5], anisotropic variants of (1.26) are treated for various mesh elements and interpolation methods. These anisotropic interpolation error estimates can be written in a general form

$$|v - \pi_K v|_{H^m(K)} \le c \sum_{\substack{i_1 + \cdots + i_d \le \mu - m \\ i_1, \ldots, i_d \ge 0}} h_{K,1}^{i_1} \ldots h_{K,d}^{i_d} \left| \frac{\partial^{\mu - m} u}{\partial x_1^{i_1} \ldots \partial x_d^{i_d}} \right|_{H^m(K)}, \tag{1.27}$$

where $h_{K,l}$ is the "directional size" of the element in the x_l direction, $l = 1, \ldots, d$, see Fig. 1.3. The estimates of type (1.27) do not require the shape regularity condition (1.25). Moreover, the derivatives of u appear on the right-hand side of (1.27) with varying multiplicative factors depending on the directions. Therefore, the size of each particular partial derivative can be compensated by suitable choice of the corresponding directional size of the mesh element. These mesh elements, with varying "directional sizes," are called *anisotropic elements*. In Sect. 1.4, we

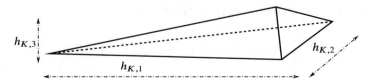

Fig. 1.3 The directional sizes of a tetrahedron

present several illustrating examples which clearly demonstrate the benefit of using anisotropic meshes.

Estimates of type (1.27) are not suitable for practical construction of anisotropic meshes. The terms in the sum on the right-hand side of (1.27) correspond only to "Cartesian" directions or their combinations. Moreover, for $p \geq 2$, the number of terms in the sum is too large in order to efficiently optimize the size and shape of mesh elements.

Interpolation estimates avoiding these drawbacks were developed in a series of papers by W. Cao [23–26]. Since the error of the interpolation by polynomials of degree p depends on $(p + 1)$-th derivatives of the interpolated function, the crucial aspect is to describe the anisotropic behavior of $\nabla^{p+1} v$ which is a $(p + 1)$ order tensor. At the core of the interpolation error estimates is a bound of type

$$|(\boldsymbol{\xi} \cdot \nabla)^{p+1} v| \leq (\boldsymbol{\xi}^{\mathsf{T}} \mathbb{U} \boldsymbol{\xi})^{(p+1)/2} \qquad \boldsymbol{\xi} \in \mathbb{R}^d, \tag{1.28}$$

where $\mathbb{U}$ is a suitable $d \times d$ positive definite matrix and symbol ^{T} denotes the transpose of the (column) vector. Specifying the matrix $\mathbb{U}$ requires one to determine the maximal higher-order directional derivatives of the interpolated function and the directional derivative(s) in the perpendicular directions(s).

Following these results, we derive in Chaps. 3 and 4 interpolation error estimates taking into account the anisotropic behavior of the interpolated functions and also the geometry (anisotropy) of mesh elements. These estimates are employed for mesh optimization.

1.4 Illustrative Examples

In this section, we present several examples demonstrating the potential of the use of anisotropic meshes for the solution of partial differential equations using Galerkin approximation. We show that, taking into account the anisotropy of the solution, we can achieve significantly smaller error of the numerical approximation using a smaller number of degrees of freedom DoF (= dimension of the space of trial functions).

Remark 1.14 Consider, for simplicity, a linear problem. A smaller number of DoF results in a smaller linear algebraic system which is, roughly speaking, cheaper to solve. However, algebraic systems arising by a discretization on anisotropic meshes have larger condition numbers in comparison with systems arising from a discretization on isotropic meshes (for the same number of DoF). Therefore, the benefit of using anisotropic meshes is not always proportional to the reduction of DoF.

1.4.1 Poisson Problem

Let $\Omega := (0, 1)^2$ be a unit square with boundary $\Gamma := \partial\Omega$. We consider the Poisson problem with Dirichlet boundary conditions:

$$-\Delta u = g \qquad \text{in } \Omega, \tag{1.29}$$

$$u = u_D \qquad \text{on } \Gamma.$$

The functions g and u_D are chosen such that the exact solution of (1.29) is

$$u(x_1, x_2) = \sin(2\pi x_1), \qquad (x_1, x_2) \in \Omega. \tag{1.30}$$

This function depends only on x_1 and it is constant in the x_2 direction. We seek its numerical approximation u_h in the space of continuous piecewise linear functions Z_{h1} (cf. (1.21)) by the conforming finite element method. Therefore, we seek $u_h \in Z_{h1}$, satisfying the identities

$$\int_\Omega \nabla u_h \cdot \nabla v_h \, dx = \int_\Omega g \, v_h \, dx \qquad \forall v_h \in Z_{h1} \cap H_0^1(\Omega) \tag{1.31}$$

and

$$\int_\Gamma u v_h \, dS = \int_\Gamma u_D v_h \, dS \qquad \forall v_h \in Z_{h1}. \tag{1.32}$$

We perform computations on two types of triangulation, see Fig. 1.4:

- *Rectangular triangular* meshes, where Ω is split uniformly by step δ in both x_1 and x_2 directions and the resulting square elements are split by diagonals into rectangular triangles. We use the values $\delta = 0.1$ and $\delta = 0.025$.
- *Anisotropic* meshes, where Ω is split by step δ only in the x_1 direction, while in the x_2 direction we have only one layer of elements. The resulting elongated rectangles are split by diagonals into anisotropic triangles. We use the values $\delta = 0.02$ and $\delta = 0.004$.

The corresponding results are presented in Table 1.1 which shows the errors $e_h = u - u_h$ measured in the $L^2(\Omega)$-norm, $H^1(\Omega)$-seminorm, and $L^\infty(\Omega)$-norm. We observe that due to the alignment of elements, the anisotropic meshes give smaller

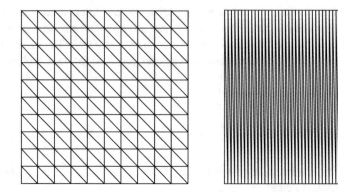

Fig. 1.4 Uniform triangular mesh with $\delta = 0.1$ (left) and anisotropic mesh with $\delta = 0.02$ (right)

Table 1.1 Errors associated with the numerical approximation of (1.30) on uniform and anisotropic meshes

Mesh	δ	$\#\mathcal{T}_h$	$\|e_h\|_{L^2(\Omega)}$	$|e_h|_{H^1(\Omega)}$	$\|e_h\|_{L^\infty(\Omega)}$
Uniform	0.100	200	1.9250E-02	7.9575E-01	4.4942E-02
Uniform	0.025	3200	1.2178E-03	1.9946E-01	2.8340E-03
Anisotropic	0.020	100	6.0954E-04	1.3247E-01	1.0912E-03
Anisotropic	0.004	500	1.1214E-04	2.7008E-02	2.9072E-04

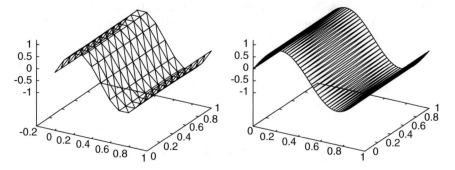

Fig. 1.5 Piecewise linear approximation of $u = \sin(2\pi x_1)$ obtained on a uniform triangular mesh with $\delta = 0.1$ (left) and anisotropic mesh with $\delta = 0.02$ (right)

errors with significantly smaller number of elements $\#\mathcal{T}_h$ in comparison with the uniform meshes. Figure 1.5 shows the P_1 approximation u_h obtained on the uniform mesh with $\delta = 0.1$ and anisotropic mesh with $\delta = 0.02$. The anisotropic mesh gives noticeably smoother approximation of u.

This example clearly demonstrates that suitable (in this simple case a priori) mesh adaptation allows us to achieve the same error tolerance using significantly smaller number of degrees of freedom.

1.4.2 Linear Advection-Diffusion Equation: Triple Layer Problem

We consider the linear convection-dominated problem

$$- \varepsilon \Delta u + b_1 \frac{\partial u}{\partial x_1} + b_2 \frac{\partial u}{\partial x_2} = 0 \quad \text{in } \Omega \tag{1.33}$$

with $\Omega := (0, 2) \times (0, 1)$, $\varepsilon = 10^{-6}$, and the advective field $(b_1, b_2) = (x_2, (1 - x_1)^2)$. We prescribe homogeneous Neumann data at the outflow part $\Gamma_N := \{2\} \times (0, 1) \cup (0, 2) \times \{1\}$ and discontinuous Dirichlet data

$$u = \begin{cases} 1 & x_1 \in (\frac{1}{8}, \frac{1}{2}), \ x_2 = 0 \\ 2 & x_1 \in (\frac{1}{2}, \frac{3}{4}), \ x_2 = 0 \\ 0 & \text{elsewhere on } \Gamma_D := \Gamma \setminus \Gamma_N. \end{cases} \tag{1.34}$$

Due to the small amount of diffusion the discontinuous profile (1.34) is basically transported along the characteristic curves leading to sharp characteristic interior layers.

Solving such a problem by conforming finite element method requires a suitable stabilization, otherwise the numerical solution is completely spoilt by spurious oscillations. We discretize (1.33) and (1.34) by the discontinuous Galerkin method (DGM) where the approximate solution is sought in the space S_{hp}, cf. (1.22). The corresponding scheme is described in Sect. 7.3.

We compare the results obtained by the symmetric interior penalty variant of DGM (cf. [37] or [45]) using P_4 polynomial approximation on

(A1) A globally fine uniform triangulation with 40,000 right-angled triangles ($h = 0.01\sqrt{2}$).
(A2) An adaptively refined triangular grid obtained from isotropic refinement, where the elements with largest value of an residual indicator are split onto 4 subelements (in this case we admit hanging nodes).
(A3) A mesh obtained using an anisotropic mesh adaptation algorithm.

These meshes are shown in Fig. 1.6 together with a detail close to the upper right corner of Ω. For both adaptive techniques, we observe a strong refinement along all three interior layers.

Further, Fig. 1.7 presents the isolines of the solution. We observe that whereas the computation obtained on the fine uniform mesh contains some nonphysical oscillations, the approximate solution obtained on the adaptively adapted grids (isotropically as well as anisotropically) are completely without any noticeable oscillations. Although, we do not know the exact solution of this problem, we can deduce from Fig. 1.7 that the accuracy of the numerical approximation obtained by (A2) and (A3) are very similar. On the other hand the costs of the computations are different. Table 1.2 shows the number of triangles $\#\mathcal{T}_h$ of the final grids and the total

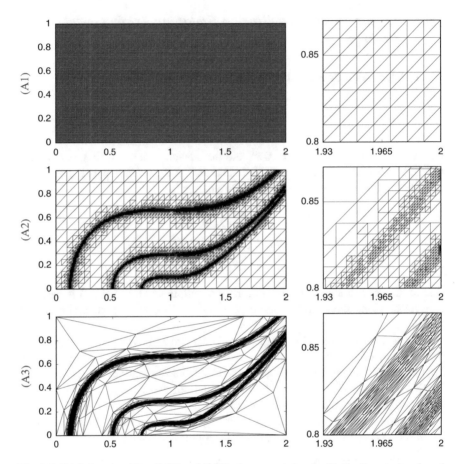

Fig. 1.6 Triple layer problem (1.33) and (1.34): the computational grids for three types of mesh adaptations. Total view (left) and detail close to the top-right corner (right)

computational time in seconds including the computations on all adaptation levels for (A2) and (A3). Obviously, the anisotropic mesh adaptation technique is superior.

1.4.3   High-Speed Flow Through a Scramjet

The last example exhibits an inviscid high-speed flow through a model of a scramjet (supersonic combustion ramjet) where two axisymmetric bodies are inserted into a wind channel, see Fig. 1.8. The problem is described by the Euler equations (cf. Sect. 7.5) which are nonlinear hyperbolic equations of the first order. At the inlet, the flow has a Mach number (= ratio between the flow speed and the local speed of

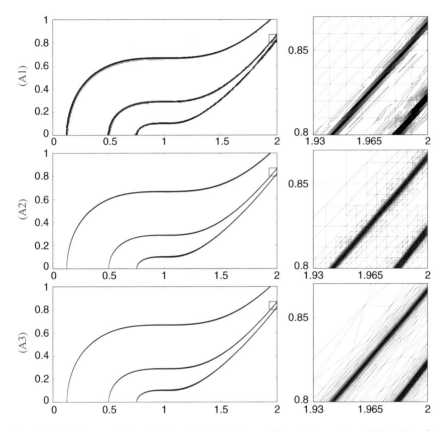

Fig. 1.7 Triple layer problem (1.33) and (1.34): isolines of the solution with equidistant $\delta = 0.1$ for three types of mesh adaptations. Total view (left) and detail close to the top-right corner (right)

Table 1.2 Triple layer problem (1.33) and (1.34): comparison of the efficiency of the mesh adaptive algorithms	Method	#$\mathscr{T}_h$	Time (s)
	(A1)	40,000	865.5
	(A2)	36,988	1360.5
	(A3)	13,430	663.7

sound) equal to 3. The presence of several interior angles gives rise to a complicated structure of shock waves (discontinuities), see, e.g., [36, 89].

We solve this problem using a finite volume method (FVM) employing a piecewise constant approximation. This is a special case of DGM with $p = 0$. FVM is monotone and, therefore, the resulting approximate solution does not suffer from spurious oscillations. However, the drawback is the excessive dissipation of the scheme, which smears the shock waves unless the mesh is sufficiently fine. Therefore, anisotropic mesh adaptation is a very efficient tool for the numerical simulation of such problems, since it generates fine grids only in the vicinity of

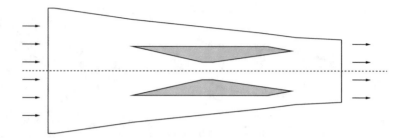

Fig. 1.8 High-speed flow through a scramjet: geometry of the domain

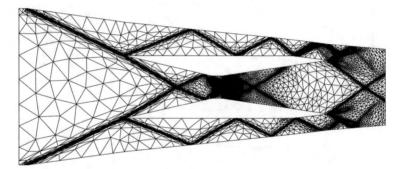

Fig. 1.9 High-speed flow through a scramjet: the anisotropically adapted mesh

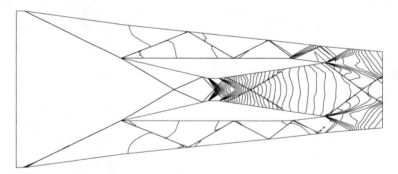

Fig. 1.10 High-speed flow through a scramjet: isolines of the Mach number achieved on the anisotropically adapted mesh

shock waves and only in the direction perpendicular to these waves. Figure 1.9 shows the mesh obtained by the anisotropic mesh adaptation algorithm and Fig. 1.10 presents the corresponding isolines of the Mach number. The complicated structure of shock waves is well captured and the shock waves are sufficiently sharp.

Chapter 2
Metric Based Mesh Representation

In this chapter, we formulate the initial problems related to the partition of the given domain Ω into a simplicial mesh. Furthermore, we recall the concept of the representation of simplicial meshes by a metric field, which is at the core of anisotropic mesh adaptation algorithms. This metric field defines a Riemannian metric over Ω, and it allows us to describe the geometry of the corresponding mesh elements.

Theorem 1.8 in the previous section asserts that the bound of the error of the approximate solution of the given problem is proportional to the approximation property of the spaces W_h where the approximate solutions are sought. In Sect. 1.3, we introduced examples of finite element spaces W_h and mentioned their approximation properties with respect to the diameter of mesh elements. Illustrative numerical examples in Sect. 1.4 show that not only the diameter of elements but also their geometry (shape, orientation) strongly influences the quality of numerical approximations. As mentioned in the Introduction, our general goal is the construction of simplicial meshes and the corresponding finite element spaces in such a way that we minimize the number of degrees of freedom (=dimension of W_h), while the approximation property and/or the error of the approximate solution is under the given tolerance.

Hereafter, we describe how the geometry of the simplicial mesh elements can be defined, which is important for the forthcoming formulation of the "optimal" mesh and the corresponding mesh adaptation algorithm.

2.1 Uniform Meshes

In Sect. 1.4, we presented numerical experiments on meshes consisting of identical rectangular triangles (Fig. 1.4, left and Fig. 1.6, top), which are easy to construct for rectangular domains. However, meshes consisting of rectangular elements are

V. Dolejší, G. May, *Anisotropic* hp-*Mesh Adaptation Methods*, Nečas Center Series, https://doi.org/10.1007/978-3-031-04279-9_2

not optimal when we want to cover a given domain Ω with the minimal number of elements whose diameter is bounded by the given value $\bar{h}$. We formulate this task more precisely.

Problem 2.1 Let $\Omega \subset \mathbb{R}^d$ be a bounded domain and $\bar{h} > 0$ be given. We seek a simplicial mesh $\mathcal{T}_h$ (cf. Definition 1.12) such that

 (i) The step size of $\mathcal{T}_h$ is $\leq \bar{h}$.
(ii) The number of elements #$\mathcal{T}_h$ is the minimal possible.

The existence of the solution of this problem is guaranteed by the following result.

Lemma 2.2 *Let $\Omega \subset \mathbb{R}^d$ be bounded polygonal domain and $\bar{h} > 0$ be given. Then, Problem 2.1 has at least one solution.*

Proof Let $\mathcal{S}$ denote the set of all meshes of Ω such that condition (i) of Problem 2.1 is satisfied. Obviously, set $\mathcal{S}$ is non-empty since there exists a mesh with arbitrary small step size. The number of triangles of any mesh from $\mathcal{S}$ is integer and positive. Then, according to Zorn's lemma [82], set $\mathcal{S}$ contains at least one minimal element, i.e., the mesh with the minimal number of simplexes.   □

On the other hand, the construction of the solution of Problem 2.1 is not, in general, an easy task. It depends on several factors, namely the size and shape of Ω. However, if $\bar{h} \ll \mathrm{diam}(\Omega)$, then the mesh, which is a hypothetical solution of Problem 2.1, is intuitively expected to be close to a mesh consisting of equilateral simplexes, with (Euclidean) length of edges approximately equal to $\bar{h}$, i.e.,

$$|\mathbf{e}| := (\mathbf{e}^\mathsf{T}\mathbf{e})^{1/2} \approx \bar{h} \qquad \forall \mathbf{e} \in \mathcal{F}_h, \tag{2.1}$$

where $\mathbf{e} = (e_1, \ldots, e_d)^\mathsf{T} \in \mathbb{R}^d$ denotes an edge of $\mathcal{T}_h$, defined by its initial and final nodes, symbol $^\mathsf{T}$ denotes the transpose of the (column) vector, $\mathbf{e}^\mathsf{T}\mathbf{e} = e_1^2 + \cdots + e_d^2 = |\mathbf{e}|^2$ is the scalar product of these vectors, and $\mathcal{F}_h$ is the set of all edges of $\mathcal{T}_h$.

Figure 2.1, left, shows a triangular mesh ($d = 2$) consisting of equilateral triangles of equal size, corresponding to some given edge length. It corresponds to the solution of Problem 2.1 when Ω is (in the limit) the whole space $\mathbb{R}^2$. In this case, relation (2.1) is valid with equality (=). Moreover, Fig. 2.1, right, shows a triangulation of a square domain generated by a mesh generator trying to fulfill relation (2.1) as best as possible, particularly in the sense of least squares. We observe that this mesh has a similar pattern, compared to that on the left, with several disturbances. It is clear that it is impossible to cover any square domain (having right angles) by equilateral triangles having angles of $\pi/3$.

The situation is much more complicated for $d = 3$, because it is not possible to cover any (nontrivial) domain by uniform tetrahedra, since any arbitrary composition of equilateral tetrahedra with a given edge length results in a hole in the tetrahedrization [20].

In virtue of this argumentation, we consider another problem whose solution is an approximation of the solution of Problem 2.1.

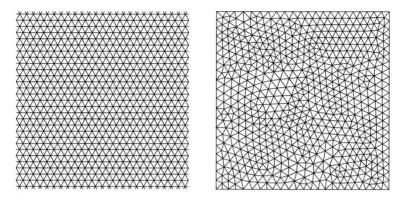

Fig. 2.1 Uniform triangulations: ideal mesh consisting of equilateral triangles (left) and a real mesh of a square domain (right)

Problem 2.3 Let $\Omega \subset \mathbb{R}^d$ be a bounded domain and let $\bar{h} \geq 0$. We seek a simplicial mesh $\mathscr{T}_h$ of Ω such that

$$|\mathbf{e}| \approx \bar{h} \qquad \forall \mathbf{e} \in \mathscr{F}_h. \tag{2.2}$$

A mesh satisfying (2.2) for $\bar{h} > 0$ is called *isotropic*.

Relation (2.2) is important also from the practical point of view of mesh generators. To begin with, the approximate inequality in (2.2) has to be given a more precise meaning. One possibility is the following least squares sense: for a given mesh $\mathscr{T}_h$ (having set of edges $\mathscr{F}_h$) and given $\bar{h} > 0$, we define the quantity

$$Q(\mathscr{T}_h) := \frac{1}{\#\mathscr{F}_h} \sum_{\mathbf{e} \in \mathscr{F}_h} \left(|\mathbf{e}| - \bar{h} \right)^2, \tag{2.3}$$

where $\#\mathscr{F}_h$ denotes the number of edges of mesh $\mathscr{T}_h$. Obviously, $Q(\mathscr{T}_h) \geq 0$ and $Q(\mathscr{T}_h) = 0$ if and only if (2.2) is valid with equality. Hence, the idea of mesh optimization is to locally adapt the given mesh (for example, by adding nodes, removing edges, and moving nodes, cf. Sect. 2.5.3) in such a way that parameter $Q(\mathscr{T}_h)$ is as small as possible. The resulting mesh is then accepted as a solution of Problem 2.3.

2.2 Uniform Meshes Under a Riemannian Metric

We generalize the concept from Sect. 2.1. Relation (2.2) defines an isotropic mesh in such a way that the standard Euclidean length of all mesh edges is approximately equal to a given constant. It is possible to consider a Riemannian metric instead of the Euclidean one and define an anisotropic mesh as a uniform mesh under the Riemannian metric. This concept was first introduced in [67], and then it has been intensively developed for more than 20 years; we mention [4, 63, 64, 73] for example.

Let Sym denote the space of real $d \times d$ symmetric positive definite matrices,

$$Sym := \left\{ M \in \mathbb{R}^{d \times d}; \ \mathbf{e}^{\mathsf{T}} M \mathbf{e} > 0 \ \forall \mathbf{e} \in \mathbb{R}^d, \ \mathbf{e} \neq 0 \right\}. \tag{2.4}$$

Moreover, let $\mathcal{M} : \Omega \to Sym$ be a mapping defining the Riemannian metric space. Let $\mathbf{a}_0, \mathbf{a}_1 \in \Omega$ be the start and end nodes of edge $\mathbf{e} \in \mathcal{F}_h$, i.e., $\mathbf{e}$ is the set of points $\{ \mathbf{a}_0 + t(\mathbf{a}_1 - \mathbf{a}_0), \ t \in [0, 1] \}$. We define the length of $\mathbf{e}$ (= distance between $\mathbf{a}_0$ and $\mathbf{a}_1$) in the Riemannian metric space $\mathcal{M}$ by

$$\|\mathbf{e}\|_{\mathcal{M}} = \|\mathbf{a}_1 - \mathbf{a}_0\|_{\mathcal{M}} := \int_0^1 \left((\mathbf{a}_1 - \mathbf{a}_0)^{\mathsf{T}} \mathcal{M}(\mathbf{a}_0 + t(\mathbf{a}_1 - \mathbf{a}_0))(\mathbf{a}_1 - \mathbf{a}_0) \right)^{1/2} dt. \tag{2.5}$$

We simply observe that if $\mathcal{M}$ is constant on $\mathbf{e}$ and equal to $M \in Sym$, then

$$\|\mathbf{e}\|_{\mathcal{M}} = (\mathbf{e}^{\mathsf{T}} M \mathbf{e})^{1/2} =: \|\mathbf{e}\|_{M}. \tag{2.6}$$

In a similar fashion, we can generalize Problem 2.3 in the following way.

Problem 2.4 Let $\Omega \subset \mathbb{R}^d$ be a bounded domain, and let $\mathcal{M} : \Omega \to Sym$ be a given Riemannian metric field. For given $\bar{h} \geq 0$, we seek a simplicial mesh $\mathcal{T}_h$ of Ω such that

$$\|\mathbf{e}\|_{\mathcal{M}} \approx \bar{h} \qquad \forall \mathbf{e} \in \mathcal{F}_h. \tag{2.7}$$

However, this problem (similar to Problem 2.3) is not well-defined since it is not clear how to interpret the approximate equality. In order to formulate a well-defined problem, we generalize the quantity $Q(\mathcal{T}_h)$ given by (2.3) and seek the best mesh in the sense of the least squares.

Problem 2.5 Let $\Omega \subset \mathbb{R}^d$ be a bounded domain, and let $\mathcal{M} : \Omega \to Sym$ be a given Riemannian metric field. For given $\bar{h} \geq 0$, we seek a simplicial mesh $\mathcal{T}_h$ of Ω, which minimizes $Q(\mathcal{T}_h)$, i.e.,

$$Q(\mathcal{T}_h) \leq Q(\mathcal{T}_h') \qquad \forall \mathcal{T}_h' \in \{\text{meshes of } \Omega\}, \tag{2.8}$$

where $Q(\mathcal{T}_h)$ is given by

$$Q(\mathcal{T}_h) := \frac{1}{\# \mathcal{F}_h} \sum_{\mathbf{e} \in \mathcal{F}_h} \left(\|\mathbf{e}\|_{\mathcal{M}} - \bar{h} \right)^2. \tag{2.9}$$

Lemma 2.6 *For the given domain $\Omega \subset \mathbb{R}^d$, metric field $\mathcal{M} : \Omega \rightarrow Sym$, and $\bar{h} > 0$, Problem 2.5 has at least one solution.*

Proof The proof is analogous to the proof of Lemma 2.2.   □

Remark 2.7 Without loss of generality, we fix the value $\bar{h}$, since, if a mesh $\mathcal{T}_h$ is the solution of Problem 2.5 for given $\mathcal{M}$ and $\bar{h}$, then $\mathcal{T}_h$ is the solution of Problem 2.5 for $\tilde{\mathcal{M}} := \frac{c}{\bar{h}} \mathcal{M}$ and $\tilde{h} := c$ for any $c > 0$. Therefore, we set $\bar{h} := \sqrt{3}$ for $d = 2$ and $\bar{h} := \sqrt{8/3}$ for $d = 3$ in the following. These values are equal to the lengths of uniform simplexes inscribed in the unit ball, cf. Lemmas 2.12 and 2.17 hereafter.

In the following, we discuss the space anisotropy induced by the Riemannian metric field $\mathcal{M}$. For a clear exposition, we first present the case $d = 2$ in Sect. 2.3 and then the more complicated case $d = 3$ in Sect. 2.4.

2.3 Space Anisotropy in Two Dimensions

We explain how the metric field $\mathcal{M} : \Omega \rightarrow Sym$ defines the anisotropy of the space, together with the corresponding geometrical interpretation. Then we derive relations between a triangle and a matrix from *Sym*.

2.3.1 Space Anisotropy Described by an Ellipse

Let $\Omega \subset \mathbb{R}^2$ be a domain, $\mathcal{M} : \Omega \rightarrow Sym$ be a given Riemannian metric field, and $\bar{x} \in \Omega$ be arbitrary but fixed. We set $\mathbb{M} := \mathcal{M}(\bar{x})$. The matrix $\mathbb{M}$ defines the space anisotropy at $\bar{x}$ in the following way. Since $\mathbb{M} \in Sym$, it is diagonalizable and has positive eigenvalues $\lambda_2 \geq \lambda_1 > 0$ and corresponding mutually orthonormal eigenvectors, v_i, $i = 1, 2$. We define the diagonal matrix

$$\Lambda_{\mathbb{M}} = \begin{pmatrix} \lambda_1 & 0 \\ 0 & \lambda_2 \end{pmatrix} \tag{2.10}$$

and the matrix $\mathbb{Q}_{\mathbb{M}}$, whose columns are the eigenvectors v_1 and v_2:

$$\mathbb{Q}_{\mathbb{M}} = \begin{pmatrix} v_1 & v_2 \end{pmatrix}. \tag{2.11}$$

Since v_1 and v_2 are orthonormal, $\mathbb{Q}_{\mathbb{M}}$ is a unitary matrix, i.e.,

$$\mathbb{Q}_{\mathbb{M}}^{\mathsf{T}} \mathbb{Q}_{\mathbb{M}} = \mathbb{Q}_{\mathbb{M}} \mathbb{Q}_{\mathbb{M}}^{\mathsf{T}} = \mathbb{I}, \tag{2.12}$$

where symbol $^\mathsf{T}$ denotes the transpose of the matrix and $\mathbb{I}$ is the identity matrix. Moreover, matrix $\mathbb{Q}_\mathbb{M}$ can be written as

$$\mathbb{Q}_\mathbb{M} = \begin{pmatrix} \cos\phi_\mathbb{M} & -\sin\phi_\mathbb{M} \\ \sin\phi_\mathbb{M} & \cos\phi_\mathbb{M} \end{pmatrix}, \tag{2.13}$$

where $\phi_\mathbb{M} \in [0, 2\pi]$. Then matrix $\mathbb{Q}_\mathbb{M}$ represents a *rotation* through angle $\phi_\mathbb{M}$ in the counterclockwise direction. It follows from (2.10) and (2.11) that $\mathbb{M}$ has the eigenvalue decomposition

$$\mathbb{M} = \mathbb{Q}_\mathbb{M}^\mathsf{T} \Lambda_\mathbb{M} \mathbb{Q}_\mathbb{M}. \tag{2.14}$$

Moreover, we define a planar domain by the relation

$$\mathscr{E}_\mathbb{M} := \left\{ x \in \mathbb{R}^2;\ (x - \bar{x})^\mathsf{T} \mathbb{M}(x - \bar{x}) \le 1 \right\}. \tag{2.15}$$

We introduce a new coordinate system $\tilde{x} = (\tilde{x}_1, \tilde{x}_2)$ with the origin at $\bar{x}$ and the axis $\tilde{x}_1$ in the direction $\phi_\mathbb{M}$ (cf. (2.13)), and hence

$$\tilde{x} = \mathbb{Q}_\mathbb{M}(x - \bar{x}). \tag{2.16}$$

In virtue of (2.10), (2.14), and (2.16), we rewrite the relation in (2.15) as

$$(x - \bar{x})^\mathsf{T}\mathbb{M}(x - \bar{x}) = (x - \bar{x})^\mathsf{T}\mathbb{Q}_\mathbb{M}^\mathsf{T}\Lambda_\mathbb{M}\mathbb{Q}_\mathbb{M}(x - \bar{x}) = \tilde{x}^\mathsf{T}\Lambda_\mathbb{M}\tilde{x} = \lambda_1\tilde{x}_1^2 + \lambda_2\tilde{x}_2^2 \le 1. \tag{2.17}$$

Therefore, relation (2.15) defines the *ellipse* $\mathscr{E}_\mathbb{M}$ centered at $\bar{x}$ whose semi-axes have lengths equal to $(\lambda_i)^{-1/2}$, $i = 1, 2$, and its orientation is $\phi_\mathbb{M}$ (=angle between the major semi-axis and the axis x_1 of the coordinate system), see Fig. 2.2.

The ellipse $\mathscr{E}_\mathbb{M}$ gives a geometric interpretation of the anisotropy of the space defined by M at $\bar{x}$. Namely, λ_1 and λ_2 give the lengthening or shortening of distances in the direction given by v_1 and the perpendicular direction v_2, respectively.

Finally, we introduce the following auxiliary lemma which will be used later.

Lemma 2.8 *Let* $\hat{\mathscr{E}} := \{\hat{x} \in \mathbb{R}^2 : |\hat{x}| \le 1\}$ *be the unit ball,* $\mathbb{M} \in Sym$, $\mathscr{E}_\mathbb{M}$ *be the corresponding ellipse* (2.15), *and* $B_\mathbb{M} : \mathbb{R}^2 \to \mathbb{R}^2$ *be the mapping given by*

$$B_\mathbb{M}(\hat{x}) := \mathbb{Q}_\mathbb{M}^\mathsf{T}\Lambda_\mathbb{M}^{-1/2}\mathbb{Q}_\psi\hat{x} + \bar{x}, \qquad \hat{x} \in \mathbb{R}^2, \tag{2.18}$$

where $\mathbb{Q}_\mathbb{M}^\mathsf{T}$ *is the transpose of matrix* (2.11), $\Lambda_\mathbb{M}^{-1/2} := \mathrm{diag}(1/\sqrt{\lambda_1}, 1/\sqrt{\lambda_2})$, λ_1 *and* λ_2 *are the eigenvalues of* $\mathbb{M}$, *and* $\mathbb{Q}_\psi$ *is a rotation matrix through arbitrary angle* $\psi \in [0, 2\pi)$. *Then* $B_\mathbb{M}$ *maps the ball* $\hat{\mathscr{E}}$ *onto the ellipse* $\mathscr{E}_\mathbb{M}$.

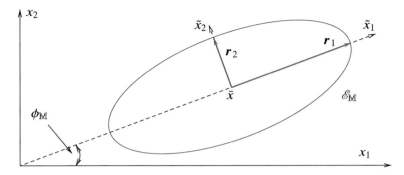

Fig. 2.2 The ellipse $\mathscr{E}_{\mathrm{M}}$ with the semi-axes defined by vectors $r_i = v_i/\sqrt{\lambda_i}$, $i = 1, 2$, the angle between v_1 and the axis x_1 is equal to ϕ_{M}; the coordinate system $(\tilde{x}_1, \tilde{x}_2)$ is defined by (2.16)

Proof Let $x = B_{\mathrm{M}}(\hat{x}) = Q_{\mathrm{M}}^{\mathsf{T}} \Lambda_{\mathrm{M}}^{-1/2} Q_\psi \hat{x} + \bar{x}$ for any $\hat{x} \in \hat{\mathscr{E}}$. Then $x - \bar{x} = Q_{\mathrm{M}}^{\mathsf{T}} \Lambda_{\mathrm{M}}^{-1/2} Q_\psi \hat{x}$ and $(x - \bar{x})^{\mathsf{T}} = \hat{x}^{\mathsf{T}} Q_\psi^{\mathsf{T}} \Lambda_{\mathrm{M}}^{-1/2} Q_{\mathrm{M}}$. By direct calculation, decomposition (2.14), and identity (2.12), one obtains

$$(x - \bar{x})^{\mathsf{T}} \mathrm{M} (x - \bar{x}) = \hat{x}^{\mathsf{T}} Q_\psi^{\mathsf{T}} \Lambda_{\mathrm{M}}^{-1/2} Q_{\mathrm{M}} \; Q_{\mathrm{M}}^{\mathsf{T}} \Lambda_{\mathrm{M}} Q_{\mathrm{M}} \; Q_{\mathrm{M}}^{\mathsf{T}} \Lambda_{\mathrm{M}}^{-1/2} Q_\psi \hat{x}$$

$$= \hat{x}^{\mathsf{T}} Q_\psi^{\mathsf{T}} \Lambda_{\mathrm{M}}^{-1/2} \Lambda_{\mathrm{M}} \Lambda_{\mathrm{M}}^{-1/2} Q_\psi \hat{x} = \hat{x}^{\mathsf{T}} Q_\psi^{\mathsf{T}} Q_\psi \hat{x} = |\hat{x}|^2 \leq 1,$$

which proves the lemma due to (2.15). $\square$

2.3.2 Geometry of a Triangle

We describe the geometry of a given triangle K employing the geometry of the corresponding ellipse. Let $\hat{K} \subset \mathbb{R}^2$ be a *reference equilateral triangle* with vertices $\hat{\mathbf{a}}_1 = (1, 0)$, $\hat{\mathbf{a}}_2 = (-\frac{1}{2}, \frac{\sqrt{3}}{2})$, and $\hat{\mathbf{a}}_3 = (-\frac{1}{2}, -\frac{\sqrt{3}}{2})$, i.e.,

$$\hat{K} := \left\{ x \in \mathbb{R}^2; \; x = \sum_{i=1}^{3} \kappa_i \hat{\mathbf{a}}_i, \quad \sum_{i=1}^{3} \kappa_i \leq 1, \quad \kappa_i \geq 0, \quad i = 1, \ldots, 3 \right\}. \tag{2.19}$$

The length of edges of K is equal to $\sqrt{3}$, while its barycenter is at $x = 0$. We choose this reference triangle because it is the triangle of maximum area inscribed into a unit ball

$$\hat{\mathscr{E}} := \{ x \in \mathbb{R}^2; \; |x| \leq 1 \}. \tag{2.20}$$

However, an arbitrary rotation of $\hat{K}$ gives an acceptable reference triangle as well.

Let K be an arbitrary but fixed triangle. Then there exists an affine mapping $F_K : \hat{K} \to K$, which maps $\hat{K}$ onto K. It is written in the form

$$F_K(\hat{x}) = \mathbb{F}_K \hat{x} + x_K, \quad \hat{x} \in \hat{K}, \tag{2.21}$$

where $x_K \in \mathbb{R}^2$ is the barycenter of K and $\mathbb{F}_K$ is a 2×2 matrix defining the size and the shape of K. In particular, if the vertices of K have the Cartesian coordinates $\mathbf{a}_1 = (a_{1,1}, a_{1,2})$, $\mathbf{a}_2 = (a_{2,1}, a_{2,2})$, and $\mathbf{a}_3 = (a_{3,1}, a_{3,2})$, then

$$\mathbb{F}_K = \begin{pmatrix} \frac{2}{3}a_{1,1} - \frac{1}{3}(a_{2,1} + a_{3,1}) & \frac{1}{\sqrt{3}}(a_{2,1} - a_{3,1}) \\ \frac{2}{3}a_{1,2} - \frac{1}{3}(a_{2,2} + a_{3,2}) & \frac{1}{\sqrt{3}}(a_{2,2} - a_{3,2}) \end{pmatrix}, \quad x_K = \begin{pmatrix} \frac{1}{3}(a_{1,1} + a_{2,1} + a_{3,1}) \\ \frac{1}{3}(a_{1,2} + a_{2,2} + a_{3,2}) \end{pmatrix}. \tag{2.22}$$

We note that matrix $\mathbb{F}_K$ is regular if the aforementioned vertices are not lying on a straight line.

Since matrix $\mathbb{F}_K$ is regular, we can apply the *singular value decomposition* and write

$$\mathbb{F}_K = \mathbb{Q}_{\phi_K} \mathbb{L}_K \mathbb{Q}_{\psi_K}^{\mathsf{T}}, \tag{2.23}$$

where $\mathbb{Q}_{\phi_K}$ and $\mathbb{Q}_{\psi_K}$ are the *rotation* through angles ϕ_K and ψ_K counterclockwise,

$$\mathbb{Q}_{\phi_K} := \begin{pmatrix} \cos\phi_K & -\sin\phi_K \\ \sin\phi_K & \cos\phi_K \end{pmatrix}, \qquad \mathbb{Q}_{\psi_K} := \begin{pmatrix} \cos\psi_K & -\sin\psi_K \\ \sin\psi_K & \cos\psi_K \end{pmatrix}, \tag{2.24}$$

and $\mathbb{L}_K = \mathrm{diag}(\ell_{K,1}, \ell_{K,2})$ is the diagonal matrix with the singular values $\ell_{K,1} \geq \ell_{K,2} > 0$.

Figure 2.3 shows the action of particular phases of the mapping $\mathbb{F}_K = \mathbb{Q}_{\phi_K} \mathbb{L}_K \mathbb{Q}_{\psi_K}^{\mathsf{T}}$ on the reference triangle $\hat{K}$ and the reference ellipse (=unit circle) $\hat{\mathcal{E}}$. First we apply a rotation through the angle ψ in the clockwise direction given by $\mathbb{Q}_{\psi_K}^{\mathsf{T}}$ (step (b)), then we perform a lengthening or shortening in the x_1 and x_2 directions by $\mathbb{L}_K$ (step (c)), and finally, we rotate through the angle ϕ in the counterclockwise direction given by $\mathbb{Q}_{\phi_K}$ (step (d)).

It is convenient to express $\mathbb{L}_K$ in a slightly different way and introduce new quantities. Namely, we define

$$\mu_K := \sqrt{\ell_{K,1}\ell_{K,2}} > 0, \qquad \sigma_K := \sqrt{\ell_{K,1}/\ell_{K,2}} \geq 1. \tag{2.25}$$

Then, we have

$$\mathbb{L}_K = \begin{pmatrix} \ell_{K,1} & 0 \\ 0 & \ell_{K,2} \end{pmatrix} = \mu_K \mathbb{S}_K, \quad \text{where } \mathbb{S}_K := \begin{pmatrix} \sigma_K & 0 \\ 0 & \sigma_K^{-1} \end{pmatrix}, \tag{2.26}$$

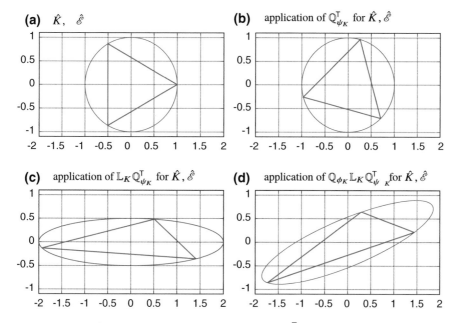

Fig. 2.3 Particular phases of the mapping $\mathbb{F}_K = \mathbb{Q}_{\phi_K} \mathbb{L}_K \mathbb{Q}_{\psi_K}^\mathsf{T}$, cf. (2.23), with $\psi_K = \pi/4$, $\phi_K = \pi/8$ and $\mathbb{L}_K = \mathrm{diag}(2, 1/2)$, applied on the reference triangle $\hat{K}$ (red) and the reference ellipse. (**a**) $\hat{K}$, $\hat{\mathscr{E}}$. (**b**)   application of $\mathbb{Q}_{\psi_K}^\mathsf{T}$ for $\hat{K}, \hat{\mathscr{E}}$. (**c**)   application of $\mathbb{L}_K \mathbb{Q}_{\psi_K}^\mathsf{T}$ for $\hat{K}, \hat{\mathscr{E}}$. (**d**)   application of $\mathbb{Q}_{\phi_K} \mathbb{L}_K \mathbb{Q}_{\psi_K}^\mathsf{T}$ for $\hat{K}, \hat{\mathscr{E}}$

and, since matrices $\mathbb{Q}_{\phi_K}$ and $\mathbb{Q}_{\psi_K}$ in (2.23) are orthogonal,

$$\det \mathbb{F}_K = \det \mathbb{L}_K = \mu_K^2. \tag{2.27}$$

The area of K is given by the substitution theorem as

$$|K| = \int_K 1 \, \mathrm{d}x = \int_{\hat{K}} \det \mathbb{F}_K \, \mathrm{d}\hat{x} = \det \mathbb{F}_K |\hat{K}| = \mu_K^2 |\hat{K}|, \tag{2.28}$$

where $|\hat{K}| = 3\sqrt{3}/4$ is the area of the reference triangle. Therefore, the size of element K is given only by the determinant of matrix $\mathbb{L}_K$. Hence, the triangles in Fig. 2.3a and b have the same area and similarly triangles in Fig. 2.3c and d.

Moreover, we establish the result which connects decompositions (2.14) and (2.23).

Lemma 2.9 *Let $\mathscr{E}_K$ be an image of the reference ellipse $\hat{\mathscr{E}}$ (cf. (2.20)) given by mapping (2.21) with its decomposition (2.23), i.e.,*

$$\mathscr{E}_K := \{ x \in \mathbb{R}^2; \ x = \mathbb{Q}_{\phi_K} \mathbb{L}_K \mathbb{Q}_{\psi_K}^\mathsf{T} \hat{x} + x_K, \ \hat{x} \in \hat{\mathscr{E}} \}. \tag{2.29}$$

Then $\mathscr{E}_K$ is the ellipse defined by (2.15) with $\bar{x} := x_K$ and

$$\mathbb{M} := \mathbb{Q}_{\phi_K}^{\mathsf{T}} (\mathbb{L}_K^{-1})^2 \mathbb{Q}_{\phi_K} = \mathbb{Q}_{\phi_K}^{\mathsf{T}} \begin{pmatrix} 1/(\ell_{K,1})^2 & 0 \\ 0 & 1/(\ell_{K,2})^2 \end{pmatrix} \mathbb{Q}_{\phi_K}. \tag{2.30}$$

Proof Comparing (2.29) with the mapping (2.18) in Lemma 2.8, we find that $\Lambda_{\mathbb{M}}^{-1/2} = \mathbb{L}_K = \mathrm{diag}(\ell_{K,1}, \ell_{K,2})$ and hence

$$\Lambda_{\mathbb{M}} = \mathrm{diag}(\lambda_1, \lambda_2) = \mathbb{L}_K^{-2} = \mathrm{diag}(1/\ell_{K,1}^2, 1/\ell_{K,2}^2), \tag{2.31}$$

where λ_i, $i = 1, 2$, are the eigenvalues of $\mathbb{M}$. This implies the lemma.   □

The ellipse $\mathscr{E}_K$ given by (2.29) can be uniquely determined by the triplet of parameters $\{\mu_K, \sigma_K, \phi_K\}$, which is called the *geometry* of $\mathscr{E}_K$. The parameter μ_K is proportional to the area of $\mathscr{E}_K$ and it is called the *size* of the ellipse. Furthermore, σ_K is equal to the ratio between the lengths of major and minor semi-axes of $\mathscr{E}_K$ and it is called the *shape*. Finally, ϕ_K is the angle between the major semi-axis of $\mathscr{E}_K$ and x_1-axis and it is called the *orientation* of the ellipse.

The same terms are used for the description of the geometry of the triangle K.

Definition 2.10 Let K be a triangle given by the affine mapping $F_K : \hat{K} \to K$, cf. (2.21)–(2.22), and let (2.23)–(2.26) be the corresponding decomposition. The value μ_K is called the *size* of K, σ_K is called the *shape* of K, and the angle ϕ_K is the *orientation* of K. The pair $\{\sigma_K, \phi_K\}$ is called the *anisotropy* of K and the triplet $\{\mu_K, \sigma_K, \phi_K\}$ is called the *geometry* of K.

Remark 2.11 The angle ψ_K from (2.23) corresponds to the so-called *distortion* of K. For $\psi_K = 0$, the triangle is acute isosceles, for $\psi_K = \pi/2$, the triangle is obtuse isosceles, and otherwise the triangle is not isosceles. Hence, there exist infinitely many triangles with the same geometry, consistent with the arbitrary value of ψ in Lemma 2.8.

2.3.3 Equilateral Triangle with Respect to Matrix $\mathbb{M} \in Sym$

We have already mentioned that a triangle, inscribed into $\hat{\mathscr{E}}$ and having the maximal possible area, is the equilateral triangle with edge length equal to $\sqrt{3}$, see Fig. 2.4, left. This triangle can be arbitrarily rotated and it corresponds to elements from an isotropic mesh. In virtue of Lemma 2.8, we can write $\hat{\mathscr{E}} = \mathscr{E}_{\mathbb{I}}$, where $\mathbb{I}$ is the identity matrix.

Similar considerations can be carried out for a general matrix $\mathbb{M} \in Sym$ and the corresponding ellipse $\mathscr{E}_{\mathbb{M}}$ given by (2.15). The goal is to find a triangle inscribed into $\mathscr{E}_{\mathbb{M}}$ and having the maximal possible area. Lemma 2.8 implies that $\mathscr{E}_{\mathbb{M}}$ is an image of $\hat{\mathscr{E}}$ by the mapping $B_{\mathbb{M}}$ given by (2.18). Moreover, if $\hat{K}$ is the equilateral triangle with center at origin and the edge length equal to $\sqrt{3}$, then $B_{\mathbb{M}}$ maps $\hat{K}$ on triangle K,

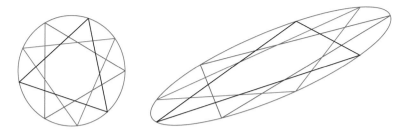

Fig. 2.4 Ellipse $\mathscr{E}_{\mathbb{M}}$ and examples of the maximal triangles $K_{\mathbb{M}}$ inscribed into $\mathscr{E}_{\mathbb{M}}$, the isotropic case (left) and the anisotropic one (right). Each triangle in the right figure is the image of the triangle of the same color from the left, obtained from mapping $B_{\mathbb{M}}$ in (2.18)

which is inscribed into $\mathscr{E}_{\mathbb{M}}$. Since the mapping $B_{\mathbb{M}}$ is linear, it increases/decreases the area of the input object by the constant factor

$$\det \mathbb{Q}_{\mathbb{M}}^{\mathsf{T}} \Lambda_{\mathbb{M}}^{-1/2} \mathbb{Q}_{\psi} = 1/\sqrt{\lambda_1 \lambda_2} = \mu_K^2,$$

where the last equality follows from Lemma 2.9 and (2.27). Since $\hat{K}$ has the maximal area among all triangles inscribed in $\hat{\mathscr{E}}$, then $K = B_{\mathbb{M}}(\hat{K})$ has the maximal area among all triangles inscribed in $\mathscr{E}_{\mathbb{M}}$. Obviously, there exist infinitely many triangles (acute angle as well as obtuse angle) with this property, cf. Fig. 2.4.

The following property is an anisotropic analogue of the length of edges of the maximal triangle inscribed into a unit ball.

Lemma 2.12 *Let* $\mathbb{M} \in Sym$ *and* $\mathscr{E}_{\mathbb{M}}$ *be given by* (2.15). *Let* $K_{\mathbb{M}}$ *be a triangle inscribed into* $\mathscr{E}_{\mathbb{M}}$ *and having the maximal possible area. Then*

$$\|\mathbf{e}_i\|_{\mathbb{M}} := (\mathbf{e}_i^{\mathsf{T}} \mathbb{M} \mathbf{e}_i)^{1/2} = \sqrt{3}, \qquad i = 1, 2, 3, \tag{2.32}$$

where $\mathbf{e}_i$, $i = 1, 2, 3$, *denotes the edges of* $K_{\mathbb{M}}$ *considered as vectors.*

Proof Let $K_{\mathbb{M}}$ be a maximal triangle inscribed into $\mathscr{E}_{\mathbb{M}}$, and then there exists a mapping that maps the reference triangle $\hat{K}$ onto $K_{\mathbb{M}}$. In virtue of Lemmas 2.8 and 2.9, this mapping can be written in the form $K_{\mathbb{M}} = B_{\mathbb{M}}(\hat{K})$, where $B_{\mathbb{M}}$ is the mapping given by (2.18). Let $\hat{\mathbf{z}}_1$ and $\hat{\mathbf{z}}_2$ be any pair of vertices of $\hat{K}$, i.e., $|\hat{\mathbf{z}}_1 - \hat{\mathbf{z}}_2| = \sqrt{3}$. Then the corresponding edge of $K_{\mathbb{M}}$ has end points $B_{\mathbb{M}}\hat{\mathbf{z}}_i$, $i = 1, 2$. Hence,

$$\mathbf{e} = B_{\mathbb{M}}\hat{\mathbf{z}}_2 - B_{\mathbb{M}}\hat{\mathbf{z}}_1 = \mathbb{Q}_{\mathbb{M}}^{\mathsf{T}} \Lambda_{\mathbb{M}}^{-1/2} \mathbb{Q}_{\psi} (\hat{\mathbf{z}}_2 - \hat{\mathbf{z}}_1)$$

is an edge of $K_{\mathbb{M}}$. This together with decomposition (2.14) and identity (2.12) implies

$$
\begin{aligned}
\|e\|_{\mathbb{M}}^2 = e^{\mathsf{T}} \mathbb{M} e &= (Q_{\mathbb{M}}^{\mathsf{T}} \Lambda_{\mathbb{M}}^{-1/2} Q_{\psi} (\hat{z}_2 - \hat{z}_1))^{\mathsf{T}} \mathbb{M} Q_{\mathbb{M}}^{\mathsf{T}} \Lambda_{\mathbb{M}}^{-1/2} Q_{\psi} (\hat{z}_2 - \hat{z}_1) \\
&= (\hat{z}_2 - \hat{z}_1)^{\mathsf{T}} Q_{\psi}^{\mathsf{T}} \Lambda_{\mathbb{M}}^{-1/2} Q_{\mathbb{M}} Q_{\mathbb{M}}^{\mathsf{T}} \Lambda_{\mathbb{M}} Q_{\mathbb{M}} Q_{\mathbb{M}}^{\mathsf{T}} \Lambda_{\mathbb{M}}^{-1/2} Q_{\psi} (\hat{z}_2 - \hat{z}_1) \\
&= (\hat{z}_2 - \hat{z}_1)^{\mathsf{T}} Q_{\psi}^{\mathsf{T}} \Lambda_{\mathbb{M}}^{-1/2} \Lambda_{\mathbb{M}} \Lambda_{\mathbb{M}}^{-1/2} Q_{\psi} (\hat{z}_2 - \hat{z}_1) = |\hat{z}_2 - \hat{z}_1|^2 = 3,
\end{aligned}
$$

which proves the lemma.

$\square$

2.3.4   Steiner Ellipse

In Sect. 2.3.3, we defined triangle $K_{\mathbb{M}}$ for a given matrix $\mathbb{M} \in Sym$. Triangle $K_{\mathbb{M}}$ has the maximal possible area and it is inscribed into $\mathscr{E}_{\mathbb{M}}$ related to $\mathbb{M}$ by (2.15). In this section, we consider the opposite task. For a given triangle K find an ellipse $\mathscr{E}_K$ which circumscribes K and has the minimal possible area.

Such an ellipse is called the *Steiner ellipse* (see, e.g., [83]). Its center coincides with the barycenter of K, and the areas of K and $\mathscr{E}_K$ satisfy the relations

$$
|K| = c_a r_{K,1} r_{K,2} = \frac{c_a}{\pi} |\mathscr{E}_K|, \tag{2.33}
$$

where $r_{K,1}$ and $r_{K,2}$ are the lengths of the semi-axes of the ellipse $\mathscr{E}_K$ and $c_a = 3\sqrt{3}/4$. The Steiner ellipse $\mathscr{E}_K$ is given by the relation

$$
\mathscr{E}_K := \left\{ x \in \mathbb{R}^2; \ x^{\mathsf{T}} \mathbb{M}_K x \le 1 \right\}, \tag{2.34}
$$

where $\mathbb{M}_K = \{m_{ij}\}_{i,j=1,2}^2 \in Sym$ and its coefficients are the solution of the following linear algebraic system:

$$
\begin{pmatrix} (a_{1,1} - a_{2,1})^2 & 2(a_{1,1} - a_{2,1})(a_{1,2} - a_{2,2}) & (a_{1,2} - a_{2,2})^2 \\ (a_{2,1} - a_{3,1})^2 & 2(a_{2,1} - a_{3,1})(a_{2,2} - a_{3,2}) & (a_{2,2} - a_{3,2})^2 \\ (a_{3,1} - a_{1,1})^2 & 2(a_{3,1} - a_{1,1})(a_{3,2} - a_{1,2}) & (a_{3,2} - a_{1,2})^2 \end{pmatrix} \begin{pmatrix} m_{11} \\ m_{12} \\ m_{22} \end{pmatrix} = \begin{pmatrix} 3 \\ 3 \\ 3 \end{pmatrix},
$$

$$
\tag{2.35}
$$

where $\mathbf{a}_1 = (a_{1,1}, a_{1,2})$, $\mathbf{a}_2 = (a_{2,1}, a_{2,2})$, and $\mathbf{a}_3 = (a_{3,1}, a_{3,2})$ are the Cartesian coordinates of the vertices of K. The first row of (2.35) gives

$$(a_{1,1} - a_{2,1})^2 m_{11} + 2(a_{1,1} - a_{2,1})(a_{1,2} - a_{2,2})m_{12} + (a_{1,2} - a_{2,2})^2 m_{22} = 3 \tag{2.36}$$

$$\Leftrightarrow \quad (\mathbf{a}_1 - \mathbf{a}_2)^T \mathbb{M}_K (\mathbf{a}_1 - \mathbf{a}_2) = 3$$

and similarly for the edges $\mathbf{a}_2 - \mathbf{a}_3$ and $\mathbf{a}_3 - \mathbf{a}_1$. This is in agreement with Lemma 2.12.

2.3.5   Summary of the Description of Space Anisotropy

The results of Sect. 2.3 can be summarized as a list of the following connections.

Matrix $\mathbb{M} \in Sym$ with the eigenvalue decomposition $\mathbb{M} = \mathbb{Q}_{\mathbb{M}}^\mathsf{T} \Lambda_{\mathbb{M}} \mathbb{Q}_{\mathbb{M}}$

$\leftrightarrow$ Ellipse $\mathscr{E}_{\mathbb{M}}$ given by (2.15)

$\leftrightarrow$ Triangle $K_{\mathbb{M}}$ inscribed into $\mathscr{E}_{\mathbb{M}}$ and having the maximal area

$\leftrightarrow$ $K_{\mathbb{M}}$ is the imagine of $\hat{K}$ by mapping $F_K(\hat{x}) = \mathbb{Q}_{\phi_K} \mathbb{L}_K \mathbb{Q}_{\psi_K}^\mathsf{T} \hat{x}$ with $\mathbb{L}_K = \Lambda_{\mathbb{M}}^{-1/2}$

$\leftrightarrow$ Steiner ellipse of K is $\mathscr{E}_{\mathbb{M}}$

We note that the analogous relations are valid also for 3D case explained below.

2.4   Space Anisotropy in Three Dimensions

Similar to Sect. 2.3, we describe the anisotropy of the space given by the metric field $M : \Omega \to Sym$ for the dimension $d = 3$. Since the 3D case is analogous, we avoid some obvious details.

2.4.1   Space Anisotropy Described by an Ellipsoid

Let $\Omega \subset \mathbb{R}^3$ be a domain, $M : \Omega \to Sym$ be a given Riemannian metric field, and $\bar{x} \in \Omega$ be arbitrary but fixed. We set $\mathbb{M} := M(\bar{x})$. The matrix $\mathbb{M}$ defines the space anisotropy at $\bar{x}$ in the following way. Since $\mathbb{M} \in Sym$, it is diagonalizable and has positive eigenvalues $\lambda_3 \geq \lambda_2 \geq \lambda_1 > 0$ and corresponding mutually orthonormal eigenvectors v_i, $i = 1, 2, 3$. Similar to (2.10) and (2.11), we define the matrices

$$\Lambda_M = \begin{pmatrix} \lambda_1 & 0 & 0 \\ 0 & \lambda_2 & 0 \\ 0 & 0 & \lambda_3 \end{pmatrix}, \qquad Q_M = \begin{pmatrix} v_1 & v_2 & v_3 \end{pmatrix}. \tag{2.37}$$

Analogously to (2.14), M has the eigenvalue decomposition

$$M = Q_M^T \Lambda_M Q_M. \tag{2.38}$$

The matrix Q_M is unitary and fulfills (2.12). According to, e.g., [114], any 3×3 unitary matrix can be decomposed into the Givens rotations

$$Q_\alpha^{(1)} = \begin{pmatrix} 1 & 0 & 0 \\ 0 & \cos\alpha & -\sin\alpha \\ 0 & \sin\alpha & \cos\alpha \end{pmatrix}, \quad Q_\beta^{(2)} = \begin{pmatrix} \cos\beta & 0 & -\sin\beta \\ 0 & 1 & 0 \\ \sin\beta & 0 & \cos\beta \end{pmatrix}, \quad Q_\gamma^{(3)} = \begin{pmatrix} \cos\gamma & -\sin\gamma & 0 \\ \sin\gamma & \cos\gamma & 0 \\ 0 & 0 & 1 \end{pmatrix},$$

around axes of the coordinate system through angles α, β, and γ. This decomposition corresponds to three degrees of freedom of the orientation in 3D, see Fig. 2.5. Therefore, a unitary matrix Q represents a rotation in $\mathbb{R}^3$, and we denote formally the rotation angles by $\boldsymbol{\phi} = (\phi_1, \phi_2, \phi_3) \in \mathscr{U}$, where

$$\mathscr{U} = [0, 2\pi] \times [0, \pi] \times [0, 2\pi]. \tag{2.39}$$

Therefore, for a triplet of orthonormal vectors v_i, $i = 1, 2, 3$, there exists a triplet $\boldsymbol{\phi} \in \mathscr{U}$. For our purpose, it is not necessary to define explicit relations among v_i, $i = 1, 2, 3$, and $\boldsymbol{\phi} \in \mathscr{U}$.

Remark 2.13 We note that for symmetric objects (as an ellipsoid), the domain $\mathscr{U}$ can be taken smaller, namely $\mathscr{U} = [0, \pi] \times [0, \pi] \times [0, \pi]$.

Fig. 2.5 Degrees of freedom of a rotation in 3D, ϕ_1 and ϕ_2 define the 3D space direction and ϕ_3 defines the rotation around this direction

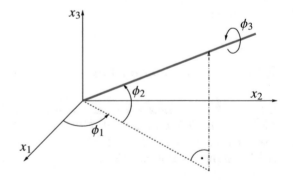

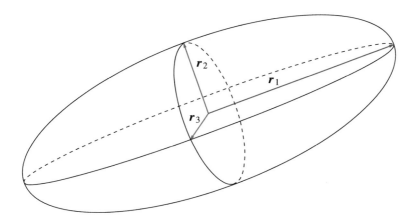

Fig. 2.6 3D ellipsoid $\mathscr{E}_{\mathbb{M}}$ with the semi-axes defined by vectors $r_i = v_i/\sqrt{\lambda_i}$, $i = 1, 2, 3$, where λ_i and v_i, $i = 1, 2, 3$ are the eigenvalues and eigenvectors of $\mathbb{M}$, respectively

Remark 2.14 In order to be precise, matrix $\mathbb{Q}_{\mathbb{M}}$ given by (2.37) can include also a reflection depending on the sign of eigenvectors. A possible consideration of reflections has no impact on the subsequent results.

Similar to (2.15), we define by

$$\mathscr{E}_{\mathbb{M}} := \left\{ x \in \mathbb{R}^3; \ (x - \bar{x})^{\mathsf{T}} \mathbb{M} (x - \bar{x}) \le 1 \right\} \tag{2.40}$$

the *ellipsoid* $\mathscr{E}_{\mathbb{M}}$ centered at $\bar{x}$ whose semi-axes have lengths $r_i = (\lambda_i)^{-1/2}$, $i = 1, 2, 3$, and its orientation of the semi-axis corresponds to directions of eigenvectors v_i, $i = 1, 2, 3$, see Fig. 2.6.

Similar to the 2D case, the ellipsoid $\mathscr{E}_{\mathbb{M}}$ gives a geometric interpretation of the anisotropy of the space given by M at $\bar{x}$. Namely, λ_i, $i = 1, 2, 3$, give the lengthening or shortening of distances in the mutually perpendicular directions given by v_i, $i = 1, 2, 3$.

2.4.2   Geometry of a Tetrahedron

Let $\hat{K} \subset \mathbb{R}^3$ be a *reference equilateral tetrahedron* with vertices $\hat{\mathbf{a}}_1 = (1, 0, 0)$, $\hat{\mathbf{a}}_2 = (-\frac{1}{3}, \sqrt{\frac{8}{9}}, 0)$, $\hat{\mathbf{a}}_3 = (-\frac{1}{3}, -\sqrt{\frac{2}{9}}, \sqrt{\frac{2}{3}})$, and $\hat{\mathbf{a}}_4 = (-\frac{1}{3}, -\sqrt{\frac{2}{9}}, -\sqrt{\frac{2}{3}})$, i.e.,

$$\hat{K} := \left\{ x \in \mathbb{R}^3; \ x = \sum_{i=1}^{4} \kappa_i \hat{\mathbf{a}}_i, \quad \sum_{i=1}^{4} \kappa_i \le 1, \ \kappa_i \ge 0, \ i = 1, \dots, 4 \right\}. \tag{2.41}$$

The length of all edges of $\hat{K}$ is equal to $\sqrt{8/3}$. This reference tetrahedron is the tetrahedron of maximum area inscribed into a unit ball with the center at the origin, denoted hereafter as

$$\hat{\mathscr{E}} := \{x \in \mathbb{R}^3;\ |x| \leq 1\}. \tag{2.42}$$

Let K be an arbitrary but fixed tetrahedron. Then there exists an affine mapping $F_K : \hat{K} \to K$, which maps $\hat{K}$ onto K, namely

$$F_K(\hat{x}) = \mathbb{F}_K \hat{x} + x_K, \quad \hat{x} \in \hat{K}, \tag{2.43}$$

where $x_K \in \mathbb{R}^3$ is the barycenter of K and $\mathbb{F}_K$ is a 3×3 matrix defining the size and the shape of K. The mapping F_K can be constructed in a manner similar to (2.22).

If the volume of K is positive, then $\mathbb{F}_K$ is a regular matrix and it can be decomposed by the *singular value decomposition* such that

$$\mathbb{F}_K = \mathbb{Q}_{\phi_K} \mathbb{L}_K \mathbb{Q}_{\psi_K}^\mathsf{T}, \tag{2.44}$$

where $\mathbb{L}_K = \mathrm{diag}(\ell_{K,1}, \ell_{K,2}, \ell_{K,3})$ is a diagonal matrix with the singular values $\ell_{K,1} \geq \ell_{K,2} \geq \ell_{K,3} > 0$. Moreover, $\mathbb{Q}_{\phi_K}$ and $\mathbb{Q}_{\psi_K}$ are *unitary matrices*, cf. (2.12), which exhibit a rotation in 3D, see Sect. 2.4.1. The subscripts $\phi_K, \psi_K \in \mathscr{U}$, cf. (2.39), formally represent the rotation angles.

Similarly, as in the 2D case, the decomposition of the mapping $\mathbb{F}_K$ in (2.44) has the following steps: first, $\mathbb{Q}_{\psi_K}^\mathsf{T}$ on $\hat{K}$ or $\hat{\mathscr{E}}$ performs the 3D rotation given by ψ_K. Then, $\mathbb{L}_K$ performs a lengthening or shortening in the x_1, x_2, and x_3 directions. Finally, $\mathbb{Q}_{\phi_K}$ performs the 3D rotation by ψ_K.

Since matrices $\mathbb{Q}_{\phi_K}$ and $\mathbb{Q}_{\psi_K}$ in (2.44) are unitary, we have $\det \mathbb{F}_K = \det \mathbb{L}_K$, and therefore the volume of tetrahedron K is given by $\mathbb{L}_K$. We define

$$\mu_K := \left(\ell_{K,1} \ell_{K,2} \ell_{K,3}\right)^{1/3}, \quad (\sigma_K, \varsigma_K) := \left(\left(\frac{\ell_{K,1}}{\ell_{K,2}}\right)^{1/3}, \left(\frac{\ell_{K,1}}{\ell_{K,3}}\right)^{1/3}\right). \tag{2.45}$$

Obviously, $\mu_K > 0$, $1 \leq \sigma_K \leq \varsigma_K$. Then, we have

$$\mathbb{L}_K = \begin{pmatrix} \ell_{K,1} & 0 & 0 \\ 0 & \ell_{K,2} & 0 \\ 0 & 0 & \ell_{K,3} \end{pmatrix} = \mu_K \mathbb{S}_K, \quad \text{where } \mathbb{S}_K := \begin{pmatrix} \sigma_K \varsigma_K & 0 & 0 \\ 0 & \varsigma_K/\sigma_K^2 & 0 \\ 0 & 0 & \sigma_K/\varsigma_K^2 \end{pmatrix}. \tag{2.46}$$

Similar to (2.28), the volume of K is equal to

$$|K| = \ell_{K,1}\ell_{K,2}\ell_{K,3}|\hat{K}| = \mu_K^3 |\hat{K}|, \tag{2.47}$$

where $|\hat{K}| = 8/(9\sqrt{3})$ is the volume of the reference tetrahedron $\hat{K}$. Furthermore, relations (2.44) and (2.46) imply that

$$\det \mathbb{F}_K = \det \mathbb{L}_K = \mu_K^3. \tag{2.48}$$

We present the 3D complement of Lemma 2.9.

Lemma 2.15 *Let $\mathscr{E}_K$ be an image of the reference ellipse $\hat{\mathscr{E}}$ (cf. (2.42)) given by mapping (2.43) with its decomposition (2.44), i.e.,*

$$\mathscr{E}_K := \{x \in \mathbb{R}^3;\ x = \mathbb{Q}_{\phi_K} \mathbb{L}_K \mathbb{Q}_{\psi_K}^\mathsf{T} \hat{x} + x_K,\ \hat{x} \in \hat{\mathscr{E}}\}. \tag{2.49}$$

Then $\mathscr{E}_K$ is the ellipsoid defined by (2.40) with $\bar{x} := x_K$ and

$$\mathbb{M} := \mathbb{Q}_{\phi_K}^\mathsf{T} (\mathbb{L}_K^{-1})^2 \mathbb{Q}_{\phi_K} = \mathbb{Q}_{\phi_K}^\mathsf{T} \begin{pmatrix} 1/(\ell_{K,1})^2 & 0 & 0 \\ 0 & 1/(\ell_{K,2})^2 & 0 \\ 0 & 0 & 1/(\ell_{K,3})^2 \end{pmatrix} \mathbb{Q}_{\phi_K}. \tag{2.50}$$

Proof The proof is completely analogous to the proof of Lemma 2.9. In particular, we have

$$\Lambda_\mathbb{M} = \mathrm{diag}(\lambda_1, \lambda_2, \lambda_3) = \mathbb{L}_K^{-2} = \mathrm{diag}(1/\ell_{K,1}^2, 1/\ell_{K,2}^2, 1/\ell_{K,3}^2), \tag{2.51}$$

where λ_i, $i = 1, 2, 3$, are the eigenvalues of $\mathbb{M}$.   □

Analogously to Definition 2.10, we define the geometry of a tetrahedron.

Definition 2.16 Let K be a tetrahedron given by the affine mapping $F_K : \hat{K} \rightarrow K$, cf. (2.43), and let (2.44), (2.45), and (2.46) be the corresponding decomposition. The value μ_K is called the *size* of K, the pair (σ_K, ς_K) is called the *shape* of K, and the triplet $\phi_K \in \mathscr{U}$ (cf. (2.39)) is the *orientation* of K. The group of six $\{\mu_K, \sigma_K, \varsigma_K, \phi_K\}$ is called the *geometry* of K. Finally, the shape and orientation are called *anisotropy* of K.

2.4.3   Equilateral Tetrahedra with Respect to Matrix $\mathbb{M} \in Sym$

Analogously to Sect. 2.3.3, we define, for a matrix $\mathbb{M} \in Sym$, the tetrahedron $K_\mathbb{M}$ which is inscribed into the ellipsoid $\mathscr{E}_\mathbb{M}$ given by (2.15) and has the maximum possible volume. There exist infinitely many tetrahedra having the same volume (acute angle as well as obtuse angle) as in the 2D case.

It can be derived that the maximal tetrahedron inscribed in the unit ball has the length of all six edges equal to $\sqrt{8/3}$. Then, the 3D variant of Lemma 2.12 is the following.

Lemma 2.17 *Let* $\mathbb{M} \in Sym$ *and* $\mathscr{E}_{\mathbb{M}}$ *be given by* (2.40). *Let* $K_{\mathbb{M}}$ *be a tetrahedron inscribed into* $\mathscr{E}_{\mathbb{M}}$ *and having the maximum possible area. Then*

$$\|\mathbf{e}_i\|_{\mathbb{M}} := (\mathbf{e}_i^{\mathsf{T}} \mathbb{M} \mathbf{e}_i)^{1/2} = \sqrt{8/3}, \qquad i = 1, \ldots, 6, \tag{2.52}$$

where $\mathbf{e}_i$, $i = 1, \ldots, 6$, *denotes the edges of* $K_{\mathbb{M}}$ *considered as vectors.*

Proof The proof is completely analogous to the two-dimensional case, following from Lemmas 2.8 and 2.12. □

Obviously, equivalences similar to those in Sect. 2.3.5 are valid also for the 3D case. These equivalences together with Lemmas 2.12 and 2.17 have a practical impact on the construction of the anisotropic mesh adaptation algorithms. In the forthcoming chapters, we derive estimates depending on the the size, shape, and orientation of mesh elements given in Definitions 2.10 and 2.16. Then, by an optimization process, we derive the optimal size, shape, and orientation of corresponding ellipses or ellipsoids. Further, we define the matrices $\mathbb{M} \in Sym$ at a set of given nodes in Ω, and, using a linear interpolation, we define the metric field $M : \Omega \to Sym$. Finally, we obtain the desired mesh solving approximately Problem 2.5. This is briefly described in the remainder of this chapter.

2.5   Mesh Generation by a Metric Field

In this section, we briefly describe how to (approximately) solve Problem 2.5. Usually, we derive the optimal geometry of mesh elements (cf. Definitions 2.10 and 2.16) at a finite set of nodes. In Sect. 2.5.1, we describe how to define the metric field M at this set of nodes. Then it is possible to perform an interpolation of this nodal metric and define M in the whole domain Ω, which is explained in Sect. 2.5.2. Furthermore, we present the main ideas of the construction of the corresponding simplicial mesh in Sect. 2.5.3.

2.5.1   Setting of the Nodal Metric

We start with the two-dimensional case; the extension to 3D is obvious. In practical computations, we derive the optimal geometry of mesh elements only in a finite set of nodes $\mathscr{N}_h$. Usually, these nodes are the vertices or barycenters of an initial mesh $\mathscr{T}_h^0$ which we need to adapt. Here, we consider the latter case. Let $\mathscr{T}_h^0$ be a simplicial mesh of Ω and x_K be barycenter of $K \in \mathscr{T}_h^0$.

In the following parts of this book, we derive several techniques whose output is the optimal element size $\mu_K^\star > 0$, the optimal element shape $\sigma_K^\star \geq 1$, and the optimal element orientation $\phi_K^\star \in [0, \pi]$ for each $K \in \mathscr{T}_h^0$. Then, in virtue of (2.25), we set the values

$$\ell_{K,1}^{\star} := \mu_K^{\star}\sigma_K^{\star}, \qquad \ell_{K,2}^{\star} := \mu_K^{\star}/\sigma_K^{\star}, \tag{2.53}$$

and using (2.30), we define the matrix

$$\mathbb{M}_K^{\star} := \begin{pmatrix} \cos\phi_K^{\star} & \sin\phi_K^{\star} \\ -\sin\phi_K^{\star} & \cos\phi_K^{\star} \end{pmatrix} \begin{pmatrix} 1/(\ell_{K,1}^{\star})^2 & 0 \\ 0 & 1/(\ell_{K,2\star})^2 \end{pmatrix} \begin{pmatrix} \cos\phi_K^{\star} & -\sin\phi_K^{\star} \\ \sin\phi_K^{\star} & \cos\phi_K^{\star} \end{pmatrix} \tag{2.54}$$

for all elements $K \in \mathcal{T}_h^0$.

For practical reasons, it is advantageous to recompute the matrices $\mathbb{M}_K^{\star}$, which are associated with the barycenters of elements $K \in \mathcal{T}_h^0$, to matrices associated with the set of vertices of $\mathcal{T}_h^0$ denoted as

$$\mathcal{N}_h := \{x \in \Omega : x \text{ is a vertex of } \mathcal{T}_h^0\}. \tag{2.55}$$

One can use an arithmetic average of elements sharing the vertex, i.e., we set

$$\mathbb{M}_x := \frac{1}{\#\mathcal{P}_x} \sum_{K \in \mathcal{P}_x} \mathbb{M}_K^{\star}, \qquad x \in \mathcal{N}_h, \tag{2.56}$$

where $\mathcal{P}_x$ denotes the set of $K \in \mathcal{T}_h^0$ sharing the vertex $x \in \mathcal{N}_h$ and $\#\mathcal{P}_x$ is the number or elements contained in the set $\mathcal{P}_x$, $x \in \mathcal{N}_h$. Obviously, $\mathbb{M}_x \in Sym$ for all $x \in \mathcal{N}_h$.

2.5.2 Interpolation of the Metric

Let $\mathcal{T}_h^0$ be an initial mesh to be adapted. We assume that for each $x \in \mathcal{N}_h$, given by (2.55), the matrix $\mathbb{M}_x \in Sym$ is given, cf. Sect. 2.5.1. Let $x \in \Omega$, and then there exists $K \in \mathcal{T}_h^0$ such that $x \in K$. If x belongs to ∂K, then there exist more such simplexes and we choose one of them. We denote the vertices of K by $\mathbf{a}_i$, $i = 1, \ldots, d+1$. Then there exist barycentric coordinates $\kappa_i \geq 0$, $i = 1, \ldots, d+1$, $\sum_{i=1}^{d+1} \kappa_i = 1$ such that

$$x = \sum_{i=1}^{d+1} \kappa_i \mathbf{a}_i. \tag{2.57}$$

In virtue of (2.57), we define $\mathcal{M}$ at x by

$$\mathcal{M}(x) := \sum_{i=1}^{d+1} \kappa_i \mathbb{M}_{\mathbf{a}_i}, \tag{2.58}$$

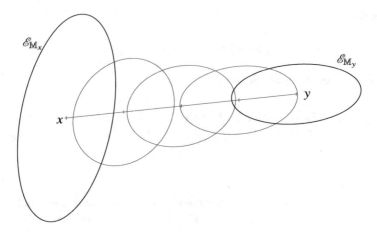

Fig. 2.7 Linear interpolation of the metric between x and y

where $\mathbb{M}_{\mathbf{a}_i}$ is the matrix given for $\mathbf{a}_i \in \mathcal{N}_h$. Obviously, $M \in Sym$ and the resulting metric field $M(x)$, $x \in \Omega$, is continuous on Ω. Figure 2.7 illustrates the interpolation of a metric given by $\mathbb{M}_x$ and $\mathbb{M}_y$ between nodes x and y, respectively, using (2.58). Obviously, relation (2.58) implies that

$$M\big(\tfrac{1}{2}(\mathbf{a}_i + \mathbf{a}_j)\big) = \tfrac{1}{2}\big(\mathbb{M}_{\mathbf{a}_i} + \mathbb{M}_{\mathbf{a}_j}\big), \tag{2.59}$$

where $\mathbf{a}_i$ and $\mathbf{a}_j$ are any vertices of $\mathcal{T}_h^0$ connected by an edge.

Using this technique, we are able to set matrix $M(x) \in Sym$ at an arbitrary node $x \in \Omega$ and therefore to define the metric field M on the whole domain Ω. However, an approximation of length of an edge in the Riemann metric space M, which is explained in the following section, significantly reduces the number of nodes $x \in \Omega$ where $M(x)$ has to be evaluated in practice.

2.5.3 Mesh Optimization

We present the main ideas of the mesh optimization corresponding to the solution of Problem 2.5. As mentioned above, we assume that an initial mesh $\mathcal{T}_h^0$ of the domain Ω, satisfying assumptions of Definitions 1.12, is given. Furthermore, matrices $\mathbb{M}_{\mathbf{a}} \in Sym$ are prescribed for each vertex $\mathbf{a}$ of $\mathcal{T}_h^0$.

The main idea of the solution of Problem 2.5 is the minimization of $Q(\mathcal{T}_h)$ given by (2.9). The evaluation of $Q(\mathcal{T}_h)$ requires the computation of the length of all edges $\mathbf{e}$ of the mesh under the metric given by M, cf. (2.5). From a practical point of view, it makes sense to approximate the integral in (2.5) by a numerical quadrature; the simplest one is the midpoint rule. Hence, let $\mathbf{e} \in \mathcal{F}_h$ be an edge of the mesh $\mathcal{T}_h$ having the end points $\mathbf{a}_1$ and $\mathbf{a}_2$. Putting $\mathbf{a}_{\mathbf{e}} := (\mathbf{a}_1 + \mathbf{a}_2)/2$ and applying the midpoint integration formula, we obtain from (2.5) the approximation

$$\|e\|_{\mathcal{M}} = \int_0^1 \left(e^T \mathcal{M}(a_1 + t(a_2 - a_1))e\right)^{1/2} dt \tag{2.60}$$

$$\approx \int_0^1 \left(e^T \mathcal{M}(a_e)e\right)^{1/2} dt = \left(e^T \mathbb{M}_e e\right)^{1/2}, \qquad e \in \mathcal{F}_h,$$

where $\mathbb{M}_e = \mathcal{M}(a_e) = \mathcal{M}((a_1 + a_2)/2)$. Moreover, in virtue of (2.59), we have $\mathbb{M}_e = (\mathbb{M}_{a_1} + \mathbb{M}_{a_2})/2$.

The mesh optimization process exhibits the construction of the sequence of meshes $\mathcal{T}_h^k$, $k = 1, 2, \ldots$, starting from the input grid $\mathcal{T}_h^0$, such that the sequence quantities $Q(\mathcal{T}_h^k)$, $k = 0, 1, 2 \ldots$, converge to the smallest possible value. The mesh $\mathcal{T}_h^{k+1}$ arises by an application of one of the following local operations on mesh $\mathcal{T}_h^k$:

- Adding a node at the center of the given edge
- Removing an edge
- Swapping of the diagonal of two neighboring elements
- Moving a node

Moreover, in order to keep the approximation of the boundary, it is necessary to distinguish between the interior (I) and boundary (B) nodes. These operations for two dimensions are shown in Fig. 2.8. They cover also the case of non-polygonal boundaries. Some of the local operations can be extended to three-dimensional case

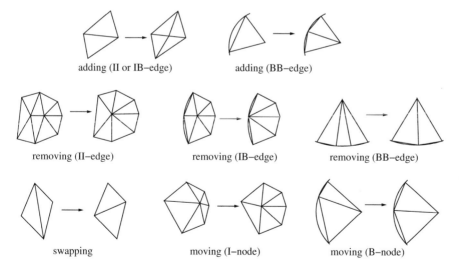

Fig. 2.8 2D local operations used in the mesh adaptation. I denotes an interior node, B denotes a boundary node, and II, IB, and BB are edges defined by the end nodes of the corresponding type

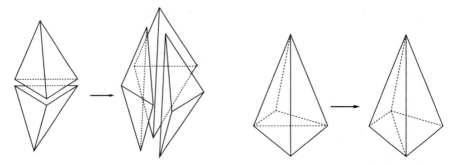

Fig. 2.9 Examples of swapping for interior tetrahedra (left) and boundary ones (right)

in a straightforward way, while others require a more sophisticated treatment, see Fig. 2.9 for two examples.

Roughly speaking, each operation is carried out if it leads to a decrease of parameter $Q(\cdot)$, given by (2.9). It means that we evaluate $Q(\cdot)$ for the current mesh and for the mesh arising from performing one of the previously mentioned local operations. If $Q(\cdot)$ is smaller, then this operation is accepted and we proceed with the mesh adaptation. The mesh optimization process is stopped if any other local operation cannot be performed. Obviously, there are many technical issues which are necessary to take into account. Namely, we have to guarantee that the arising mesh is still conforming and that the elements do not collapse, i.e., $|K|/h_K^d$ has to be bounded from below. More details can be found in [39, 42]. These ideas form the base of the code ANGENER [41]. Several animations showing the process of mesh adaptation can be viewed in [40]. Similar techniques are used, e.g., in [79, 85].

Finally, we illustrate the mesh optimization process by a two-dimensional example in Fig. 2.10. We consider a unit square $(0, 1) \times (0, 1)$ and prescribe the matrices $\mathbb{M}_{\mathbf{a}} \in Sym$ at a finite set of nodes $\mathbf{a} \in \mathcal{N}_h$. They are shown in Fig. 2.10, left. The matrices $\mathbb{M}_{\mathbf{a}}$, $\mathbf{a} \in \mathcal{N}_h$, are represented by the ellipses given by (2.15). Fig. 2.10, right shows the resulting triangular mesh arising from the mesh optimization process outlined in this section. The size, shape, and orientation of the ellipses and the corresponding elements are obvious.

We summarize the brief exposition of this chapter.

(1) A simplicial mesh of Ω can be defined by the given metric field $\mathcal{M} : \Omega \rightarrow Sym$ as the solution of Problem 2.5.
(2) Problem 2.5 can be solved iteratively by the anisotropic mesh adaptation process consisting of applications of local operations, cf. Sect. 2.5.3.
(3) The mapping $\mathcal{M}$ can be given only by matrices $\mathbb{M}_{\mathbf{a}} \in Sym$ at a finite set of nodes $\mathbf{a} \in \mathcal{N}_h \subset \Omega$, i.e., we set $\mathcal{M}(x) := \mathbb{M}(x)$ for $x \in \mathcal{N}_h$ and $\mathcal{M}(x)$ at $x \notin \mathcal{N}_h$ is obtained by a suitable interpolation.

Therefore, in order to construct an "optimal" mesh, where the optimality can be considered from various points of view, we set only matrices $\mathbb{M}(\mathbf{a})$ at $\mathbf{a} \in \mathcal{N}_h$,

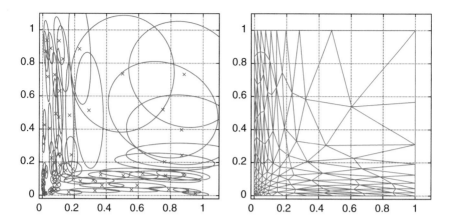

Fig. 2.10 Example of a metric $\mathcal{M}$ represented by ellipses at nodes $\mathbf{a} \in \mathcal{N}_h$ (blue crosses) (left) and the corresponding triangular grid arising by the mesh optimization process (right)

where $\mathcal{N}_h$ is typically the set of barycenters (or vertices) of the initial mesh. This approach is followed in the rest of this monograph.

Chapter 3
Interpolation Error Estimates for Two Dimensions

We formulate the fundamental theoretical results which are later employed for the anisotropic mesh adaptation method. First, we recall the geometry terms of a mesh triangle K discussed in the previous chapter. Further, we define an interpolation of a sufficiently smooth function u on element K as a polynomial function having the same value and partial derivatives as the original function at the barycenter of K. Moreover, we derive estimates of the difference between u and its interpolation (=interpolation error estimates) in several norms. These estimates take into account the geometry of mesh element K. Finally, we derive the optimal shape of a triangle with given barycenter, minimizing the interpolation error estimates.

3.1 Geometry of a Triangle

In order to have this chapter self-contained, we recall the basic terms and relations from Sect. 2.3.2. Let $\hat{K}$ be the reference triangle given by (2.19). For a triangle K there exists an affine mapping $F_K : \hat{K} \to K$ (cf. (2.21)) written in the form

$$F_K(\hat{x}) = \mathbb{F}_K \hat{x} + x_K, \quad \hat{x} \in \hat{K}, \tag{3.1}$$

where x_K is the barycenter of K and $\mathbb{F}_K$ is a regular matrix. The *singular value decomposition* of $\mathbb{F}_K$ reads

$$\mathbb{F}_K = \mathbb{Q}_{\phi_K} \mathbb{L}_K \mathbb{Q}_{\psi_K}^{\mathsf{T}}, \tag{3.2}$$

where $\mathbb{Q}_{\phi_K}$ and $\mathbb{Q}_{\psi_K}$ are rotation matrices through angles ϕ_K and ψ_K, respectively, (cf. (2.24)) and $\mathbb{L}_K = \mathrm{diag}(\ell_{K,1}, \ell_{K,2})$ is a diagonal matrix with the singular values $\ell_{K,1} \geq \ell_{K,2} > 0$.

© The Author(s), under exclusive license to Springer Nature Switzerland AG 2022
V. Dolejší, G. May, *Anisotropic hp-Mesh Adaptation Methods*, Nečas Center Series,
https://doi.org/10.1007/978-3-031-04279-9_3

Moreover, setting $\lambda_i := 1/(\ell_{K,i})^2$, $i = 1, 2$, $\Lambda_{\mathbb{M}} := \mathrm{diag}(\lambda_1, \lambda_2)$ and

$$\mathbb{M} := \mathbb{Q}_{\phi_K}^{\mathsf{T}} \Lambda_{\mathbb{M}} \mathbb{Q}_{\phi_K}, \tag{3.3}$$

then K is the maximal triangle inscribed into ellipse $\mathscr{E}_{\mathbb{M}}$ generated by matrix $\mathbb{M}$ using (2.15).

We recall the terms describing the geometry of triangle, cf. Definition 2.10 and relations (2.26) and (2.27).

Definition 3.1 Let K be a triangle given by mapping (3.1) with decomposition (3.2), singular values $\ell_{K,1} \geq \ell_{K,2} > 0$, and angle $\phi_K \in [0, \pi)$. Moreover, let $\lambda_i := 1/(\ell_{K,i})^2$, $i = 1, 2$, and $\Lambda_{\mathbb{M}} := \mathrm{diag}(\lambda_1, \lambda_2)$ be as in (3.3). Then

- $\mu_K := \sqrt{\ell_{K,1}\ell_{K,2}} = (\lambda_1\lambda_2)^{-1/4} > 0$ is called the *size* of K.
- $\sigma_K := \sqrt{\ell_{K,1}/\ell_{K,2}} = (\lambda_2/\lambda_1)^{1/4} \geq 1$ is called the *shape* of K.
- The angle ϕ_K is the *orientation* of K.
- The pair $\{\sigma_K, \phi_K\}$ is called the *anisotropy* of K.
- The triplet $\{\mu_K, \sigma_K, \phi_K\}$ is called the *geometry* of K.
- Using the definition of μ_K and σ_K we have

$$\mathbb{L}_K = \begin{pmatrix} \ell_{K,1} & 0 \\ 0 & \ell_{K,2} \end{pmatrix} = \mu_K \mathbb{S}_K, \quad \text{where } \mathbb{S}_K := \begin{pmatrix} \sigma_K & 0 \\ 0 & \sigma_K^{-1} \end{pmatrix} \tag{3.4}$$

and

$$\det \mathbb{F}_K = \det \mathbb{L}_K = \mu_K^2. \tag{3.5}$$

The geometry of K, given by $\{\mu_K, \sigma_K, \phi_K\}$, is important for the forthcoming interpolation error estimates which depend explicitly on these parameters. Later, we employ these estimates and define, by a mesh optimization process, a new geometry of mesh element. Using the technique from Sect. 2.5, we are then able to construct a new mesh.

3.2   Interpolation Error Function and Its Anisotropic Bound

The aim of this section is to derive estimates of the interpolation error taking into account the geometry of a triangle introduced in Sect. 3.1. For simplicity, we assume that the function to be interpolated is sufficiently regular, namely that it belongs to the space $C^\infty := C^\infty(\overline{\Omega})$.

3.2.1 Interpolation Operator

Let $u \in C^{\infty}$ be a given function, $\bar{x} \in \Omega$ be a given node, and $p \in \mathbb{N}$ be an integer representing the polynomial approximation degree. By $P^p \Omega$ we denote the space of polynomial functions of the total degree at most p on Ω, cf. (1.20).

We define an interpolation operator $\Pi_{\bar{x},p} : C^{\infty} \to P^p(\Omega)$ such that $\Pi_{\bar{x},p} u$ is the polynomial function of degree p on Ω which has the same values of all partial derivatives up to order p at $\bar{x}$ as the function u.

Definition 3.2 Let $\bar{x} = (\bar{x}_1, \bar{x}_2) \in \Omega$ and $p \in \mathbb{N}$ be given. The *interpolation operator* $\Pi_{\bar{x},p} : C^{\infty} \to P^p(\Omega)$ of degree p at $\bar{x}$ is given by

$$\frac{\partial^k \left(\Pi_{\bar{x},p} u(\bar{x}) \right)}{\partial x_1^l \partial x_2^{k-l}} = \frac{\partial^k u(\bar{x})}{\partial x_1^l \partial x_2^{k-l}} \qquad \forall l = 0, \dots, k \quad \forall k = 0, \dots, p. \tag{3.6}$$

Lemma 3.3 *Let $u \in C^{\infty}$, $\bar{x} \in \Omega$, and $p \in \mathbb{N}$ be given. Then the interpolation $\Pi_{\bar{x},p} u$ in virtue of Definition 3.2 exists and satisfies*

$$\Pi_{\bar{x},p} u(x) = \sum_{k=0}^{p} \frac{1}{k!} \left(\sum_{l=0}^{k} \binom{k}{l} \frac{\partial^k u(\bar{x})}{\partial x_1^l \partial x_2^{k-l}} (x_1 - \bar{x}_1)^l (x_2 - \bar{x}_2)^{k-l} \right), \qquad x \in \Omega, \tag{3.7}$$

where $\binom{k}{l} = \frac{k!}{l!(k-l)!}$.

Proof Obviously, the function given by the right-hand side of (3.7) exists since $u \in C^{\infty}$. Moreover, it is a polynomial function of degree at most p and their partial derivatives satisfy (3.6). □

3.2.2 Interpolation Error Function

In this section, we introduce the *interpolation error function* which approximates the difference between a function u and its interpolation given by Definition 3.2. Let $u \in C^{\infty}$ be a given function, $\bar{x} \in \Omega$ and $p \in \mathbb{N}$ be an integer. Using the Taylor expansion of degree $p + 1$ at $\bar{x}$, we have

$$u(x) = \sum_{k=0}^{p+1} \frac{1}{k!} \left(\sum_{l=0}^{k} \binom{k}{l} \frac{\partial^k u(\bar{x})}{\partial x_1^l \partial x_2^{k-l}} (x_1 - \bar{x}_1)^l (x_2 - \bar{x}_2)^{k-l} \right) + O(|x - \bar{x}|^{p+2}), \ x \in \Omega. \tag{3.8}$$

Since $\Pi_{\bar{x},p}u$ satisfies (3.7), then we obtain from (3.8) the relation

$$u(x) - \Pi_{\bar{x},p}u(x) = w_{\bar{x},p}^{\text{int}}(x) + O(|x - \bar{x}|^{p+2}), \tag{3.9}$$

where

$$w_{\bar{x},p}^{\text{int}}(x) := \frac{1}{(p+1)!} \sum_{l=0}^{p+1} \left[\binom{p+1}{l} \frac{\partial^{p+1} u(\bar{x})}{\partial x_1^l \partial x_2^{p+1-l}} (x_1 - \bar{x}_1)^l (x_2 - \bar{x}_2)^{p+1-l} \right]. \tag{3.10}$$

Definition 3.4 Let $u \in C^\infty$, $\bar{x} \in \Omega$, and $p \in \mathbb{N}$ be given. Then the polynomial function $w_{\bar{x},p}^{\text{int}}$ of degree $p+1$ defined by (3.10) is called the *interpolation error function* of degree p, located at $\bar{x}$, of function u.

Remark 3.5 We denote by

$$d^{p+1}u(\bar{x};\xi) := \frac{1}{(p+1)!} \sum_{l=0}^{p+1} \binom{p+1}{l} \frac{\partial^{p+1} u(x)}{\partial x_1^l \partial x_2^{p+1-l}} \xi_1^l \xi_2^{k-l}, \tag{3.11}$$

the $(p+1)$th-(scaled) *directional derivative* of $u \in C^\infty$ along the direction $\xi = (\xi_1, \xi_2) \in \mathbb{R}^2$ at $\bar{x} \in \Omega$. Then the interpolation error function is equal to this derivation, namely

$$d^{p+1}u(\bar{x}; x - \bar{x}) = w_{\bar{x},p}^{\text{int}}(x), \quad x \in \Omega. \tag{3.12}$$

3.2.3 s-Homogeneous Functions

The interpolation error function $w_{\bar{x},p}^{\text{int}}$ from Definition 3.4 belongs to the following type of functions:

Definition 3.6 Let $\bar{x} = (\bar{x}_1, \bar{x}_2) \in \Omega$ and $s \in \mathbb{N}$ be given. We say that a polynomial function $z : \Omega \to \mathbb{R}$ is an *s-homogeneous function located at $\bar{x}$* if

$$z(x) = \sum_{l=0}^{s} \alpha_l (x_1 - \bar{x}_1)^l (x_2 - \bar{x}_2)^{s-l}, \quad x = (x_1, x_2) \in \Omega, \tag{3.13}$$

where $\alpha_l \in \mathbb{R}$, $l = 0, \ldots, s$ are the coefficients.

Putting $\xi := x - \bar{x}$ and expressing ξ in polar coordinates (r, ψ), the s-homogeneous function z located at $\bar{x}$ given by (3.13) has the form

$$z(x) = r^s \sum_{l=0}^{s} \alpha_l \cos^l \psi \, \sin^{s-l} \psi, \quad x = \xi + \bar{x}, \ \xi = \xi(r, \psi). \tag{3.14}$$

In virtue of Definition 3.6, function $w_{\bar{x},p}^{\text{int}}$ from (3.10) is $(p+1)$-homogeneous with the coefficients

$$\alpha_l = \frac{1}{(p+1)!} \binom{p+1}{l} \frac{\partial^{p+1} u(x)}{\partial x_1^l \partial x_2^{p+1-l}}, \qquad l = 0, \dots, p+1. \tag{3.15}$$

Hereafter, we consider s-function for $s \geq 2$, the case $s = 2$ corresponds to the lowest considered polynomial approximation degree $p = 1$. The forthcoming estimates are motivated by a *level set function*. We denote by

$$B_1 := \{\xi; \; \xi = (\xi_1, \xi_2) \in \mathbb{R}^2, \; \xi_1^2 + \xi_2^2 = 1\} \tag{3.16}$$

the unit sphere in $\mathbb{R}^2$.

Let z be an s-homogeneous function located at $\bar{x} \in \Omega$ (cf. 3.13). We define the closed curve

$$\theta^s(z) := \left\{ x \in \mathbb{R}^2; \; x = \bar{x} + |z(\bar{x} + \xi)|\xi \quad \forall \xi \in B_1 \right\} \tag{3.17}$$

and the corresponding domain bounded by $\theta^s(z)$ is denoted by $\bar{\theta}^s(z)$. Obviously, $z(\bar{x}) = 0$ and z is monotone with respect to its polar coordinate r, cf. (3.14). Then the curve $\theta^s(z)$ describes how the value of function z is increasing along each particular direction $x - \bar{x}$. Hence, we can say that $\theta^s(z)$ describes the *anisotropy of function z*. Particularly, for each direction $\xi \in B_1$, the absolute value of $z(\bar{x} + \xi)$ is equal to the length of the vector pointing from $\bar{x}$ in the direction ξ and ending on the curve $\theta^s(z)$. This is illustrated in Fig. 3.1.

Since the interpolation error function $w_{\bar{x},p}^{\text{int}}(x)$ from (3.10) is a $(p+1)$-homogeneous function, the corresponding curve $\theta^{p+1}(w_{\bar{x},p}^{\text{int}})$ defined by (3.17)

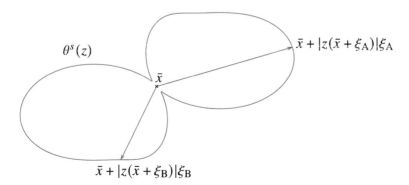

Fig. 3.1 A curve $\theta^s(z)$, directions $\xi_A, \xi_B \in B_1$. The values $|z(\bar{x} + \xi_A)|$ and $|z(\bar{x} + \xi_B)|$ are equal to the length of the colored vectors. Obviously $|z(\bar{x} + \xi_A)| > |z(\bar{x} + \xi_B)|$

determines the size of $|w^{int}_{\bar{x},p}(\bar{x} + \xi)|$ for all $\xi \in B_1$. Moreover, it is equal to the size of the scaled directional derivatives $|d^{p+1}u(\bar{x}; \xi)|$, cf. (3.11).

Example 3.7 We consider the polynomial function

$$u(x_1, x_2) = 50 + 10x_1 + 20x_2 + 6x_1^2 + 2x_1x_2 + 24x_2^2 \tag{3.18}$$
$$+ 12x_1^4 + 3x_1^3x_2 + x_1^2x_2^2 + 6x_1x_2^3 + 24x_2^4$$
$$+ 18x_1^6 + 12x_1^5x_2 - 1.92x_1^4x_2^2 + 3.6x_1^3x_2^3 + 14.4x_1^2x_2^4 + 2.4x_1x_2^5 + 90x_2^6,$$

for $(x_1, x_2) \in [-1, 1]^2$, see Fig. 3.2, top left. By a direct differentiation, we evaluate the corresponding weighted directional derivatives of u at $\bar{x} = 0$ (cf. (3.11)):

$$d^2u(\bar{x}; \xi) = (3\xi_1^2 + 2\xi_1\xi_2 + 12\xi_2^2), \tag{3.19}$$
$$d^4u(\bar{x}; \xi) = (5\xi_1^4 + 5\xi_1^3\xi_2 + 2.5\xi_1^2\xi_2^2 + 10\xi_1\xi_2^3 + 10\xi_2^4)/10,$$
$$d^6u(\bar{x}; \xi) = (2.5\xi_1^6 + 10\xi_1^5\xi_2 - 4\xi_1^4\xi_2^2 + 10\xi_1^3\xi_2^3 + 30\xi_1^2\xi_2^4 + 2\xi_1\xi_2^5 + 12.5\xi_2^6)/100$$

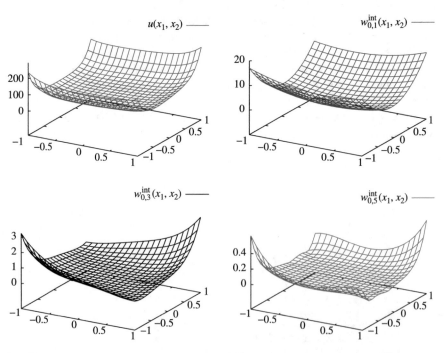

Fig. 3.2 The function $u(x_1, x_2)$ (top left) from (3.18) and the corresponding interpolation error functions $w^{int}_{0,1}$ (top right), $w^{int}_{0,3}$ (bottom left), $w^{int}_{0,5}$ (bottom right) at $\bar{x} = 0$ for $(x_1, x_2) = (\xi_1, \xi_2) \in [-1, 1]^2$

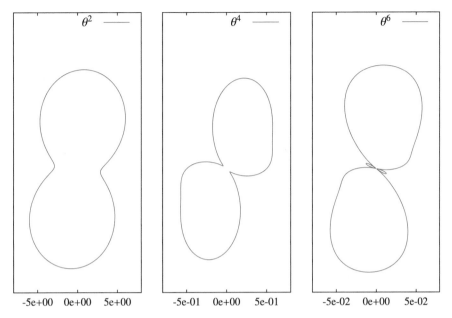

Fig. 3.3 The curves θ^2, θ^4, and θ^6 defined by (3.17) corresponding to the interpolation error functions $w_{0,p}^{\text{int}}$, $p = 1, 3, 5$ from Example 3.7

for $\xi = (\xi_1, \xi_2) \in \mathbb{R}^2$. In virtue of (3.12), the interpolation error functions $w_{\bar{x},p}^{\text{int}}$ given by (3.10) at $\bar{x} = 0$ fulfill

$$w_{0,p}^{\text{int}}(\xi) = \mathrm{d}^{p+1}u(0; \xi), \quad \xi \in \mathbb{R}^2, \ p = 1, 3, 5, \tag{3.20}$$

see Fig. 3.2. Obviously, $w_{0,p}^{\text{int}}$ are $(p + 1)$-homogeneous functions, $p = 1, 3, 5$, Fig. 3.3 shows the corresponding curves θ^2, θ^4, and θ^6 given by (3.17) (we omit the argument, for simplicity). Obviously, these curves are closed and have several symmetries.

Relation (3.13) implies that any s-function, $s \geq 1$, is given by $s + 1$ parameters, i.e., the interpolation error function $w_{\bar{x},p}^{\text{int}}$ is given by $p + 2$ parameters. On the other hand, the geometry of a triangle is described by three parameters (size, shape, and orientation), see Definition 3.1. Therefore, for a mesh optimization process, it would be advantageous to describe the anisotropy of the interpolation error function $w_{\bar{x},p}^{\text{int}}$ also only by three parameters. It can be done directly for $p = 1$ (linear approximation). This is shown in Sect. 3.2.4. Then we extend this approach for $p > 1$ in Sect. 3.2.5.

3.2.4 Anisotropic Bound of Interpolation Error Function for $p = 1$

Let $\bar{x} \in \Omega$, $u \in C^\infty$, and $p = 1$ be given. Let $d^2 u(\bar{x}; \xi)$, $\xi \in \mathbb{R}^2$ be the second order (scaled) directional derivative of u at $\bar{x}$, given by (3.11). In virtue of Definition 3.6, $d^2 u(\bar{x}; \cdot)$ is a 2-homogeneous function.

Let B_1 be the unit sphere, cf. (3.16). We define the values $A_1 \geq 0$, $\xi_1 \in B_1$, $\varphi_1 \in [0, 2\pi)$, $A_1^\perp \geq 0$, $\xi_1^\perp \in B_1$, and $\rho_1 \geq 1$ by

$$A_1 := \max_{\xi \in B_1} |d^2 u(\bar{x}; \xi)|, \tag{3.21a}$$

$$\xi_1 := \arg\max_{\xi \in B_1} |d^2 u(\bar{x}; \xi)|, \tag{3.21b}$$

$$\varphi_1 \in [0, 2\pi) \text{ such that } (\cos\varphi_1, \sin\varphi_1)^\mathsf{T} = \xi_1, \tag{3.21c}$$

$$A_1^\perp := |d^2 u(\bar{x}; \xi_1^\perp)|, \text{ where } \xi_1^\perp \in B_1, \ \xi_1^\perp \cdot \xi_1 = 0, \tag{3.21d}$$

$$\rho_1 := \frac{A_1}{A_1^\perp}, \tag{3.21e}$$

where $a \cdot b = (a_1 b_1 + a_2 b_2)$, $a, b \in \mathbb{R}^2$ is the scalar product in $\mathbb{R}^2$. It means that A_1 is the maximal value of the second order scaled directional derivative of u at $\bar{x}$, ξ_1 is the direction which maximizes this derivative, φ_1 is the angle corresponding to ξ_1, $\xi_1^\perp$ is the direction perpendicular to ξ_1, $A_1^\perp$ is the second order scaled directional derivative of u along the direction $\xi_1^\perp$ at $\bar{x}$, and ρ_1 is the ratio between A_1 and $A_1^\perp$.

Let us define a matrix $\mathbb{D}_{\rho_1}$ by

$$\mathbb{D}_{\rho_1} := \begin{pmatrix} 1 & 0 \\ 0 & (\rho_1)^{-1} \end{pmatrix}, \tag{3.22}$$

where ρ_1 is given by (3.21e). Then, we have the following property.

Lemma 3.8 *Let $\bar{x} \in \Omega$, $u \in C^\infty$, and $p = 1$ be given. We set $A_1 \geq 0$, $\varphi_1 \in [0, 2\pi)$ and ρ_1 by (3.21a), (3.21c), and (3.21e), respectively. Then*

$$\left| w_{\bar{x},1}^{\text{int}}(x) \right| = \left| w_{\bar{x},1}^{\text{int}}(\bar{x} + \zeta) \right| \leq A_1 \zeta^\mathsf{T} \mathbb{Q}_{\varphi_1} \mathbb{D}_{\rho_1} \mathbb{Q}_{\varphi_1}^\mathsf{T} \zeta \quad \forall \zeta = x - \bar{x}, \ x \in \Omega, \tag{3.23}$$

where $w_{\bar{x},1}^{\text{int}}$ is given by (3.10), $\mathbb{Q}_{\varphi_1}$ is the rotation (2.24), and $\mathbb{D}_{\rho_1}$ is defined by (3.22).

Proof Obviously, both sides of (3.23) are 2-homogeneous functions located at $\bar{x}$ with respect to ζ, i.e.,

$$w_{\bar{x},1}^{\text{int}}(\bar{x} + \beta\zeta) = \beta^2 w_{\bar{x},1}^{\text{int}}(\bar{x} + \zeta) \qquad \forall \zeta \in \mathbb{R}^2 \, \forall \beta > 0, \tag{3.24}$$

$$A_1(\beta\zeta)^\mathsf{T} \mathbb{Q}_{\varphi_1} \mathbb{D}_{\rho_1} \mathbb{Q}_{\varphi_1}^\mathsf{T} (\beta\zeta) = \beta^2 A_1 \zeta^\mathsf{T} \mathbb{Q}_{\varphi_1} \mathbb{D}_{\rho_1} \mathbb{Q}_{\varphi_1}^\mathsf{T} \zeta \quad \forall \zeta \in \mathbb{R}^2 \, \forall \beta > 0.$$

Therefore, it is enough to prove (3.23) for all $\zeta \in B_1$. From (3.10), we have

$$w^{int}_{\bar{x},1}(\bar{x}+\zeta) = \frac{1}{2}\sum_{i,j=1}^{2} \frac{\partial^2 u(\bar{x})}{\partial x_i \partial x_j}\zeta_i\zeta_j = \tfrac{1}{2}\zeta^{\mathsf{T}}\mathbb{H}\zeta \quad \forall \zeta = (\zeta_1,\zeta_2), \tag{3.25}$$

where

$$\mathbb{H} := \left\{\frac{\partial^2 u(\bar{x})}{\partial x_i \partial x_j}\right\}^{2}_{i,j=1} \tag{3.26}$$

is the Hessian matrix. Since $\mathbb{H}$ is symmetric, we decompose it in the form

$$\mathbb{H} = \mathbb{Q}^{\mathsf{T}}_{\phi}\mathrm{diag}(\lambda_1,\lambda_2)\mathbb{Q}_{\phi} = \mathbb{Q}^{\mathsf{T}}_{\phi}\begin{pmatrix} \lambda_1 & 0 \\ 0 & \lambda_2 \end{pmatrix}\mathbb{Q}_{\phi}, \tag{3.27}$$

where λ_1, λ_2 are the real eigenvalues of $\mathbb{H}$ and $\mathbb{Q}_{\phi}$ is the rotation through angle $\phi \in [0,2\pi)$ given by (2.24). The eigenvalues are indexed such that $|\lambda_1| \geq |\lambda_2|$. The columns of $\mathbb{Q}_{\phi}$ are the eigenvectors corresponding to λ_1 and λ_2, hence the eigenvector corresponding to λ_1 is $(\cos\phi, \sin\phi)^{\mathsf{T}}$.

Moreover, we express $\zeta \in B_1$ as a function of the corresponding angle, i.e., $\zeta = \zeta(\alpha) = (\cos\alpha, \sin\alpha)^{\mathsf{T}}, \alpha \in [0,2\pi)$. By a direct computation, we have $\mathbb{Q}_{\phi}\zeta = (\cos(\phi+\alpha), \sin(\phi+\alpha))^{\mathsf{T}}$. Then, using (3.25)–(3.27), we obtain

$$w^{int}_{\bar{x},1}(\bar{x}+\zeta) = \tfrac{1}{2}\zeta^{\mathsf{T}}\mathbb{H}\zeta = \tfrac{1}{2}\zeta^{\mathsf{T}}\mathbb{Q}^{\mathsf{T}}_{\phi}\mathrm{diag}(\lambda_1,\lambda_2)\mathbb{Q}_{\phi}\zeta \tag{3.28}$$

$$= \tfrac{1}{2}(\cos(\phi+\alpha), \sin(\phi+\alpha))\begin{pmatrix} \lambda_1 & 0 \\ 0 & \lambda_2 \end{pmatrix}\begin{pmatrix} \cos(\phi+\alpha) \\ \sin(\phi+\alpha) \end{pmatrix}$$

$$= \tfrac{1}{2}\left(\lambda_1\cos^2(\phi+\alpha) + \lambda_2\sin^2(\phi+\alpha)\right).$$

Since $|\lambda_1| \geq \lambda_2$, then

$$\max_{\alpha\in[0,2\pi)}|w^{int}_{\bar{x},1}(\bar{x}+\zeta(\alpha))| = \frac{|\lambda_1|}{2}, \quad \arg\max_{\alpha\in(0,2\pi)}|w^{int}_{\bar{x},1}(\bar{x}+\zeta(\alpha))| = -\phi, \tag{3.29}$$

which together with (3.12) and (3.21a)–(3.21c) gives

$$A_1 = \max_{\xi\in B_1}|\mathrm{d}^2 u(\bar{x};\xi)| = \max_{\alpha\in[0,2\pi)}|w^{int}_{\bar{x},1}(\bar{x}+\zeta(\alpha))| = \frac{|\lambda_1|}{2}, \tag{3.30}$$

$$\xi_1 = \arg\max_{\xi\in B_1}|\mathrm{d}^2 u(\bar{x};\xi)| \quad\Rightarrow\quad \varphi_1 = \arg\max_{\alpha\in(0,2\pi)}|w^{int}_{\bar{x},1}(\bar{x}+\zeta(\alpha))| = -\phi.$$

Moreover, let $\phi^\perp := -\phi + \pi/2$ then the vector $(\cos\phi^\perp, \sin\phi^\perp)$ is perpendicular to the vector $(\cos(-\phi), \sin(-\phi))$. The relation (3.28) implies

$$|w_{\bar{x},1}^{\text{int}}(\bar{x} + \zeta(\phi^\perp))| = \frac{|\lambda_2|}{2}. \tag{3.31}$$

Using (3.12), (3.21d) and (3.21e), and (3.31), we obtain

$$A_1/\rho_1 = A_1^\perp = |d^2u(\bar{x}; \xi_1^\perp)| = |w_{\bar{x},1}^{\text{int}}(\bar{x} + \zeta(\phi^\perp))| = \frac{|\lambda_2|}{2}. \tag{3.32}$$

Due to identity $Q_{\varphi_1}^\mathsf{T}\zeta = (\cos(\alpha - \varphi_1), \sin(\alpha - \varphi_1))^\mathsf{T}$, $\zeta \in B_1$, the right-hand side of (3.23) can be expressed by

$$A_1\zeta^\mathsf{T}Q_{\varphi_1}D_{\rho_1}Q_{\varphi_1}^\mathsf{T}\zeta = (\cos(\alpha - \varphi_1), \sin(\alpha - \varphi_1)) \begin{pmatrix} A_1 & 0 \\ 0 & A_1/\rho_1 \end{pmatrix} \begin{pmatrix} \cos(\alpha - \varphi_1) \\ \sin(\alpha - \varphi_1) \end{pmatrix}$$

$$= A_1 \cos^2(\alpha - \varphi_1) + A_1/\rho_1 \sin^2(\alpha - \varphi_1). \tag{3.33}$$

Inserting the identities $A_1 = \frac{|\lambda_1|}{2}$, $A_1/\rho_1 = \frac{|\lambda_2|}{2}$, and $\varphi_1 = -\phi$ (following from (3.30) and (3.32)) in (3.33), we obtain the identity

$$A_1\zeta^\mathsf{T}Q_{\varphi_1}D_{\rho_1}Q_{\varphi_1}^\mathsf{T}\zeta = \frac{|\lambda_1|}{2}\cos^2(\alpha + \phi) + \frac{|\lambda_2|}{2}\sin^2(\alpha + \phi), \tag{3.34}$$

which together with (3.28) proves (3.23).   □

The relation (3.23) expresses the interpolation error function $w_{\bar{x},p}^{\text{int}}$ for $p = 1$ using the quantities A_1, ρ_1, and φ_1 denoting the size, the shape, and the orientation of the interpolation error function, respectively.

Remark 3.9 If the eigenvalues of the Hessian matrix λ_1 and λ_2 have the same sign, then (3.23) is valid with equality. Moreover, if these eigenvalues are positive (the Hessian matrix is positive definite), then (3.23) is valid in the form $w_{\bar{x},1}^{\text{int}}(\bar{x} + \zeta) = A_1\zeta^\mathsf{T}Q_{\varphi_1}D_{\rho_1}Q_{\varphi_1}^\mathsf{T}\zeta$. This case frequently appears in the literature.

3.2.5 Anisotropic Bound of Interpolation Error Function for $p > 1$

The aim is to derive an estimate of the interpolation error function $w_{\bar{x},p}^{\text{int}}$ like Lemma 3.8 also for $p > 1$. Motivated by [24, 26], we intent to derive the *anisotropic estimate* in the form

$$|w_{\bar{x},p}^{\text{int}}(x)| \leq A_w \left(\zeta^\mathsf{T}Q_{\varphi_w}D_{\rho_w}^{[p+1]}Q_{\varphi_w}^\mathsf{T}\zeta\right)^{\frac{p+1}{2}} \forall \zeta = x - \bar{x}, \ x \in \Omega, \tag{3.35}$$

where $\mathbb{Q}_{\varphi_w}$ is the rotation through angle φ_w (2.24) and $\mathbb{D}_{\rho_w}^{[q]}$ is the matrix given by

$$\mathbb{D}_{\rho}^{[q]} := \begin{pmatrix} 1 & 0 \\ 0 & \rho^{-2/q} \end{pmatrix}, \qquad \rho \geq 1,\ q \geq 1. \tag{3.36}$$

The symbol $[q]$ denotes a superscript and the value q corresponds to the power $(= -2/q)$ of the second diagonal entry of $\mathbb{D}_{\rho}^{[q]}$, $q \geq 1$. The values $A_w \geq 0$, $\rho_w \geq 1$, and $\varphi_w \in [0, 2\pi)$ in (3.35) represent the size, the shape, and the orientation of the interpolation error function $w_{\bar{x},p}^{\text{int}}$, which have to be defined. It is important that the right-hand side of (3.35) is a $(p+1)$-homogeneous function, in the sense of Definition 3.6, like $w_{\bar{x},p}^{\text{int}}$.

In the forthcoming Sect. 3.2.5.1, we define quantities A_w, ρ_w, and φ_w by a formal extension of (3.21). We demonstrate by an example that these values do not lead to an upper bound on the interpolation function (3.35). Therefore, in Sect. 3.2.5.2 we propose a modification of quantities A_w, ρ_w, and φ_w in such a way that the inequality (3.35) is valid.

3.2.5.1   First Non-guaranteed Estimate

First, we formally extend (3.21) for $p > 1$. Let $\bar{x} \in \Omega$, $u \in C^\infty$, and $p \in \mathbb{N}$ be given. We define the values $\tilde{A}_p \geq 0$, $\tilde{\xi}_p \in B_1$, $\tilde{\varphi}_p \in [0, 2\pi)$, $\tilde{A}_p^\perp \geq 0$, and $\tilde{\rho}_p \geq 1$ by

$$\tilde{A}_p := \max_{\xi \in B_1} |d^{p+1}u(\bar{x}; \xi)|, \tag{3.37a}$$

$$\tilde{\xi}_p := \arg\max_{\xi \in B_1} |d^{p+1}u(\bar{x}; \xi)|, \tag{3.37b}$$

$$\tilde{\varphi}_p \in [0, 2\pi) \text{ such that } (\cos\tilde{\varphi}_p, \sin\tilde{\varphi}_p)^\mathsf{T} = \tilde{\xi}_p, \tag{3.37c}$$

$$\tilde{A}_p^\perp := |d^{p+1}u(\bar{x}; \tilde{\xi}_p^\perp)|, \text{ where } \xi_p^\perp \in B_1,\ \tilde{\xi}_p^\perp \cdot \tilde{\xi}_p = 0, \tag{3.37d}$$

$$\tilde{\rho}_p := \frac{\tilde{A}_p}{\tilde{A}_p^\perp}. \tag{3.37e}$$

Hence, $\tilde{A}_p$ is the maximal value of the $(p+1)$th-order scaled directional derivative of u at $\bar{x}$, $\tilde{\xi}_p$ is the direction which maximizes this derivative, $\tilde{\varphi}_p$ is the angle corresponding to $\tilde{\xi}_p$, $\tilde{A}_p^\perp$ is the $(p+1)$th-order scaled directional derivative of u along the direction perpendicular to $\tilde{\xi}_p$, and $\tilde{\rho}_p$ is the ratio between $\tilde{A}_p$ and $\tilde{A}_p^\perp$.

However, numerical experiments show (see Example 3.10 below) that the estimate

$$\left| w_{\bar{x},p}^{\text{int}}(\bar{x} + \zeta) \right| \leq \tilde{A}_p \left(\zeta^\mathsf{T} \mathbb{Q}_{\tilde{\varphi}_p} \mathbb{D}_{\tilde{\rho}_p}^{[p+1]} \mathbb{Q}_{\tilde{\varphi}_p}^\mathsf{T} \zeta \right)^{\frac{p+1}{2}}, \qquad \forall \zeta \in \mathbb{R}^2 \tag{3.38}$$

is not valid for the values $\tilde{A}_p$, $\tilde{\varphi}_p$, and $\tilde{\rho}_p$ defined by (3.37a), (3.37c), and (3.37e), respectively.

In order to show that (3.38) is not valid in general, we introduce, similarly as in (3.17), a closed curve representing the right-hand side of (3.38). Namely, let $\bar{x} \in \Omega$, $p \in \mathbb{N}$, $A \geq 0$, $\rho \geq 1$, and $\varphi \in [0, 2\pi)$ be given, we define the curve

$$\vartheta^{p+1}(A, \rho, \varphi) := \left\{ x \in \mathbb{R}^2; \; x = \bar{x} + A\big(\zeta^\mathsf{T} \mathbb{Q}_\varphi \mathbb{D}_\rho^{[p+1]} \mathbb{Q}_\varphi^\mathsf{T} \zeta\big)^{\frac{p+1}{2}} \zeta \;\; \forall \zeta \in B_1 \right\},$$

(3.39)

where $\mathbb{Q}_\varphi$ and $\mathbb{D}_\rho^{[p+1]}$ are given by (2.24) and (3.36), respectively. Obviously, the distance between $\bar{x}$ and $x \in \vartheta^{p+1}$ is equal to

$$|x - \bar{x}| = A\big(\zeta^\mathsf{T} \mathbb{Q}_\varphi \mathbb{D}_\rho^{[p+1]} \mathbb{Q}_\varphi^\mathsf{T} \zeta\big)^{\frac{p+1}{2}} \quad \text{with } \zeta = (x - \bar{x})|x - \bar{x}|^{-1}.$$

Further,

$$A\big((x - \bar{x})^\mathsf{T} \mathbb{Q}_\varphi \mathbb{D}_\rho^{[p+1]} \mathbb{Q}_\varphi^\mathsf{T} (x - \bar{x})\big)^{\frac{p+1}{2}}, \quad x \in \mathbb{R}^2$$

is a $(p + 1)$-homogeneous function, cf. Definition 3.6. Therefore, for the symbol ϑ, we use the superscript $p+1$. Moreover, we denoted by $\bar{\vartheta}^{p+1}$ the domain bounded by ϑ^{p+1}.

Figure 3.4 shows the dependence of the graph of curve $\vartheta^{p+1}(A, \rho, \varphi)$ on its arguments. An increase of A enlarges the curve in all directions by the same multiplicative factor. On the other hand, decreasing ρ does not change the maximal distance from the origin, but the curve ϑ^{p+1} approaches a circle with radius A. A modification of φ only rotates the object and it is not shown in the figure.

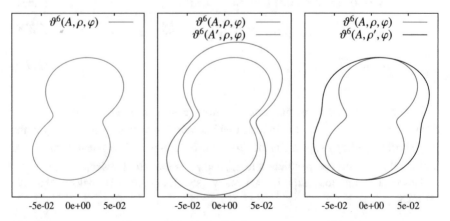

Fig. 3.4 The curves $\vartheta^{p+1}(A, \rho, \varphi)$ (red), $\vartheta^{p+1}(A', \rho, \varphi)$ (blue), $\vartheta^{p+1}(A, \rho', \varphi)$ (black) with $A' = 1.25A$, $\rho' = 0.5\rho$, and $p = 5$

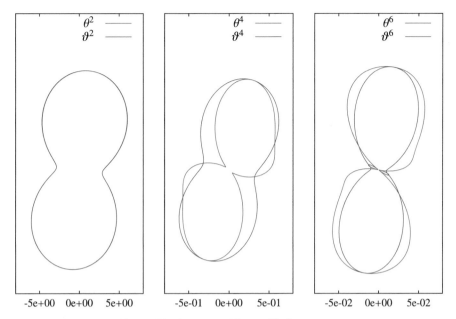

Fig. 3.5 The curves $\theta^{p+1} = \theta^{p+1}(w_{0,p}^{\mathrm{int}})$ and $\vartheta^{p+1} = \vartheta^{p+1}(\tilde{A}_p, \tilde{\rho}_p, \tilde{\varphi}_p)$, $p = 1, 3, 5$ for function u from Example 3.10

Using the curves θ^{p+1} and ϑ^{p+1} given by (3.17) and (3.39), respectively, we explain graphically the meaning of inequality (3.38). Since $\theta^{p+1} = \theta^{p+1}(w_{\bar{x},p}^{\mathrm{int}})$ corresponds to the left-hand side of (3.38) and $\vartheta^{p+1} = \vartheta^{p+1}(\tilde{A}_p, \tilde{\rho}_p, \tilde{\varphi}_p)$ corresponds to the right-hand side of (3.38), the inequality (3.38) is valid for any $\zeta \in \mathbb{R}^2$ if and only if $\bar{\theta}^{p+1} \subset \bar{\vartheta}^{p+1}$. The following example shows that this is not true in general and, therefore, estimate (3.38) does not hold.

Example 3.10 We consider again the function u from Example 3.7, relation (3.18), and $\bar{x} = (0, 0)$. Figure 3.5 shows the curves $\theta^{p+1}(w_{0,p}^{\mathrm{int}})$, $p = 1, 3, 5$ (cf. (3.17)) and the curves $\vartheta^{p+1} = \vartheta^{p+1}(\tilde{A}_p, \tilde{\rho}_p, \tilde{\varphi}_p)$, $p = 1, 3, 5$, where $\tilde{A}_p, \tilde{\varphi}_p$, and $\tilde{\rho}_p$ are given by (3.37a), (3.37c), and (3.37e), respectively. We observe that $\bar{\theta}^2 = \bar{\vartheta}^2$ which is in agreement with Lemma 3.8. However, $\bar{\theta}^p \not\subset \bar{\vartheta}^p$ for $p = 4, 6$ which violates inequality (3.38).

As mentioned above, in Sect. 3.2.5.2 we propose a modification of quantities A_w, ρ_w, and φ_w in such a way that the inequality (3.35) is valid. This modification requires an evaluation of the size of the area of $\bar{\vartheta}^{p+1}$.

Lemma 3.11 *Let $A > 0$, $\rho \geq 1$, and $\varphi \in [0, 2\pi)$ be given, let $\vartheta^{p+1} = \vartheta^{p+1}(A, \rho, \varphi)$ be the curve defined by (3.39) and $\bar{\vartheta}^{p+1}$ the domain bounded by*

ϑ^{p+1}. *Its area satisfies*

$$|\bar{\vartheta}^{p+1}(A, \rho, \phi)| = A^2 \sum_{i=0}^{p+1} c_p^{(i)} \rho^{-\frac{2i}{p+1}}, \tag{3.40}$$

where

$$c_p^{(i)} = \frac{1}{2}\binom{p+1}{i} \int_0^{2\pi} (\sin \phi)^{2i} (\cos \phi)^{2(p+1-i)} d\phi, \quad i = 0, \ldots, p+1. \tag{3.41}$$

Proof We transform $\bar{\vartheta}^{p+1}$ into the coordinate system $(\tilde{x}_1, \tilde{x}_2)$ with origin at $\bar{x}$ and the axis $\tilde{x}_1$ parallel with the direction $(\cos \varphi, \sin \varphi)^{\mathsf{T}}$. Then $\bar{\vartheta}^{p+1}$ is parametrized by

$$\begin{aligned}
\tilde{x}_1 &= A\, t \cos \phi \left(\cos^2 \phi + \rho^{-\frac{2}{p+1}} \sin^2 \phi \right)^{\frac{p+1}{2}}, \\
\tilde{x}_2 &= A\, t \sin \phi \left(\cos^2 \phi + \rho^{-\frac{2}{p+1}} \sin^2 \phi \right)^{\frac{p+1}{2}},
\end{aligned} \qquad t \in [0, 1],\ \phi \in [0, 2\pi). \tag{3.42}$$

Obviously, if $\rho = 1$, then $\bar{\vartheta}$ reduces to a circle with radius A. Using the integration by substitution and parametrization (3.42), we have

$$|\bar{\vartheta}^{p+1}(A, \rho, \phi)| = \int_{t=0}^1 \int_{\phi=0}^{2\pi} \det \frac{D(\tilde{x}_1, \tilde{x}_2)}{D(t, \phi)} d\phi\, dt, \tag{3.43}$$

where the integrand in (3.43) is the Jacobian of mapping (3.42). We write relations (3.42) in the form

$$\tilde{x}_1 = t \cos(\phi) P(\phi), \tag{3.44}$$

$$\tilde{x}_2 = t \sin(\phi) P(\phi), \quad \text{with } P(\phi) := A\left(\cos^2 \phi + \rho^{-\frac{2}{p+1}} \sin^2 \phi \right)^{\frac{p+1}{2}}.$$

Direct differentiation of (3.44) and evaluation of the determinant give

$$\begin{aligned}
\det \frac{D(\tilde{x}_1, \tilde{x}_2)}{D(t, \phi)} &= \det \begin{pmatrix} \cos(\phi) P(\phi) & -t \sin(\phi) P(\phi) + t \cos(\phi) P'(\phi) \\ \sin(\phi) P(\phi) & t \cos(\phi) P(\phi) + t \sin(\phi) P'(\phi) \end{pmatrix} \\
&= t \cos^2(\phi) P(\phi)^2 + t \cos(\phi) P(\phi) \sin(\phi) P'(\phi) \\
&\quad + t \sin^2(\phi) P(\phi)^2 - t \sin(\phi) P(\phi) \cos(\phi) P'(\phi) = t\, P(\phi)^2.
\end{aligned} \tag{3.45}$$

Table 3.1 Values of coefficients $c_p^{(i)}$, $i = 0, \ldots, p + 1$ from (3.41) for $p = 1, \ldots 7$

p	$c_p^{(0)}$	$c_p^{(1)}$	$c_p^{(2)}$	$c_p^{(3)}$	$c_p^{(4)}$	$c_p^{(5)}$	$c_p^{(6)}$	$c_p^{(7)}$	$c_p^{(8)}$
1	$\frac{3\pi}{8}$	$\frac{2\pi}{8}$	$\frac{3\pi}{8}$						
2	$\frac{5\pi}{16}$	$\frac{3\pi}{16}$	$\frac{3\pi}{16}$	$\frac{5\pi}{16}$					
3	$\frac{35\pi}{128}$	$\frac{20\pi}{128}$	$\frac{18\pi}{128}$	$\frac{20\pi}{128}$	$\frac{35\pi}{128}$				
4	$\frac{63\pi}{256}$	$\frac{35\pi}{256}$	$\frac{30\pi}{256}$	$\frac{30\pi}{256}$	$\frac{35\pi}{256}$	$\frac{63\pi}{256}$			
5	$\frac{231\pi}{1024}$	$\frac{126\pi}{1024}$	$\frac{105\pi}{1024}$	$\frac{100\pi}{1024}$	$\frac{105\pi}{1024}$	$\frac{126\pi}{1024}$	$\frac{231\pi}{1024}$		
6	$\frac{429\pi}{2048}$	$\frac{231\pi}{2048}$	$\frac{189\pi}{2048}$	$\frac{175\pi}{2048}$	$\frac{175\pi}{2048}$	$\frac{189\pi}{2048}$	$\frac{231\pi}{2048}$	$\frac{429\pi}{2048}$	
7	$\frac{6435\pi}{32768}$	$\frac{3432\pi}{32768}$	$\frac{2772\pi}{32768}$	$\frac{2520\pi}{32768}$	$\frac{2450\pi}{32768}$	$\frac{2520\pi}{32768}$	$\frac{2772\pi}{32768}$	$\frac{3432\pi}{32768}$	$\frac{6435\pi}{32768}$

Hence, (3.43)–(3.45), Fubini's theorem and formula $(a + b)^n = \sum_{i=0}^{n} \binom{n}{i} a^i b^{n-i}$, $a, b \in \mathbb{R}$, $n \in \mathbb{N}$, imply

$$|\bar{\vartheta}^{p+1}(A, \rho, \phi)| = \int_{t=0}^{1} \int_{\phi=0}^{2\pi} t \, A^2 (\cos^2 \phi + \rho^{-\frac{2}{p+1}} \sin^2 \phi)^{p+1} d\phi \, dt \qquad (3.46)$$

$$= \frac{1}{2} A^2 \int_{0}^{2\pi} (\cos^2 \phi + \rho^{-\frac{2}{p+1}} \sin^2 \phi)^{p+1} d\phi$$

$$= \frac{1}{2} A^2 \int_{0}^{2\pi} \sum_{i=0}^{p+1} \binom{p+1}{i} (\cos \phi)^{2(p+1-i)} \rho^{-\frac{2i}{p+1}} (\sin \phi)^{2i} d\phi$$

$$= A^2 \sum_{i=0}^{p+1} c_p^{(i)} \rho^{-\frac{2i}{p+1}},$$

where $c_p^{(i)}$, $i = 0, \ldots, p + 1$ are given by (3.41). Therefore, (3.40) is proved. □

The coefficients $c_p^{(i)}$, $i = 0, \ldots, p + 1$ introduced in (3.41) can be evaluated analytically. Examples are given in Table 3.1 for $p = 1, \ldots, 7$.

Finally, we note how to find values $\tilde{A}_p$ and $\tilde{\xi}_p$ given by (3.37a) and (3.37b). In principle the evaluation of the maximum can be carried out analytically. The directional derivative $d^{p+1} u(\bar{x}; \xi)$, $\xi \in B_1$ given by (3.11) is a trigonometric-polynomial function of a real variable $\psi \in [0, 2\pi]$, cf. (3.14). Then it has several local extrema (depending on p) that can be compared. However, it requires finding the roots of a polynomial, which has to be carried out iteratively for large p.

A more direct way is to evaluate the directional derivative $|d^{p+1} u(\bar{x}; \xi)|$ only for a finite set of $\xi \in B_1$, i.e., $\xi \in \Xi := \{\cos(i\pi/n), \sin(i\pi/n), \ i = 1, \ldots, n\}$, where n is a suitably chosen integer. Here we have employed the symmetry of the directional derivative such that the derivative for ψ and $\psi + \pi$ is the same. For example, the choice $N = 200$ gives (in most cases) an accuracy with respect to the correct direction of less than 1%. This technique is sketched in Algorithm 3.1. For a more sophisticated technique, we refer to [33].

Algorithm 3.1: Evaluation of values $\tilde{A}_p$ and $\tilde{\xi}_p$ given by (3.37a) and (3.37b)

---

inputs: $p \in \mathbb{N}, u \in C^\infty, \bar{x} \in \Omega$
set $n \in \mathbb{N}$ (= the number of tested directions)
set $\tilde{A}_p := 0, \tilde{\xi}_p = 0$
for $i = 0, 1, \ldots, n$ **do**
$\quad \xi := (\cos(i\pi/n), \sin(i\pi/n))$
$\quad A := d^{p+1}u(\bar{x}, \xi)$
$\quad$ **if** $A > \tilde{A}_p$ **then**
$\quad\quad \tilde{A}_p := A, \tilde{\xi}_p := \xi$
$\quad$ **end if**
end for

3.2.5.2 Second Guaranteed Estimate

Example 3.10 demonstrates that estimate (3.38) is not valid for $p > 1$ since $\bar{\theta}^p \not\subset \bar{\vartheta}^p$. We propose a modification of the definitions of $\tilde{A}_p$, $\tilde{\varphi}_p$, and $\tilde{\rho}_p$ such that resulting estimate (3.35) is true. For example, keeping $\tilde{A}_p$ and setting $\tilde{\rho}_p = 1$, the corresponding curve $\vartheta^{p+1}(\tilde{A}_p, 1, \tilde{\varphi}_p)$ is a minimal circle which circumscribes θ^{p+1}. Then estimate (3.35) is obviously valid but we lose the anisotropy of $w_{\bar{x},p}^{\text{int}}$. Therefore, the idea is to define $\tilde{A}_p$ and $\tilde{\rho}_p$ such that $\bar{\theta}^p \subset \bar{\vartheta}^p$ and $\bar{\vartheta}^p$ is as small as possible.

In principle it would be possible to modify all three parameters $\tilde{A}_p$, $\tilde{\varphi}_p$, and $\tilde{\rho}_p$. However, it would require an optimization over a three-dimensional set of parameters. In order to simplify the task we fix the direction $\tilde{\varphi}_p$ using (3.37b) and (3.37c). Further, we define the set of pairs (A, ρ) which fulfill estimate (3.35).

Definition 3.12 Let $\bar{x} \in \Omega$, $u \in C^\infty$, and $p \in \mathbb{N}$ be given and $w_{\bar{x},p}^{\text{int}}$ be the corresponding interpolation error function (3.10). We define the *set of admissible anisotropies* of $w_{\bar{x},p}^{\text{int}}$ by

$$R := \left\{ (A, \rho); \ A \geq 0, \ \rho \geq 1 : \left| w_{\bar{x},p}^{\text{int}}(\bar{x} + \zeta) \right| \leq A \left(\zeta^\mathsf{T} Q_{\varphi_w} D_\rho^{[p+1]} Q_{\varphi_w}^\mathsf{T} \zeta \right)^{\frac{p+1}{2}} \forall \zeta \in \mathbb{R}^2 \right\}, \tag{3.47}$$

where $\varphi_w = \tilde{\varphi}_p$ given by (3.37b) and (3.37c).

Lemma 3.13 *The set R given by (3.47) is nonempty and simply connected. Moreover,*

(a) *For given A, the set R is bounded with respect to ρ.*
(b) *For given ρ, the set R is bounded from below with respect to A.*

Figure 3.4 shows examples of the set R.

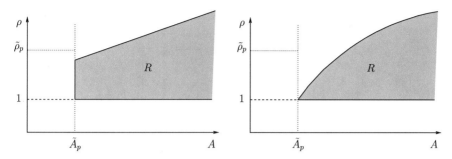

Fig. 3.6 Examples of the set R given by (3.47). The values $\tilde{A}_p$ and $\tilde{\rho}_p$ are given by (3.37a) and (3.37e), respectively

Proof Both sides of the inequality in (3.47) are $(p+1)$-homogeneous functions.

(i)   The set $R \subset \mathbb{R}^2$ is nonempty since $(\tilde{A}_p, 1) \in R$, where $\tilde{A}_p$ is given by (3.37a). This follows from (3.11), (3.12), (3.37a), and the fact that $\zeta^{\mathsf{T}} \mathbb{Q}_{\varphi_w} \mathbb{D}_\rho^{[p+1]} \mathbb{Q}_{\varphi_w}^{\mathsf{T}} \zeta = |\zeta|^2$ for $\rho = 1$.

(ii) Further, $(A, \rho) \notin R$ for any $A < \tilde{A}_p$ since the inequality in (3.47) does not hold for $\zeta = (\cos \tilde{\varphi}_p, \sin \tilde{\varphi}_p)$, hence R is bounded from below w.r.t. A.

(iii) Let $A \geq \tilde{A}_p$ be arbitrary, then $(A, 1) \in R$ due to item (i).

(iv) Let $A \geq \tilde{A}_p$ be fixed, taking into account the properties of the term on the right-hand side of the inequality in (3.47) explained by Fig. 3.4, we deduce that if we increase ρ starting from the value $\rho = 1$ we achieve a limit value $\bar{\rho}$ such that $(A, \rho) \notin R$ for $\rho > \bar{\rho}$, i.e., R is bounded w.r.t. ρ.

This relations imply that the set R looks like that in Fig. 3.6 which implies that R is a simply connected domain. $\square$

Obviously, if we choose any pair (A_w, ρ_w) from R and φ_w is given by (3.37c), then the estimate (3.35) is valid and the corresponding set $\vartheta^{p+1}(A_w, \rho_w, \varphi_w)$, given by (3.39), satisfies $\theta^p \subset \vartheta^{p+1}(A_w, \rho_w, \varphi_w)$. On the other hand, in order not to "overestimate" $w_{\bar{x},p}^{\text{int}}$, it is desirable to choose the pair (A_w, ρ_w) from R such that the corresponding set $\bar{\vartheta}^{p+1}(A_w, \rho_w, \varphi_w)$ has the minimal possible area. Therefore, we define the values A_w, ρ_w, and φ_w (cf. (3.37c)) by

$$\varphi_w \in [0, 2\pi) \text{ such that } (\cos \varphi_w, \sin \varphi_w)^{\mathsf{T}} = \arg \max_{\xi \in B_1} |\mathrm{d}^{p+1} u(\bar{x}; \xi)| \qquad (3.48)$$

$$(A_w, \rho_w) := \arg \min_{A, \rho \in R} |\bar{\vartheta}^{p+1}(A, \rho, \varphi_w)|,$$

where R and ϑ^{p+1} are given by (3.47) and (3.39), respectively, and $|\bar{\vartheta}^{p+1}(A, \rho, \varphi_w)|$ denotes the area of $\bar{\vartheta}^{p+1}(A, \rho, \varphi_w)$. This area can be expressed analytically using Lemma 3.11. The existence of A_w and ρ_w follows from the fact that R is simply

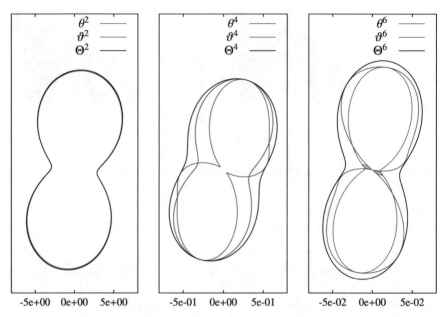

Fig. 3.7 The curves $\theta^{p+1} = \theta^{p+1}(w_{0,p}^{int})$, $\vartheta^{p+1} = \vartheta^{p+1}(\tilde{A}_p, \tilde{\rho}_p, \tilde{\varphi}_p)$, and $\Theta^{p+1} = \vartheta^{p+1}(A_w, \rho_w, \varphi_w)$ for Example 3.14

connected, bounded from below w.r.t. A and bounded w.r.t. ρ (cf. Lemma 3.13), and the fact that $|\bar{\vartheta}^{p+1}(A, \rho, \varphi_w)| \to \infty$ for $A \to \infty$.

Using (3.39) we define the curve Θ^p corresponding to the triplet $\{A_w, \rho_w, \varphi_w\}$ given by (3.48), i.e.,

$$\Theta^p := \vartheta^{p+1}(A_w, \rho_w, \varphi_w) \tag{3.49}$$

and the domain bounded by Θ^p is denoted by $\bar{\Theta}^p$. The following example demonstrates the validity of estimate (3.35).

Example 3.14 We consider again the function u from Example 3.7, relation (3.18) and $\bar{x} = (0, 0)$. Figure 3.7 shows the curve θ^p (given by (3.17)), the curve $\vartheta^p = \vartheta^{p+1}(\tilde{A}_p, \tilde{\rho}_p, \varphi_w)$ with the parameters $\tilde{A}_p$, $\tilde{\varphi}_p$, and $\tilde{\rho}_p$ defined by (3.37a), (3.37c), and (3.37e), respectively, and the curve $\Theta^p := \vartheta^{p+1}(A_w, \rho_w, \varphi_w)$ given by (3.49) with the optimal parameters A_w, φ_w, and ρ_w defined by (3.48) for $p = 1, 3, 5$. We observe that $\bar{\theta}^{p+1} \subset \bar{\Theta}^{p+1}$ for each p. Moreover, the estimate (3.35) does not overestimate $w_{\bar{x},p}^{int}$ since there are points that lie on both θ^{p+1} and Θ^{p+1}.

Remark 3.15 The values A_w, ρ_w, and φ_w given by (3.48) can be evaluated approximately by the following iterative algorithm.

1. We find φ_w by seeking the maximum of $|d^{p+1}u(\bar{x}; \xi)|$ over the set

$$\xi \in \Xi := \{\cos(i\pi/n), \sin(i\pi/n), \ i = 1, \ldots, n\},$$

where n is a suitable chosen integer, cf. Sect. 3.2.5.1.
2. For $l = 0, 1, \ldots$

   a. We set $A_p^{(l)} := \tilde{A}_p \gamma^l$, $l = 0, 1, \ldots$, where $\tilde{A}_p$ is given by (3.37a) and $\gamma > 1$ is a chosen constant.
   b. We find the maximal value $\rho_p^{(l)} \geq 1$ such that

   $$\theta^p \subset \vartheta^{p+1}(A_p^{(l)}, \rho_p^{(l)}, \varphi_w) \iff (3.35) \text{ is valid.}$$

   The value $\rho_p^{(l)}$ always exists since $\theta^p \subset \vartheta^{p+1}(A, 1, \varphi_w)$ for any $A \geq \tilde{A}_p$, cf. Lemma 3.13. Again, it is sufficient to test (3.35) for $\zeta \in \Xi$.
   c. Using Lemma 3.11, we evaluate area of $|\bar{\Theta}_l| := |\bar{\vartheta}^{p+1}(A_p^{(l)}, \rho_p^{(l)}, \varphi_w)|$.
   d. If $|\bar{\Theta}_l| > |\bar{\Theta}_{l-1}|$ we exit the loop with $A_w := A_p^{(l-1)}$ and $\rho_w := \rho_p^{(l-1)}$.

The relation (3.40) implies that the area of ϑ depends monotonously on A and ρ and, therefore, the iterative process always terminates.

Remark 3.16 We note that there exist also other choices of the set R instead of (3.47). The set R can consist of one, two, or three parameters from the set $\{A, \rho, \varphi\}$. Then the definition (3.48) of the values A_w, ρ_w, and φ_w has to be modified in the appropriate way. Obviously, a higher number of parameters leads to a sharper estimate. However, the most general case, where all three parameters are considered, appears too complicated for an accurate and fast evaluation, similar to that explained above for the case of two parameters A and ρ. Nevertheless, from the symmetry of $w_{\bar{x},p}^{int}(x)$ with respect to the direction of $x - \bar{x}$, we expect that an enrichment of the set R in (3.47) by the third parameter φ does not improve essentially the sharpness of estimate (3.35).

A direct consequence of (3.47) and (3.48) is the following fundamental *anisotropic estimate* of the interpolation error function.

Theorem 3.17 *Let $\bar{x} \in \Omega$, $u \in C^\infty$, and $p \in \mathbb{N}$ be given and $w_{\bar{x},p}^{int}$ be the corresponding interpolation error function (3.10). Let $A_w \geq 0$, $\rho_w \geq 1$, and $\varphi_w \in [0, 2\pi)$ be defined by (3.48). Then*

$$\left| w_{\bar{x},p}^{int}(x) \right| \leq A_w \left((x - \bar{x})^\mathsf{T} Q_{\varphi_w} \mathbb{D}_{\rho_w}^{[p+1]} Q_{\varphi_w}^\mathsf{T} (x - \bar{x}) \right)^{\frac{p+1}{2}} \quad \forall x \in \Omega, \qquad (3.50)$$

where Q_{φ_w} is the rotation (2.24), and $\mathbb{D}_{\rho_w}^{[p+1]}$ is defined by (3.36). Moreover, the estimate (3.50) is sharp in the sense that there exists $x \in \Omega$ such that

$$\left| w_{\bar{x},p}^{int}(x) \right| = A_w \left((x - \bar{x})^\mathsf{T} Q_{\varphi_w} \mathbb{D}_{\rho_w}^{[p+1]} Q_{\varphi_w}^\mathsf{T} (x - \bar{x}) \right)^{\frac{p+1}{2}}.$$

Finally, employing Theorem 3.17, we define the anisotropic bound of the interpolation error function $w_{\bar{x},p}^{\text{int}}$.

Definition 3.18 Let $\bar{x} \in \Omega$, $u \in C^\infty$, and $p \in \mathbb{N}$ be given. The triple $\{A_w, \rho_w, \varphi_w\}$, $A_w \geq 0$, $\varphi_w \geq 1$, $\varphi_w \in [0, \pi)$, defined by (3.48) is called the *anisotropic bound of the interpolation error function* $w_{\bar{x},p}^{\text{int}}$ at $\bar{x}$.

The importance of estimate (3.50) is the following: whereas the interpolation error function (as well as θ^p) depends on all partial derivatives of order $p + 1$, the right-hand side of (3.50) (as well as $\Theta^p = \vartheta^{p+1}(A_w, \rho_w, \varphi_w)$) depends only on three parameters A_w, φ_w, and ρ_w for arbitrary $p \in \mathbb{N}$. Since the geometry of a triangle (cf. Definition 3.1) is given also by three parameters we can directly derive a relation for the optimal geometry of mesh elements.

Furthermore, from a practical point of view, it is important that the anisotropic bound parameters $\{A_w, \varphi_w, \rho_w\}$ can be evaluated by employing the values of partial derivatives of u at $\bar{x}$ and (3.48). Moreover, estimate (3.50) is independent of a generic constant.

3.3 Interpolation Error Estimates Including the Triangle Geometry

The goal of this section is to derive estimates of the interpolation error function $w_{\bar{x},p}^{\text{int}}$ located at the barycenter a triangle $K \in \mathcal{T}_h$ in several norms. These estimates have to take into account the geometry of K. For simplicity, we consider functions from the space $C^\infty(K)$.

Problem 3.19 We assume that

- K is a triangle with the geometry $\{\mu_K, \sigma_K, \phi_K\}$, cf. Definition 3.1.
- x_K is the barycenter of K.
- $u \in C^\infty(K)$ is a function.
- $p \geq 1$ is an integer.
- $w_{x_K,p}^{\text{int}}$ is the corresponding interpolation error function located at x_K cf. (3.10).

The aim is to derive a bound

$$\left\| w_{x_K,p}^{\text{int}} \right\|_{X(K)} \leq \mathcal{G}, \tag{3.51}$$

where $\|\cdot\|_{X(K)}$ denotes a Sobolev norm over K or ∂K and the bound $\mathcal{G}$ satisfies the following requirements:

- $\mathcal{G}$ is computable without unknown constants.
- $\mathcal{G}$ depends on $\{\mu_K, \sigma_K, \phi_K\}$.

In the following, we derive these estimates employing Theorem 3.17 and a direct evaluation of the corresponding norms. The integral over K is transformed to an integral over the reference triangle $\hat{K}$ (cf. (2.19)) and then we estimate this integral by an integration of the same integrand over the corresponding circumscribed circle $\hat{\mathscr{E}}$, cf. (2.20).

We present several relations which are employed in the subsequent error estimates. Let $\mathbb{S}_K = \text{diag}(\sigma_K, \sigma_K^{-1})$ and $\mathbb{D}_{\rho_w}^{[s_2]} = \text{diag}(1, \rho_w^{-2/s_2})$ be the diagonal matrices given by (3.4) and (3.36), respectively, for $\sigma_K \geq 1$, $\rho_w \geq 1$, and $s_2 \geq 1$. Further, let $\mathbb{Q}_{\varphi_w}$ and $\mathbb{Q}_{\phi_K}$ be the rotation matrices through angles φ_w and ϕ_K, respectively, cf. (2.24). We define the symmetric 2×2 matrix

$$\mathbb{G} := \mathbb{S}_K \mathbb{Q}_{\phi_K}^{\mathsf{T}} \mathbb{Q}_{\varphi_w} \mathbb{D}_{\rho_w}^{[s_2]} \mathbb{Q}_{\varphi_w}^{\mathsf{T}} \mathbb{Q}_{\phi_K} \mathbb{S}_K \qquad (3.52)$$

and denote by $\mathbb{G}_{ij}$, $i, j = 1, 2$ its entries. Direct evaluation of $\mathbb{G}$ gives

$$\mathbb{G} = \begin{pmatrix} \sigma_K^2 (\cos^2 \tau_K + \rho_w^{-2/s_2} \sin^2 \tau_K) & -\sin \tau_K \cos \tau_K (1 - \rho_w^{-2/s_2}) \\ -\sin \tau_K \cos \tau_K (1 - \rho_w^{-2/s_2}) & \sigma_K^{-2} (\sin^2 \tau_K + \rho_w^{-2/s_2} \cos^2 \tau_K) \end{pmatrix}, \qquad (3.53)$$

where we used identity $\mathbb{Q}_{\varphi_w}^{\mathsf{T}} \mathbb{Q}_{\phi_K} = \mathbb{Q}_{\phi_K - \varphi_w} =: \mathbb{Q}_{\tau_K}$ and $\tau_K := \phi_K - \varphi_w$. Further, we evaluate the determinant of $\mathbb{G}$ by

$$\det \mathbb{G} = (\cos^2 \tau_K + \rho_w^{-2/s_2} \sin^2 \tau_K)(\sin^2 \tau_K + \rho_w^{-2/s_2} \cos^2 \tau_K) \qquad (3.54)$$

$$- \left(\sin \tau_K \cos \tau_K (1 - \rho_w^{-2/s_2}) \right)^2 = \rho_w^{-2/s_2}.$$

Obviously, $\mathbb{G}$ is symmetric and positively definite.

Let $\hat{x} = (r \cos \psi, r \sin \psi)^{\mathsf{T}}$, where (r, ψ) are the polar coordinates. Then the multiplication of $\mathbb{G}$ by $\hat{x}$ from the left and from the right fulfills

$$\hat{x}^{\mathsf{T}} \mathbb{G} \hat{x} = r^2 g(\psi) \geq 0 \qquad \forall r \geq 0 \; \forall \psi \in [0, 2\pi], \qquad (3.55)$$

where the scalar function $g(\psi) = g(s_2, \rho_w, \varphi_w; \sigma_K, \phi_K; \psi)$ is given by

$$g(\psi) := \mathbb{G}_{11} \cos^2 \psi + 2\mathbb{G}_{12} \sin \psi \cos \psi + \mathbb{G}_{22} \sin^2 \psi \qquad (3.56)$$

$$= \sigma_K^2 \left(\cos^2(\phi_K - \varphi_w) + \rho_w^{-2/s_2} \sin^2(\phi_K - \varphi_w) \right) \cos^2 \psi$$

$$- 2 \sin(\phi_K - \varphi_w) \cos(\phi_K - \varphi_w) \left(1 - \rho_w^{-2/s_2} \right) \sin \psi \cos \psi$$

$$+ \sigma_K^{-2} \left(\sin^2(\phi_K - \varphi_w) + \rho_w^{-2/s_2} \cos^2(\phi_K - \varphi_w) \right) \sin^2 \psi, \quad \psi \in [0, 2\pi].$$

Finally, we define the scalar functions

$$G(s_1, s_2, \rho_w, \varphi_w; \sigma_K, \phi_K) := \int_0^{2\pi} \left(g(s_2, \rho_w, \varphi_w; \sigma_K, \phi_K; \psi) \right)^{s_1} d\psi, \qquad (3.57)$$

$$\overline{G}(s_1, s_2, \rho_w, \varphi_w; \sigma_K, \phi_K) := \max_{\psi \in [0, 2\pi]} \left(g(s_2, \rho_w, \varphi_w; \sigma_K, \phi_K; \psi) \right)^{s_1}, \qquad (3.58)$$

$$\text{for } s_1, s_2 \geq 1, \ \rho_w, \sigma_K \geq 1, \ \varphi_w, \phi_K \in [0, 2\pi],$$

where g is given in (3.56). The variables of functions G and $\overline{G}$ are given by the matrices $\mathbb{S}_K$, $\mathbb{D}_{\rho_w}^{[s_2]}$, $\mathbb{Q}_{\varphi_w}$, and $\mathbb{Q}_{\phi_K}$ appearing in (3.52). The first variable of G and $\overline{G}$ is the power s_1 of function g and the second variable defines the power in the last entry of matrix $\mathbb{D}_{\rho_w}^{[s_2]}$. The third and fourth variables of G and $\overline{G}$ correspond to the parameters of the anisotropic bound (cf. Definition 3.18) and the last two variables correspond to the shape and orientation of the triangle. Later, we employ the following properties of G and g.

Lemma 3.20 *The functions G and g are π-periodic with respect to their argument ϕ_K. Moreover, G and g tend to infinity for $\sigma_K \to \infty$.*

Proof Both assertions easily follow from the definition of G and g and relations (3.53) and (3.56). □

We note that G can be evaluated numerically by a suitable quadrature rule. On the other hand the value of $\overline{G}$ can be evaluated analytically. Particularly, if $\mathbb{G}_{11} = \mathbb{G}_{22}$, then the maximum is achieved for $\psi = \pi/4$ or $\psi = 3\pi/4$ (depending of the sign of $\mathbb{G}_{12}$), otherwise the maximum is attained for $\psi = \frac{1}{2} \arctan(4/(\mathbb{G}_{11} - \mathbb{G}_{22}))$.

3.3.1 Estimate in the $L^q(K)$-Norm, $1 \leq q < \infty$

First, we derive the estimate in the $L^q(K)$-norm, $1 \leq q < \infty$. A similar technique is used later for estimates in other norms. We recall that the interpolation error function $w_{\bar{x},p}^{int}$, located at any node $\bar{x}$, is a polynomial function and, therefore, can be extended to $\mathbb{R}^2$.

Lemma 3.21 *Let $q \in [1, \infty)$, K be a triangle with the geometry $\{\mu_K, \sigma_K, \phi_K\}$ (cf. Definition 3.1), $u \in C^\infty(K)$, $p \in \mathbb{N}$, $w_{x_K,p}^{int}$ be the corresponding interpolation error function located at the barycenter x_K defined by (3.10) and let $\{A_w, \rho_w, \varphi_w\}$ be the anisotropic bound of $w_{x_K,p}^{int}$ given by Definition 3.18. Then*

$$\left\| w_{x_K,p}^{int} \right\|_{L^q(K)}^q \leq \frac{1}{q(p+1)+2} A_w^q \mu_K^{q(p+1)+2} \, G\left(\frac{q(p+1)}{2}, p+1, \rho_w, \varphi_w; \sigma_K, \phi_K \right),$$

$$(3.59)$$

where G is given by (3.57).

Proof Let $\hat{K}$ be the reference triangle and F_K be the affine function (3.1) mapping $\hat{K}$ onto K with the corresponding Jacobian matrix $\mathbb{F}_K$. Moreover, for a function $f(x) : K \to \mathbb{R}$, we define the corresponding "reference" function $\hat{f} : \hat{K} \to \mathbb{R}$ by

$$\hat{f}(\hat{x}) := f(F_K\hat{x}) = f(x), \qquad \hat{x} \in \hat{K}. \tag{3.60}$$

Using the substitution theorem, relation (3.5), we have

$$\int_K f(x) \, dx = \int_{\hat{K}} \hat{f}(\hat{x}) |\det \mathbb{F}_K| \, d\hat{x} = \mu_K^2 \int_{\hat{K}} \hat{f}(\hat{x}) \, d\hat{x}. \tag{3.61}$$

Furthermore, (3.1)–(3.4) imply

$$x = F_K\hat{x} + x_K = \mathbb{Q}_{\phi_K} \mathbb{L}_K \mathbb{Q}_{\psi_K}^{\mathsf{T}} \hat{x} + x_K = \mu_K \mathbb{Q}_{\phi_K} \mathbb{S}_K \mathbb{Q}_{\psi_K}^{\mathsf{T}} \hat{x} + x_K, \tag{3.62}$$

and consequently

$$x - x_K = \mu_K \mathbb{Q}_{\phi_K} \mathbb{S}_K \mathbb{Q}_{\psi_K}^{\mathsf{T}} \hat{x} \quad \text{and} \quad (x - x_K)^{\mathsf{T}} = \mu_K \hat{x}^{\mathsf{T}} \mathbb{Q}_{\psi_K} \mathbb{S}_K \mathbb{Q}_{\phi_K}^{\mathsf{T}}. \tag{3.63}$$

Using the definition of the $L^q(K)$-norm, the fundamental anisotropic estimate (3.50), relations (3.61) and (3.63), we obtain

$$\left\| w_{x_K,p}^{\text{int}} \right\|_{L^q(K)}^q = \int_K \left| w_{x_K,p}^{\text{int}}(x) \right|^q \, dx \tag{3.64}$$

$$\leq \int_K A_w^q \left((x - x_K)^{\mathsf{T}} \mathbb{Q}_{\varphi_w} \mathbb{D}_{\rho_w}^{[p+1]} \mathbb{Q}_{\varphi_w}^{\mathsf{T}} (x - x_K) \right)^{\frac{q(p+1)}{2}} dx$$

$$= A_w^q \mu_K^{q(p+1)+2} \int_{\hat{K}} \left(\hat{x}^{\mathsf{T}} \mathbb{Q}_{\psi_K} \mathbb{G} \mathbb{Q}_{\psi_K}^{\mathsf{T}} \hat{x} \right)^{\frac{q(p+1)}{2}} d\hat{x},$$

where matrix $\mathbb{G}$ is given by (cf. (3.52))

$$\mathbb{G} := \mathbb{S}_K \mathbb{Q}_{\phi_K}^{\mathsf{T}} \mathbb{Q}_{\varphi_w} \mathbb{D}_{\rho_w}^{[p+1]} \mathbb{Q}_{\varphi_w}^{\mathsf{T}} \mathbb{Q}_{\phi_K} \mathbb{S}_K. \tag{3.65}$$

Let $\hat{\mathscr{E}} = \{\hat{x} \in \mathbb{R}^2 : |\hat{x}| \leq 1\}$ be the unit ball, cf. (2.20), which circumscribes the reference triangle $\hat{K}$, see Fig. 2.3a. Since the integrand in the last integral (3.64) is non-negative, we replace there the domain of integration $\hat{K}$ by $\hat{\mathscr{E}}$ and obtain

$$\left\| w_{x_K,p}^{\text{int}} \right\|_{L^q(K)}^q \leq A_w^q \mu_K^{q(p+1)+2} \int_{\hat{\mathscr{E}}} \left(\hat{x}^{\mathsf{T}} \mathbb{G} \hat{x} \right)^{\frac{q(p+1)}{2}} d\hat{x}, \tag{3.66}$$

where we have used the fact that $\mathbb{Q}_{\psi_K}$ is a rotation matrix and thus the transformation $\hat{x} \to \mathbb{Q}_{\psi_K}^{\mathsf{T}} \hat{x}$ maps $\hat{\mathscr{E}}$ onto itself. We evaluate the integral in (3.66) using the polar coordinates (r, ψ). Thanks to (3.55) we obtain

$$\left\| w^{\text{int}}_{x_K, p} \right\|^q_{L^q(K)} \leq A^q_w \mu^{q(p+1)+2}_K \int_0^1 r^{q(p+1)+1} \left(\int_0^{2\pi} g(\psi)^{\frac{q(p+1)}{2}} \, d\psi \right) dr.$$

$$(3.67)$$

Since $g(\psi)$ is independent of r, we change the order of integration and the equality

$$\int_0^1 r^{q(p+1)+1} \, dr = (q(p+1)+2)^{-1},$$

$$(3.68)$$

together with the definition of G by (3.57) yields (3.59).   □

Later, we employ the special case corresponding to $q = 2$.

Corollary 3.22 *Let K be a triangle with the geometry $\{\mu_K, \sigma_K, \phi_K\}$ (cf. Definition 3.1), $u \in C^\infty(K)$, $p \in \mathbb{N}$, $w^{\text{int}}_{x_K, p}$ be the corresponding interpolation error function located at the barycenter x_K defined by (3.10) and let $\{A_w, \rho_w, \varphi_w\}$ be the anisotropic bound of $w^{\text{int}}_{x_K, p}$ given by Definition 3.18. Then*

$$\left\| w^{\text{int}}_{x_K, p} \right\|^2_{L^2(K)} \leq \tfrac{1}{2p+4} A^2_w \mu^{2p+4}_K G(p+1, p+1, \rho_w, \varphi_w; \sigma_K, \phi_K),$$

$$(3.69)$$

where G is given by (3.57).

3.3.2   Estimate in the $L^\infty(K)$-Norm

Using a similar technique as in Sect. 3.3.1, we derive the estimate in the L^∞-norm.

Lemma 3.23 *Let K be a triangle with the geometry $\{\mu_K, \sigma_K, \phi_K\}$ (cf. Definition 3.1), $u \in C^\infty(K)$, $p \in \mathbb{N}$, $w^{\text{int}}_{x_K, p}$ be the corresponding interpolation error function located at the barycenter x_K defined by (3.10) and let $\{A_w, \rho_w, \varphi_w\}$ be the anisotropic bound of $w^{\text{int}}_{x_K, p}$ given by Definition 3.18. Then*

$$\left\| w^{\text{int}}_{x_K, p} \right\|_{L^\infty(K)} \leq A_w \mu^{p+1}_K \overline{G} \left(\tfrac{p+1}{2}, p+1, \rho_w, \varphi_w; \sigma_K, \phi_K \right),$$

$$(3.70)$$

where $\overline{G}$ is given by (3.58).

Proof Similar to the proof of Lemma 3.21, we employ estimate (3.50) and relations (3.60), (3.63) and obtain

$$\left\|w_{x_K,p}^{int}\right\|_{L^\infty(K)} = \max_{x\in K}|w_{x_K,p}^{int}(x)| \tag{3.71}$$

$$\leq A_w \max_{x\in K}\left((x-x_K)^\mathsf{T}Q_{\varphi_w}D_{\rho_w}^{[p+1]}Q_{\varphi_w}^\mathsf{T}(x-x_K)\right)^{\frac{p+1}{2}}$$

$$= A_w\mu_K^{p+1}\max_{\hat{x}\in\hat{K}}\left(\hat{x}^\mathsf{T}Q_{\psi_K}GQ_{\psi_K}^\mathsf{T}\hat{x}\right)^{\frac{p+1}{2}},$$

where the matrix $\mathbb{G}$ is given by (3.52). Moreover, we employ the inequality $\max_{\hat{x}\in\hat{K}} f(x) \leq \max_{\hat{x}\in\hat{\mathscr{E}}} f(x)$ which is valid for any function $f(x) \geq 0$ (since $\hat{K} \subset \hat{\mathscr{E}}$). Further, we use again the fact, similar to the proof of Lemma 3.21, that Q_{ψ_K} is a rotation matrix that maps $\hat{\mathscr{E}}$ onto itself. Thus (3.71) with (3.55) reads

$$\left\|w_{x_K,p}^{int}\right\|_{L^\infty(K)} \leq A_w\mu_K^{p+1}\max_{\hat{x}\in\hat{\mathscr{E}}}\left(\hat{x}^\mathsf{T}Q_{\psi_K}GQ_{\psi_K}^\mathsf{T}\hat{x}\right)^{\frac{p+1}{2}} \tag{3.72}$$

$$\leq A_w\mu_K^{p+1}\max_{\hat{x}\in\hat{\mathscr{E}}}\left(\hat{x}^\mathsf{T}G\hat{x}\right)^{\frac{p+1}{2}} = A_w\mu_K^{p+1}\max_{\substack{r\in[0,1]\\\psi\in[0,2\pi]}}\left(r^2g(\psi)\right)^{\frac{p+1}{2}},$$

which together with (3.58) proofs (3.70). $\qquad\square$

3.3.3 Estimate in the $H^1(K)$-Seminorm

The next goal is the estimate of the interpolation error function $w_{x_K,p}^{int}$ located at the barycenter $x_K = (x_{K,1}, x_{K,2})$ in the $H^1(K)$-seminorm. In Sect. 3.2.3, we showed that interpolation error function is a $(p + 1)$-homogeneous function (cf. Definition 3.6) because it can be written in the form

$$w_{x_K,p}^{int}(x) = \sum_{l=0}^{p+1}\alpha_l(x_1-x_{K,1})^l(x_2-x_{K,2})^{p+1-l}, \quad x = (x_1, x_2) \in K, \tag{3.73}$$

where

$$\alpha_l = \frac{1}{(p+1)!}\binom{p+1}{l}\frac{\partial^{p+1}u(x_K)}{\partial x_1^l\partial x_2^{p+1-l}}, \quad l = 0, \dots, p+1. \tag{3.74}$$

Taking into account (3.73)–(3.74), we evaluate the square of the magnitude of the gradient of $w_{x_K,p}^{\text{int}}$. Setting $\xi_i = x_i - x_{K,i}$, $i = 1, 2$, we have

$$
|\nabla w_{\bar{x},p}^{\text{int}}(x)|^2 = \left(\frac{\partial}{\partial x_1} \sum_{l=0}^{p+1} \alpha_l \xi_1^l \xi_2^{p+1-l} \right)^2 + \left(\frac{\partial}{\partial x_2} \sum_{l=0}^{p+1} \alpha_l \xi_1^l \xi_2^{p+1-l} \right)^2 \tag{3.75}
$$

$$
= \left(\sum_{l=1}^{p+1} l\,\alpha_l \xi_1^{l-1} \xi_2^{p+1-l} \right)^2 + \left(\sum_{l=0}^{p} (p+1-l)\alpha_l \xi_1^l \xi_2^{p-l} \right)^2
$$

$$
= \left(\sum_{l=0}^{p} \beta_l^{(1)} \xi_1^l \xi_2^{p-l} \right)^2 + \left(\sum_{l=0}^{p} \beta_l^{(2)} \xi_1^l \xi_2^{p-l} \right)^2,
$$

where

$$
\beta_l^{(1)} := (l+1)\,\alpha_{l+1}, \qquad \beta_l^{(2)} := (p+1-l)\,\alpha_l, \qquad l = 0, \ldots, p. \tag{3.76}
$$

In order to simplify the last terms in (3.75), we introduce the following Lemma.

Lemma 3.24 *Let $\beta_l \in \mathbb{R}$ for $l = 0, \ldots, p$. Then*

$$
\left(\sum_{l=0}^{p} \beta_l \, \xi_1^l \, \xi_2^{p-l} \right)^2 = \sum_{i=0}^{2p} \gamma_i \, \xi_1^i \, \xi_2^{2p-i}, \tag{3.77}
$$

where

$$
\gamma_i = \sum_{j=1}^{i} \beta_j \beta_{i-j}, \qquad \gamma_{2p-i} = \sum_{j=0}^{i} \beta_{p-j}\beta_{p-(i-j)}, \qquad i = 0, \ldots, p.
$$

Proof The identity can be derived by a direct computation.   □

A direct consequence of (3.73), (3.75), and Lemma 3.24 is the following:

Lemma 3.25 *Let $w_{x_K,p}^{\text{int}}$ be the interpolation error function* (3.10) *located at x_K written in the form* (3.73) *and* (3.74). *Then*

$$
|\nabla w_{x_K,p}^{\text{int}}(x)|^2 = \sum_{i=0}^{2p} \gamma_i (x_1 - x_{K,1})^i (x_2 - x_{K,2})^{2p-i}, \tag{3.78}
$$

where

$$\gamma_i = \sum_{j=1}^{i}(\beta_j^{(1)}\beta_{i-j}^{(1)} + \beta_j^{(2)}\beta_{i-j}^{(2)}),$$

$$\gamma_{2k-i} = \sum_{j=1}^{i}(\beta_{p-j}^{(1)}\beta_{p-(i-j)}^{(1)} + \beta_{p-j}^{(2)}\beta_{p-(i-j)}^{(2)}), \qquad i = 0, \dots, p,$$

and $\beta_l^{(1)}$, $\beta_l^{(2)}$, $l = 0, \dots, p$ are given by (3.76).

Lemma 3.25 implies that $|\nabla w_{x_K,p}^{\text{int}}(x)|^2$ is $2p$-homogeneous function in the sense of Definition 3.6. A direct consequence of Theorem 3.17 is the following result:

Corollary 3.26 *Let $w_{x_K,p}^{\text{int}}$ be the interpolation error function* (3.73) *and* (3.74). *Then $|\nabla w_{x_K,p}^{\text{int}}(x)|^2$ is $2p$-homogeneous function and there exist values $A_{w'} \geq 0$, $\rho_{w'} \geq 1$ and $\varphi_{w'} \in [0, 2\pi)$ such that*

$$\left|\nabla w_{x_K,p}^{\text{int}}(x)\right|^2 \leq A_{w'}\left((x-x_K)^{\mathsf{T}}Q_{\varphi_{w'}}\mathbb{D}_{\rho_{w'}}^{[2p]}Q_{\varphi_{w'}}^{\mathsf{T}}(x-x_K)\right)^p \quad \forall x \in \Omega, \qquad (3.79)$$

where $Q_{\varphi_{w'}}$ is a rotation matrix and $\mathbb{D}_{\rho_{w'}}^{[2p]} = \mathrm{diag}(1, \rho_{w'}^{-1/p})$, cf. (3.36).

Proof The triplet $\{A_{w'}, \rho_{w'}, \varphi_{w'}\}$ can be defined by relations analogous to (3.48) where we replace $(p+1)$-homogeneous function $\mathrm{d}^{p+1}u(x_K; \cdot)$ by the function $|\nabla w_{x_K,p}^{\text{int}}(x)|^2$, which is $2p$-homogeneous according to Lemma 3.25. $\square$

We note that the values $\{A_{w'}, \rho_{w'}, \varphi_{w'}\}$ can be computed using an analogue of the algorithm given in Remark 3.15. Now, we formulate the corresponding error estimate in the H^1-seminorm.

Lemma 3.27 *Let K be a triangle with the geometry $\{\mu_K, \sigma_K, \phi_K\}$ (cf. Definition 3.1), $u \in C^\infty(K)$, $p \in \mathbb{N}$, $w_{x_K,p}^{\text{int}}$ be the corresponding interpolation error function located at the barycenter x_K defined by* (3.10) *and let $|\nabla w_{x_K,p}^{\text{int}}(x)|^2$ have the anisotropic bound $\{A_{w'}, \rho_{w'}, \varphi_{w'}\}$ given by* (3.79). *Then*

$$\left\|\nabla w_{x_K,p}^{\text{int}}\right\|_{L^2(K)}^2 \leq \frac{1}{2p+2}A_{w'}\mu_K^{2p+2}G(p, 2p, \rho_{w'}, \varphi_{w'}; \sigma_K, \phi_K), \qquad (3.80)$$

where G is defined by (3.57).

Proof We proceed in a manner similar to the proof of Lemma 3.21. Using (3.63) and bound (3.79), we have

$$\left\|\nabla w_{x_K,p}^{\text{int}}\right\|_{L^2(K)}^2 \leq \int_K \left|\nabla w_{x_K,p}^{\text{int}}(x)\right|^2 \mathrm{d}x \tag{3.81}$$

$$\leq \int_K A_{w'} \left((x-x_K)^{\mathsf{T}} \mathbb{Q}_{\varphi_{w'}} \mathbb{D}_{\rho_{w'}}^{[2p]} \mathbb{Q}_{\varphi_{w'}}^{\mathsf{T}} (x-x_K)\right)^p \mathrm{d}x$$

$$= A_{w'} \mu_K^{2p+2} \int_{\hat{K}} \left(\hat{x}^{\mathsf{T}} \mathbb{Q}_{\psi_K} \mathbb{G} \mathbb{Q}_{\psi_K}^{\mathsf{T}} \hat{x}\right)^p \mathrm{d}\hat{x},$$

where

$$\mathbb{G} := \mathbb{S}_K \mathbb{Q}_{\phi_K}^{\mathsf{T}} \mathbb{Q}_{\varphi_{w'}} \mathbb{D}_{\rho_{w'}}^{[2p]} \mathbb{Q}_{\varphi_{w'}}^{\mathsf{T}} \mathbb{Q}_{\phi_K} \mathbb{S}_K. \tag{3.82}$$

Finally, employing inequality $\int_{\hat{K}} f(x)\,\mathrm{d}x \leq \int_{\hat{\mathcal{E}}} f(x)\,\mathrm{d}x$ which is valid for any function $f(x) \geq 0$, we obtain, similar to (3.66) and (3.67),

$$\left\|\nabla w_{x_K,p}^{\text{int}}\right\|_{L^2(K)}^2 \leq A_{w'} \mu_K^{2p+2} \int_{\hat{\mathcal{E}}} \left(\hat{x}^{\mathsf{T}} \mathbb{G} \hat{x}\right)^p \mathrm{d}\hat{x} \tag{3.83}$$

$$\leq A_{w'} \mu_K^{2p+2} \int_0^1 r^{2p+1} \left(\int_0^{2\pi} g(\psi)^p \,\mathrm{d}\psi\right) \mathrm{d}r.$$

Finally, integration w.r.t. r, as in (3.68), together with (3.57) implies estimate (3.80).
□

Remark 3.28 It may look inconsistent that the estimate in the $L^2(K)$-norm (3.69) contains the square of size of the magnitude A_w^2, whereas the estimate in the $H^1(K)$-semi-norm (3.80) contains the magnitude $A_{w'}$ to the power one. This is caused by a different meaning of A_w and $A_{w'}$. The value A_w corresponds to the anisotropic bound of $w_{x_K,p}^{\text{int}}$, whereas $A_{w'}$ corresponds to the anisotropic bound of $|\nabla w_{x_K,p}^{\text{int}}|^2$.

3.3.4 Estimate in the $L^2(\partial K)$-Norm

In Chap. 8, we employ also estimates evaluating the interpolation error over the boundary of element K.

Lemma 3.29 *Let K be a triangle with the geometry $\{\mu_K, \sigma_K, \phi_K\}$ (cf. Definition 3.1), $u \in C^\infty(K)$, $p \in \mathbb{N}$, $w_{x_K,p}^{\text{int}}$ be the corresponding interpolation error function located at the barycenter x_K defined by (3.10) and let $\{A_w, \rho_w, \varphi_w\}$ be* *the anisotropic bound of $w_{x_K,p}^{\text{int}}$ given by Definition 3.18. Then*

$$\left\| w_{x_K,p}^{\text{int}} \right\|_{\partial K}^2 \le A_w^2 \mu_K^{2p+3} \sigma_K \, G\left(p+1, p+1, \rho_w, \varphi_w; \sigma_K, \phi_K\right), \tag{3.84}$$

where the function G is given by (3.57).

Proof Let $\hat{K}$ be the reference triangle, F_K be the affine function (3.1) mapping $\hat{K}$ onto K and let $\mathbb{F}_K$ denote the corresponding Jacobian matrix. Moreover, for a function $f(x) : K \to \mathbb{R}$, we define the corresponding "reference" function $\hat{f} : \hat{K} \to \mathbb{R}$ by (3.60). Using path integration, we have

$$\int_{\partial K} f(x) \, \mathrm{d}S = \int_{\partial \hat{K}} \hat{f}(\hat{x}) |\mathbb{F}_K \cdot t| \, \mathrm{d}\hat{S} \le \int_{\partial \hat{K}} \hat{f}(\hat{x}) \|\mathbb{F}_K\|_2 \, \mathrm{d}\hat{S}, \tag{3.85}$$

where t is the unit tangent vector to $\partial \hat{K}$, $\mathbb{F}_K$ is the matrix from (3.1), and $\|\mathbb{F}_K\|_2$ is the induced Euclidean norm. Employing (3.2)–(3.4), we have

$$\|\mathbb{F}_K\|_2 = \mathrm{Sp}(\mathbb{F}_K^\mathsf{T}\mathbb{F}_K)^{1/2} = \mathrm{Sp}(\mathbb{L}_K) = \mu_K \sigma_K, \tag{3.86}$$

where $\mathrm{Sp}(\cdot)$ denotes the spectral radius of its argument. Using estimate (3.50), relation (3.85), (3.86), (3.63) and the similar manipulation as in (3.64), we obtain

$$\left\| w_{x_K,p}^{\text{int}} \right\|_{\partial K}^2 = \int_{\partial K} \left| w_{x_K,p}^{\text{int}}(x) \right|^2 \mathrm{d}S \tag{3.87}$$

$$\le A_w^2 \int_{\partial K} \left((x - \bar{x})^\mathsf{T} Q_{\varphi_w} D_{\rho_w}^{[p+1]} Q_{\varphi_w}^\mathsf{T} (x - \bar{x}) \right)^{p+1} \mathrm{d}S$$

$$\le A_w^2 \mu_K^{2p+3} \sigma_K \int_{\partial \hat{K}} \left(\hat{x}^\mathsf{T} Q_{\psi_K} G Q_{\psi_K}^\mathsf{T} \hat{x} \right)^{p+1} \mathrm{d}\hat{x},$$

where $G = \mathbb{S}_K Q_{\phi_K}^\mathsf{T} Q_{\varphi_w} D_{\rho_w}^{[p+1]} Q_{\varphi_w}^\mathsf{T} Q_{\phi_K} \mathbb{S}_K$, cf. (3.65).

Obviously, the argument of the last integrand in (3.87) is increasing with respect to $\hat{x}$ since

$$(t_1 \hat{x}^\mathsf{T}) Q_{\psi_K} G Q_{\psi_K}^\mathsf{T} (t_1 \hat{x}) \le (t_2 \hat{x})^\mathsf{T} Q_{\psi_K} G Q_{\psi_K}^\mathsf{T} (t_2 \hat{x}) \qquad \forall t_1 \le t_2. \tag{3.88}$$

Obviously, for each $\hat{x}_K \in \partial \hat{K}$ there exists $t \ge 1$ such that $\hat{x}_E = t\hat{x}_K \in \partial\hat{\mathscr{E}}$, see Fig. 3.8. Therefore, due to (3.88) and the fact that $|\partial\hat{K}| < |\partial\hat{\mathscr{E}}|$, we replace the domain of integration in (3.87), namely

$$\left\| w_{x_K,p}^{\text{int}} \right\|_{\partial K}^2 \le A_w^2 \mu_K^{2p+3} \sigma_K \int_{\partial\hat{\mathscr{E}}} \left(\hat{x}^\mathsf{T} Q_{\psi_K} G Q_{\psi_K}^\mathsf{T} \hat{x} \right)^{p+1} \mathrm{d}\hat{x}. \tag{3.89}$$

Fig. 3.8 The reference
triangle $\hat{K}$ and reference
ellipse $\hat{\mathcal{E}}$ and the points
$\hat{x}_K \in \partial \hat{K}$ and $\hat{x}_E \in \partial \hat{\mathcal{E}}$

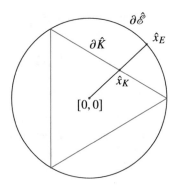

A formal rotation of the system through ψ_K implies that the last integral over the
unit sphere can be evaluated as

$$
\int_{\partial\hat{\mathcal{E}}} \left(\hat{x}^\mathsf{T} \mathsf{G} \hat{x} \right)^{p+1} d\hat{x} = \int_0^{2\pi} g(t)^{p+1} \, dt,
\tag{3.90}
$$

where $g(t)$, $t \in [0, 2\pi]$ is given by (3.56). Finally, (3.89), (3.90), and (3.57)
imply (3.84). $\square$

3.3.5   Estimate in the $H^1(\partial K)$-Seminorm

Combining the results from Sects. 3.3.3 and 3.3.4, we establish the estimate in
the $H^1(\partial K)$-seminorm, which is employed in Chap. 8. We recall that for a given
$u \in C^\infty(K)$, the square of the magnitude of its gradient $|\nabla w_{\bar{x},p}^{\text{int}}(x)|^2$ is a $2p$-
homogeneous function and, therefore, there exists the corresponding anisotropic
bound, see Corollary 3.26. Then we have the following estimate.

Lemma 3.30 *Let K be a triangle with the geometry $\{\mu_K, \sigma_K, \phi_K\}$ (cf. Definition 3.1), $u \in C^\infty(K)$, $p \in \mathbb{N}$, $w_{\bar{x},p}^{\text{int}}$ be the corresponding interpolation
error function (3.73) and (3.74) and let $|\nabla w_{\bar{x},p}^{\text{int}}(x)|^2$ have the anisotropic bound
$\{A_{w'}, \rho_{w'}, \varphi_{w'}\}$ given by (3.79). Then*

$$
\left\| \nabla w_{\bar{x},p}^{\text{int}} \right\|_{\partial K}^2 \le A_{w'} \mu_K^{2p+1} \sigma_K \, G(p, 2p, \rho_{w'}, \varphi_{w'}; \sigma_K, \phi_K),
\tag{3.91}
$$

where G is defined by (3.57).

Proof Using the definition of the norm, we have

$$
\left\| \nabla w_{\bar{x},p}^{\text{int}} \right\|_{\partial K}^2 = \int_{\partial K} |\nabla w_{\bar{x},p}^{\text{int}}|^2 \, dS.
\tag{3.92}
$$

Using the same procedure as in the proof of Lemma 3.29, we apply estimate (3.79) on the right-hand side of (3.92) and transform the integral over $\partial\hat{K}$. This integral is bounded by the integral over $\partial\hat{\mathcal{E}}$ and using (3.57), we obtain (3.91).   □

3.4   Numerical Verification

The goal of this section is the numerical verification of the interpolation error estimates from Sect. 3.3. We consider the following quartet of functions

$$w_2(x_1, x_2) := 2x_1^2 + x_2^2, \tag{3.93}$$

$$w_3(x_1, x_2) := 10x_1^3 + 2x_1^2x_2 + x_1y_1^2 - 0.1x_2^3,$$

$$w_4(x_1, x_2) := 0.5x_1^4 + 0.5x_1^3x_2 + 0.25x_1^2x_2^2 + x_1x_2^3 - x_2^4,$$

$$w_5(x_1, x_2) := 2x_1^5 + 2x_1^3x_2^2 + x_1^2x_2^3 + 3x_1x_2^4 - 4x_2^5.$$

Obviously, each w_k, $k = 2, 3, 4, 5$ is a k-homogeneous function, in the sense of Definition 3.6, located at the origin. We have chosen these functions randomly but we avoided trivial examples.

Moreover, we consider the set of acute isosceles triangles $\{K_\sigma^\phi\}$ with the barycenter at the origin, having constant size and varying anisotropy (shape and orientation). Namely, each K_σ^ϕ is given by (3.1) and (3.2) with

$$(\sigma, \phi) \in \Sigma \times \Phi, \quad \Sigma := \left\{(1.2)^j; \ j = 0, 1, 2, \ldots, 17\right\}, \tag{3.94}$$

$$\Phi := \left\{\tfrac{i}{10}\pi; \ i = 0, 1, 2, \ldots, 20\right\},$$

$x_K = 0$, $\mu_K = 1$ and $\psi_K = \pi/2$. The maximal considered shape is $\sigma = 1.2^{17} \approx 22$. Some examples of K_σ^ϕ are shown in Fig. 3.9.

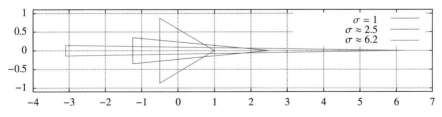

Fig. 3.9 Triangles defined by (3.94) for $\sigma = 1$, $\sigma = (1.2)^5 \approx 2.5$ and $\sigma = (1.2)^{10} \approx 6.2$

For each function w_k, $k = 2, \ldots, 5$ and each K_σ^ϕ, $\{\sigma, \phi\} \in \Sigma \times \Phi$, we evaluate the left-hand sides of the estimates (3.69), (3.70), (3.80), (3.84), and (3.91), i.e.,

$$\|w_k\|_{L^2(K_\sigma^\phi)}, \quad \|w_k\|_{L^\infty(K_\sigma^\phi)}, \quad \|\nabla w_k\|_{L^2(K_\sigma^\phi)}, \tag{3.95}$$

$$\|w_k\|_{L^2(\partial K_\sigma^\phi)}, \quad \|\nabla w_k\|_{L^2(\partial K_\sigma^\phi)},$$

respectively. Similarly, we evaluate right-hand sides of (3.69), (3.70), (3.80), (3.84), and (3.91) and denote them by

$$E(w_k)_{L^2(K_\sigma^\phi)}, \quad E(w_k)_{L^\infty(K_\sigma^\phi)}, \quad E(\nabla w_k)_{L^2(K_\sigma^\phi)}, \tag{3.96}$$

$$E(w_k)_{L^2(\partial K_\sigma^\phi)}, \quad E(\nabla w_k)_{L^2(\partial K_\sigma^\phi)},$$

respectively. Finally, we define "maximal" and "minimal" effectivity indices corresponding to the aforementioned estimates as functions of the shape σ, where the maxima and minima are taken over $\phi \in \Phi$. Hence,

$$M_{L^2(K)}^{w_k}(\sigma) = \max_{\phi \in \Phi} \frac{E(w_k)_{L^2(K_\sigma^\phi)}}{\|w_k\|_{L^2(K_\sigma^\phi)}}, \quad m_{L^2(K)}^{w_k}(\sigma) = \min_{\phi \in \Phi} \frac{E(w_k)_{L^2(K_\sigma^\phi)}}{\|w_k\|_{L^2(K_\sigma^\phi)}}, \tag{3.97}$$

$$M_{H^1(K)}^{w_k}(\sigma) = \max_{\phi \in \Phi} \frac{E(\nabla w_k)_{L^2(K_\sigma^\phi)}}{\|\nabla w_k\|_{L^2(K_\sigma^\phi)}}, \quad m_{H^1(K)}^{w_k}(\sigma) = \min_{\phi \in \Phi} \frac{E(\nabla w_k)_{L^2(K_\sigma^\phi)}}{\|\nabla w_k\|_{L^2(K_\sigma^\phi)}},$$

$$M_{L^\infty(K)}^{w_k}(\sigma) = \max_{\phi \in \Phi} \frac{E(w_k)_{L^\infty(K_\sigma^\phi)}}{\|w_k\|_{L^\infty(K_\sigma^\phi)}}, \quad m_{L^\infty(K)}^{w_k}(\sigma) = \min_{\phi \in \Phi} \frac{E(w_k)_{L^\infty(K_\sigma^\phi)}}{\|w_k\|_{L^\infty(K_\sigma^\phi)}},$$

$$M_{L^2(\partial K)}^{w_k}(\sigma) = \max_{\phi \in \Phi} \frac{E(w_k)_{L^2(\partial K_\sigma^\phi)}}{\|w_k\|_{L^2(\partial K_\sigma^\phi)}}, \quad m_{L^2(\partial K)}^{w_k}(\sigma) = \min_{\phi \in \Phi} \frac{E(w_k)_{L^2(\partial K_\sigma^\phi)}}{\|w_k\|_{L^2(\partial K_\sigma^\phi)}},$$

$$M_{H^1(\partial K)}^{w_k}(\sigma) = \max_{\phi \in \Phi} \frac{E(\nabla w_k)_{L^2(\partial K_\sigma^\phi)}}{\|\nabla w_k\|_{L^2(\partial K_\sigma^\phi)}}, \quad m_{H^1(\partial K)}^{w_k}(\sigma) = \min_{\phi \in \Phi} \frac{E(\nabla w_k)_{L^2(\partial K_\sigma^\phi)}}{\|\nabla w_k\|_{L^2(\partial K_\sigma^\phi)}}.$$

Remark 3.31 We note that the L^2- norms and the H^1-seminorms in (3.95) are computed by high-order numerical quadratures. Since the integrands are regular, the values of these norms are sufficiently accurate. On the other hand, the L^∞-norm is evaluated by seeking the maximum only in a finite list of integration nodes which is not always sufficient. Therefore, the L^∞-results suffer from integration errors but they are enough to illustrate the theoretical results.

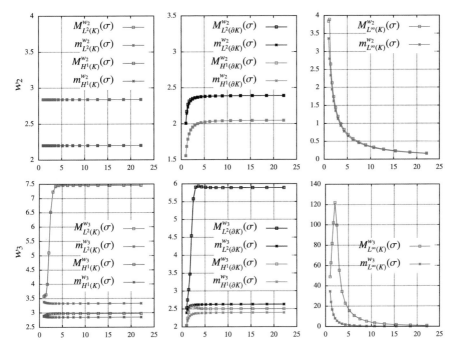

Fig. 3.10 The dependence of quantities defined by (3.97) on the shape σ for functions w_2 (first line) and w_3 (second line) from (3.93)

The computed values defined by (3.97) are shown in Figs. 3.10 and 3.11 as function of σ. We observe that the effectivity indices (3.97) are "asymptotically constant" for all $w_k, k = 2, \ldots, 5$ which is a favorable property. On the other hand, these indices are relatively high. This is caused by two reasons.

(i) In the derivation of the estimates, we replaced the integration over $\hat{K}$ by the integration over $\hat{\mathscr{E}}$, cf. the proofs of Lemmas 3.21, 3.23, and 3.27. The area of $\hat{\mathscr{E}}$ is much larger than the area of $\hat{K}$ and the integrand achieves the highest values on $\hat{\mathscr{E}} \setminus \hat{K}$. Similar argumentation is true for the "boundary" norms, where we replaced the integration over $\partial \hat{K}$ by the integration over $\partial \hat{\mathscr{E}}$, cf. proofs of Lemmas 3.29 and 3.30.

(ii) We observe that the effectivity indices are higher for higher polynomial degrees which follows from the anisotropic bound (3.50) given by Theorem 3.17. Figure 3.7 shows that the overestimation is higher for higher polynomial degrees. In virtue of Remark 3.9, the bound (3.50) is valid with equality for 2-homogeneous functions (piecewise linear approximation). This is the reason that the maximal and minimal indices are (almost) the same for the function w_2. The large difference for the L^∞-norm is due to the evaluation of the norm, see Remark 3.31.

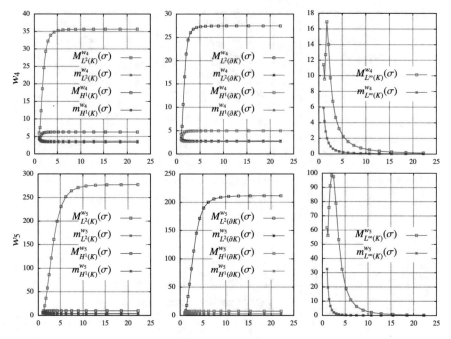

Fig. 3.11 The dependence of quantities defined by (3.97) on the shape σ for functions w_4 (first line) and w_5 (second line) from (3.93)

The large effectivity indices are not problematic for practical computations since these estimates are used only for the local setting and/or optimization of the shape of elements. On the other hand, these estimates are used neither for the setting of the elements sizes nor for the estimation of the global error itself.

Chapter 4
Interpolation Error Estimates for Three Dimensions

We extend the theoretical results to the three-dimensional case. In the same spirit as for the two-dimensional case, we recall the geometry of a tetrahedron K and define the interpolation error function and the corresponding error estimates. The extension is relatively straightforward but technically cumbersome. Therefore, we avoid some technical details that are intuitively understandable and can be derived by readers.

4.1 Geometry of a Tetrahedron

Let $\hat{K}$ be the reference tetrahedron given by (2.41). For a tetrahedron K, there exists an affine mapping $F_K : \hat{K} \to K$ (cf. (2.43)) written in the form

$$F_K(\hat{x}) = \mathbb{F}_K \hat{x} + x_K, \quad \hat{x} \in \hat{K}, \tag{4.1}$$

where x_K is the barycenter of K and $\mathbb{F}_K$ is a regular matrix. The *singular value decomposition* of $\mathbb{F}_K$ reads

$$\mathbb{F}_K = \mathbb{Q}_{\boldsymbol{\phi}_K} \mathbb{L}_K \mathbb{Q}_{\boldsymbol{\psi}_K}^{\mathsf{T}}, \tag{4.2}$$

where $\mathbb{L}_K = \mathrm{diag}(\ell_{K,1}, \ell_{K,2}, \ell_{K,3})$ is a diagonal matrix with the singular values $\ell_{K,1} \geq \ell_{K,2} \geq \ell_{K,3} > 0$, $\mathbb{Q}_{\boldsymbol{\phi}_K}$ and $\mathbb{Q}_{\boldsymbol{\psi}_K}$ are *unitary matrices*, cf. (2.12), which represent a rotation in 3D, see Sect. 2.4.1, with

$$\boldsymbol{\phi}_K, \boldsymbol{\psi}_K \in \mathscr{U} := [0, 2\pi] \times [0, \pi] \times [0, 2\pi]. \tag{4.3}$$

We recall the terms describing the geometry of tetrahedron, cf. Definition 2.16.

© The Author(s), under exclusive license to Springer Nature Switzerland AG 2022
V. Dolejší, G. May, *Anisotropic hp-Mesh Adaptation Methods*, Nečas Center Series,
https://doi.org/10.1007/978-3-031-04279-9_4

Definition 4.1 Let K be a tetrahedron given by mapping (4.1) with decomposition (4.2) with the singular values $\ell_{K,1} \geq \ell_{K,2} \geq \ell_{K,3} > 0$ and $\boldsymbol{\phi}_K \in \mathcal{U}$. Then

- $\mu_K := \left(\ell_{K,1}\,\ell_{K,2}\,\ell_{K,3}\right)^{1/3} > 0$ is called the *size* of K.
- The pair $(\sigma_K, \varsigma_K) := \left(\left(\ell_{K,1}/\ell_{K,2}\right)^{1/3}, \left(\ell_{K,1}/\ell_{K,3}\right)^{1/3}\right)$ is called the *shape* of K.
- The triplet $\boldsymbol{\phi}_K \in \mathcal{U}$ is the *orientation* of K.
- The shape and orientation are called the *anisotropy* of K.
- The group of six $\{\mu_K,\ \sigma_K,\ \varsigma_K,\ \boldsymbol{\phi}_K\}$ is called the *geometry* of K.

Using these definitions, we have

$$\mathbb{L}_K = \begin{pmatrix} \ell_{K,1} & 0 & 0 \\ 0 & \ell_{K,2} & 0 \\ 0 & 0 & \ell_{K,3} \end{pmatrix} = \mu_K \mathbb{S}_K, \quad \text{where } \mathbb{S}_K := \begin{pmatrix} \sigma_K\varsigma_K & 0 & 0 \\ 0 & \varsigma_K/\sigma_K^2 & 0 \\ 0 & 0 & \sigma_K/\varsigma_K^2 \end{pmatrix}$$

$$(4.4)$$

and

$$\det \mathbb{F}_K = \det \mathbb{L}_K = \mu_K^3. \tag{4.5}$$

4.2   Interpolation Error Function and Its Anisotropic Bound

In contrast to the geometry of element given in Sect. 4.1, the extension of the interpolation error function and its anisotropic bound from 2D to 3D is more straightforward. We deal with sufficiently regular functions from the space $C^\infty := C^\infty(\overline{\Omega})$, $\Omega \subset \mathbb{R}^3$.

4.2.1   Interpolation Error Function

Let $u \in C^\infty$ be a given function, $\bar{x} \in \Omega$ be a given node, and $p \in \mathbb{N}$ be an integer representing the polynomial approximation degree.

We define an interpolation operator $\Pi_{\bar{x},p} : C^\infty \to P^p(\Omega)$ such that $\Pi_{\bar{x},p} u$ is a polynomial function of degree p on Ω, whose partial derivatives up to order p coincide with those of u at $\bar{x}$.

Definition 4.2 Let $\bar{x} = (\bar{x}_1, \bar{x}_2, \bar{x}_3) \in \Omega$ and $p \in \mathbb{N}$ be given. The *interpolation operator* $\Pi_{\bar{x},p} : C^\infty \to P^p(\Omega)$ of degree p at $\bar{x}$ is given by

$$\frac{\partial^k \left(\Pi_{\bar{x},p} u(\bar{x})\right)}{\partial x_1^l \partial x_2^j \partial x_3^i} = \frac{\partial^k u(\bar{x})}{\partial x_1^l \partial x_2^j \partial x_3^i}, \qquad l, j, i = 0, \ldots, k,\ l + j + i \leq k. \tag{4.6}$$

The interpolation $\Pi_{\bar{x},p}u$ satisfying (4.6) for given $u \in C^\infty$, $\bar{x} \in \Omega$, and $p \in \mathbb{N}$ reads

$$\Pi_{\bar{x},p}u(x) = \sum_{k=0}^{p} \frac{1}{k!} \left(\sum_{l,j,i=0}^{l+j+i\leq k} \frac{\partial^k u(\bar{x})}{\partial x_1^l \partial x_2^j \partial x_3^i} (x_1 - \bar{x}_1)^l (x_2 - \bar{x}_2)^j (x_3 - \bar{x}_3)^i \right).$$

(4.7)

We introduce the *interpolation error function* $w_{\bar{x},p}^{\text{int}}$, which approximates the difference between u and its interpolation, given by Definition 4.2, as

$$u(x) - \Pi_{\bar{x},p}u(x) = w_{\bar{x},p}^{\text{int}}(x) + O(|x - \bar{x}|^{p+2}).$$

(4.8)

Using the Taylor expansion we derive, similarly as in (3.8)–(3.10), the relation

$$w_{\bar{x},p}^{\text{int}}(x) = \frac{1}{(p+1)!} \sum_{l,j,i=0}^{l+j+i\leq p+1} \left[\frac{\partial^{p+1} u(\bar{x})}{\partial x_1^l \partial x_2^j \partial x_3^i} (x_1 - \bar{x}_1)^l (x_2 - \bar{x}_2)^j (x_3 - \bar{x}_3)^i \right].$$

(4.9)

Definition 4.3 Let $u \in C^\infty$, $\bar{x} \in \Omega$, and $p \in \mathbb{N}$ be given. Then the polynomial function $w_{\bar{x},p}^{\text{int}}$ of degree $p + 1$ defined by (4.9) is called the *interpolation error function* of degree p located at $\bar{x}$.

Remark 4.4 Let $\xi = (\xi_1, \xi_2, \xi_3) \in \mathbb{R}^3$, and we denote by

$$d^{p+1}u(\bar{x}; \xi) := \frac{1}{(p+1)!} \sum_{l,j,i=0}^{l+j+i\leq p+1} \frac{\partial^{p+1} u(x)}{\partial x_1^l \partial x_2^j \partial x_3^i} \xi_1^l \xi_2^j \xi_3^i, \quad \xi \in \mathbb{R}^3,$$

(4.10)

the $(p + 1)$th-(scaled) *directional derivative* of $u \in C^\infty$ along the direction $\xi = (\xi_1, \xi_2, \xi_3) \in \mathbb{R}^3$ at $\bar{x} \in \Omega$. Then the interpolation error function can be expressed in terms of this derivative, namely

$$d^{p+1}u(\bar{x}; x - \bar{x}) = w_{\bar{x},p}^{\text{int}}(x), \quad x \in \Omega.$$

(4.11)

4.2.2   Anisotropic Bound of Interpolation Error Function

The interpolation error function $w_{\bar{x},p}^{\text{int}}$ (4.9) depends on all partial derivatives of u of degree $p+1$. Similarly, as for the two-dimensional case, it is advantageous to define the anisotropic bound of $w_{\bar{x},p}^{\text{int}}$ by the same number of parameters as the geometry of the tetrahedron, i.e., the six parameters identified in Definition 4.1. In particular, the aim is to estimate $w_{\bar{x},p}^{\text{int}}$ by the term

$$\left| w_{\bar{x},p}^{\text{int}}(x) \right| \leq A_w \left(\zeta^{\mathsf{T}} \mathbb{Q}_{\boldsymbol{\varphi}_w} \mathbb{D}_{\rho_p,\varrho_p}^{[p+1]} \mathbb{Q}_{\boldsymbol{\varphi}_w}^{\mathsf{T}} \zeta \right)^{\frac{p+1}{2}} \quad \forall \zeta = x - \bar{x}, \; x \in \Omega, \tag{4.12}$$

where $A_w \geq 0$, $\mathbb{Q}_{\boldsymbol{\varphi}_w}$ is a unitary matrix (=rotation matrix in 3D) characterized by triplet $\boldsymbol{\varphi}_w \in \mathscr{U}$ (cf. (4.3)), and $\mathbb{D}_{\rho,\varrho}^{[q]}$ is the matrix given by

$$\mathbb{D}_{\rho,\varrho}^{[q]} := \begin{pmatrix} 1 & 0 & 0 \\ 0 & \rho^{-2/q} & 0 \\ 0 & 0 & \varrho^{-2/q} \end{pmatrix}, \quad \rho \geq 1, \; \varrho \geq 1, \; q \geq 2. \tag{4.13}$$

The symbol $[q]$ is a superscript and the value q corresponds to the power ($= -2/q$) of the second and third diagonal entries of $\mathbb{D}_{\rho,\varrho}^{[q]}$, $q \geq 2$.

Similar to Sect. 3.2.3, it is possible to extend the term s-homogeneous function to 3D. However, for the purpose of the exposition, it is sufficient to define, as in (3.17), the closed surface for the given interpolation error function $w_{\bar{x},p}^{\text{int}}$ of degree $p + 1$ located at $\bar{x}$ by

$$\theta^{p+1}(w_{\bar{x},p}^{\text{int}}) := \left\{ x \in \mathbb{R}^3; \; x = \bar{x} + |w_{\bar{x},p}^{\text{int}}(\bar{x} + \xi)|\xi \quad \forall \xi \in B_1 \right\}, \tag{4.14}$$

where

$$B_1 := \{\xi; \; \xi = (\xi_1, \xi_2, \xi_3) \in \mathbb{R}^3, \; \xi_1^2 + \xi_2^2 + \xi_3^2 = 1\} \tag{4.15}$$

is the three-dimensional unit ball.

Figure 4.1, left, shows an example of a surface for $p = 5$. We observe a complicated structure and several symmetries.

In order to estimate the interpolation error functions, we start with an extension of the results from Sect. 3.2.5.1 to 3D. Let $\bar{x} \in \Omega$, $u \in C^\infty$, and $p \in \mathbb{N}$ be given. Taking into account the scaled directional derivative of u at $\bar{x}$ given by (4.10), we define the values $\tilde{A}_p, \tilde{A}_p^{(2)}, \tilde{A}_p^{(3)} \geq 0$, $\tilde{\xi}_p^{(1)}, \tilde{\xi}_p^{(2)}, \tilde{\xi}_p^{(3)} \in B_1$, and $\tilde{\rho}_p, \tilde{\varrho}_p \geq 1$ using the formulas

$$\tilde{\xi}_p^{(1)} := \arg\max_{\xi \in B_1} |\mathrm{d}^{p+1} u(\bar{x}; \xi)|, \qquad\qquad \tilde{A}_p := |\mathrm{d}^{p+1} u(\bar{x}; \tilde{\xi}_p^{(1)})|, \tag{4.16}$$

$$\tilde{\xi}_p^{(2)} := \arg\max_{\xi \in B_1, \xi \perp \tilde{\xi}_p^{(1)}} |\mathrm{d}^{p+1} u(\bar{x}; \xi)|, \qquad \tilde{A}_p^{(2)} := |\mathrm{d}^{p+1} u(\bar{x}; \tilde{\xi}_p^{(2)})|,$$

$$\tilde{\xi}_p^{(3)} := \tilde{\xi}_p^{(1)} \times \tilde{\xi}_p^{(2)}, \qquad\qquad\qquad \tilde{A}_p^{(3)} := |\mathrm{d}^{p+1} u(\bar{x}; \tilde{\xi}_p^{(3)})|,$$

$$\tilde{\rho}_p := \frac{\tilde{A}_p}{\tilde{A}_p^{(2)}}, \qquad \tilde{\varrho}_p := \frac{\tilde{A}_p}{\tilde{A}_p^{(3)}},$$

where $\xi \perp \tilde{\xi}_p$ means that vectors ξ and $\tilde{\xi}_p$ are perpendicular and $\tilde{\xi}_p^{(1)} \times \tilde{\xi}_p^{(2)}$ denotes the vector product returning the unit vector which is perpendicular to $\tilde{\xi}_p^{(1)}$ and $\tilde{\xi}_p^{(2)}$. Hence, $\tilde{A}_p$ is the maximal value of the $(p+1)$th-order scaled directional derivative of u at $\bar{x}$, $\tilde{\xi}_p$ is the direction which maximizes this derivative, $\tilde{A}_p^{(2)}$ is the maximal $(p+1)$th-order scaled directional derivative of u along the direction perpendicular to $\tilde{\xi}_p$, $\tilde{\xi}_p^{(2)}$ is the direction where the value $\tilde{A}_p^{(2)}$ is achieved, and $\tilde{A}_p^{(3)}$ is the $(p+1)$th-order scaled directional derivative of u along the direction $\tilde{\xi}_p^{(3)} := \tilde{\xi}_p^{(1)} \times \tilde{\xi}_p^{(2)}$. Finally, $\tilde{\rho}_p$ and $\tilde{\varrho}_p$ are ratios between $\tilde{A}_p$, $\tilde{A}_p^{(2)}$, and $\tilde{A}_p^{(3)}$.

Employing the orthonormal vectors $\tilde{\xi}_p^{(1)}$, $\tilde{\xi}_p^{(2)}$, and $\tilde{\xi}_p^{(3)}$ from (4.16), we define the unitary matrix $\mathbb{Q}_{\tilde{\varphi}_p}$ having these vectors as columns, i.e.,

$$\mathbb{Q}_{\tilde{\varphi}_p} := \left(\tilde{\xi}_p^{(1)} \ \tilde{\xi}_p^{(2)} \ \tilde{\xi}_p^{(3)} \right) \in \mathbb{R}^{3 \times 3}, \tag{4.17}$$

where $\tilde{\varphi}_p \in \mathcal{U}$ denotes formally the triplet of values defining the orientation of these vectors, cf. (4.3).

Similar to (3.38), we introduce a formal estimate

$$\left| w_{\bar{x},p}^{\text{int}}(\bar{x} + \zeta) \right| \leq \tilde{A}_p \left(\zeta^\mathsf{T} \mathbb{Q}_{\tilde{\varphi}_p} \mathbb{D}_{\tilde{\rho}_p, \tilde{\varrho}_p}^{[p+1]} \mathbb{Q}_{\tilde{\varphi}_p}^\mathsf{T} \zeta \right)^{\frac{p+1}{2}}, \quad \zeta \in \mathbb{R}^3, \tag{4.18}$$

where $\tilde{A}_p$, $\tilde{\rho}_p$, and $\tilde{\varrho}_p$ are given by (4.16), matrix $\mathbb{Q}_{\tilde{\varphi}_p}$ is defined by (4.17), and $\mathbb{D}_{\tilde{\rho}_p, \tilde{\varrho}_p}^{[p+1]}$ is the diagonal matrix from (4.13). Numerical experiments in Example 3.10 showed that this estimate is *not valid* for two-dimensional case, so it cannot be true for 3D.

The remedy is a modification of parameters $\tilde{A}_p \geq 0$, $\tilde{\rho}_p \geq 1$, and $\tilde{\varrho}_p \geq 1$ in such a way that the estimate (4.18) holds true. Therefore, similar to (3.39), we define the closed surface

$$\vartheta^{p+1}(A, \rho, \varrho, \varphi) := \left\{ x \in \mathbb{R}^3; \ x = \bar{x} + A \left(\zeta^\mathsf{T} \mathbb{Q}_\varphi \mathbb{D}_{\rho, \varrho}^{[p+1]} \mathbb{Q}_\varphi^\mathsf{T} \zeta \right)^{\frac{p+1}{2}} \zeta \quad \forall \zeta \in B_1 \right\}, \tag{4.19}$$

where $\mathbb{Q}_\varphi$, $\varphi \in \mathcal{U}$ is a unitary matrix and $\mathbb{D}_{\rho,\varrho}^{[p+1]}$ is given by (4.13).

We proceed similarly as in the two-dimensional case (3.47) and (3.48). By $\bar{\theta}$ and $\bar{\vartheta}$, we denote the domains circumscribed by the closed surfaces θ and ϑ defined by (4.14) and (4.19), respectively. We keep the orientation $\tilde{\varphi}_p$, and we define $A_w \geq 0$, $\rho_w \geq 1$, and $\varrho_w \geq 0$ such that

$$\varphi_w := \tilde{\varphi}_p \quad \text{cf. (4.16)} - \text{(4.17)}, \tag{4.20a}$$

$$\bar{\theta}^{p+1}(w_{\bar{x},p}^{\text{int}}) \subset \bar{\vartheta}^{p+1}(A_w, \rho_w, \varrho_w, \varphi_w), \tag{4.20b}$$

$$\bar{\vartheta}^{p+1}(A_w, \rho_w, \varrho_w, \varphi_w) \quad \text{has the minimal possible volume.} \tag{4.20c}$$

A direct consequence of (4.20b) – (4.20c) is the following *anisotropic estimate* of the interpolation error function.

Theorem 4.5 *Let $\bar{x} \in \Omega$, $u \in C^\infty$, and $p \in \mathbb{N}$ be given. Moreover, let $w_{\bar{x},p}^{\text{int}}$ be the corresponding interpolation error function (4.9). Let $A_w \geq 0$, $\rho_w \geq 1$, $\varrho_w \geq 1$, and $\boldsymbol{\varphi}_w \in \mathcal{U}$ be given by (4.20). Then*

$$\left| w_{\bar{x},p}^{\text{int}}(x) \right| \leq A_w \left((x - \bar{x})^\mathsf{T} \mathbb{Q}_{\boldsymbol{\varphi}_w} \mathbb{D}_{\rho_w,\varrho_w}^{[p+1]} \mathbb{Q}_{\boldsymbol{\varphi}_w}^\mathsf{T} (x - \bar{x}) \right)^{\frac{p+1}{2}} \quad \forall x \in \Omega, \qquad (4.21)$$

where $\mathbb{Q}_{\boldsymbol{\varphi}_w}$ is the unitary matrix (4.17), and $\mathbb{D}_{\rho_w,\varrho_w}^{[p+1]}$ is defined by (4.13). Moreover, the estimate (4.21) is sharp in the sense that there exists $x \in \Omega$ such that

$$\left| w_{\bar{x},p}^{\text{int}}(x) \right| = A_w \left((x - \bar{x})^\mathsf{T} \mathbb{Q}_{\boldsymbol{\varphi}_w} \mathbb{D}_{\rho_w,\varrho_w}^{[p+1]} \mathbb{Q}_{\boldsymbol{\varphi}_w}^\mathsf{T} (x - \bar{x}) \right)^{\frac{p+1}{2}}.$$

Remark 4.6 The non-guaranteed estimate (4.18), using (4.16), implies that $\tilde{\rho}_p \leq \tilde{\varrho}_p$. Without loss of generality, we assume the same property also for the guaranteed version of the estimate given by (4.21), i.e., $\rho_w \leq \varrho_w$.

Finally, we define the anisotropic bound of the interpolation error function.

Definition 4.7 Let $\bar{x} \in \Omega \subset \mathbb{R}^3$, $u \in C^\infty$, and $p \in \mathbb{N}$ be given. The parameters $\{A_w, \rho_w, \varrho_w, \boldsymbol{\varphi}_w\}$, $A_w \geq 0$, $\rho_w \geq 1$, $\varrho_w \geq 0$, and $\boldsymbol{\varphi}_w \in \mathcal{U}$ defined by (4.20) are called the *anisotropic bound of the interpolation error function* $w_{\bar{x},p}^{\text{int}}$ at $\bar{x}$.

Figure 4.1, right, shows an example of a surface ϑ^{p+1} ($p = 5$) illustrating the anisotropic bound of the interpolation error function. The analogy to the two-dimensional case is obvious.

Remark 4.8 The anisotropic bound of the given interpolation error function can be evaluated by a modification of the 2D algorithm described in Remark 3.15. Since we keep $\boldsymbol{\varphi}_w := \tilde{\boldsymbol{\varphi}}_p$ given by (4.16)–(4.17), the minimization problem (4.20b)–(4.20c) reduces to the necessity to find only three parameters A_w, ρ_w, and ϱ_w. Employing

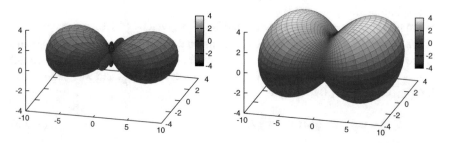

Fig. 4.1 A 3D example of the surface $\theta^{p+1}(w_{\bar{x},p}^{\text{int}})$ illustrating interpolation error function for $p = 5$, cf. (4.14) (left) and the surface ϑ^{p+1} given by (4.20) (right)

the values $\tilde{A}_p$, $\tilde{\rho}_p$, and $\tilde{\varrho}_p$ from (4.16) as the initial guesses, one is able to find an approximation A_w, ρ_w, and ϱ_w at reasonable computational costs. Obviously, it is possible to optimize also the orientation $\boldsymbol{\varphi}_w \in \mathcal{U}$, but this task is more time consuming.

4.3   Interpolation Error Estimates Including the Tetrahedra Geometry

We extend some of the results from Sect. 3.3 to the three-dimensional case, in particular, the estimates of interpolation error functions, which take into account the geometry of a tetrahedron K.

Problem 4.9 We assume that

- K is a tetrahedron with the geometry $\{\mu_K, \sigma_K, \varsigma_K, \boldsymbol{\phi}_K\}$, cf. Definition 4.1.
- x_K is the barycenter of K.
- $u \in C^\infty(K)$ is a function.
- $p \geq 1$ is an integer.
- $w_{x_K,p}^{\text{int}}$ is the corresponding interpolation error function located at x_K, cf. (4.9).

The aim is to derive a bound

$$\left\| w_{x_K,p}^{\text{int}} \right\|_{X(K)} \leq \mathcal{G}, \tag{4.22}$$

where $\|\cdot\|_{X(K)}$ denotes a Sobolev norm over K or ∂K, such that the bound $\mathcal{G}$ satisfies the following requirements:

- $\mathcal{G}$ is computable without unknown constants.
- $\mathcal{G}$ depends on $\{\mu_K, \sigma_K, \varsigma_K, \boldsymbol{\phi}_K\}$.

We use the same approach as in Sect. 3.3. We employ Theorem 4.5, the integral over K is transformed to an integral over the reference tetrahedra $\hat{K}$ (cf. (2.41)), and then we estimate this integral by an integration of the same integrand over the corresponding circumscribed ball $\hat{\mathscr{E}}$, cf. (2.42).

We present several relations that are employed in the subsequent error estimates. Let

$$\mathbb{S}_K = \text{diag}\left(\sigma_K \varsigma_K, \frac{\varsigma_K}{\sigma_K^2}, \frac{\sigma_K}{\varsigma_K^2}\right) \quad \text{and} \quad \mathbb{D}_{\rho_w,\varrho_w}^{[s_2]} = \text{diag}\left(1, \rho_w^{-2/s_2}, \varrho_w^{-2/s_2}\right) \tag{4.23}$$

be the diagonal matrices given by (4.4) and (4.13), respectively, and $s_2 \geq 1$. Furthermore, let $\mathbb{Q}_{\boldsymbol{\varphi}_w}$ and $\mathbb{Q}_{\boldsymbol{\phi}_K}$ be the 3D rotation matrices corresponding to $\boldsymbol{\varphi}_w \in \mathcal{U}$ and $\boldsymbol{\phi}_K \in \mathcal{U}$, respectively, cf. (4.17). As in (3.52), we define the symmetric 3×3

matrix

$$G := \mathbb{S}_K Q_{\phi_K}^{\mathsf{T}} Q_{\varphi_w} D_{\rho_w,\varrho_w}^{[s_2]} Q_{\varphi_w}^{\mathsf{T}} Q_{\phi_K} \mathbb{S}_K \tag{4.24}$$

and denote by G_{ij}, $i, j = 1, 2, 3$, the entries of matrix G from (4.24). Furthermore, since matrices Q in (4.24) are unitary (det $Q = 1$), det $\mathbb{S}_K = 1$, we have

$$\det G = \det D_{\rho_w,\varrho_w}^{[s_2]} = (\rho_w \varrho_w)^{-2/s_2}. \tag{4.25}$$

We express $\hat{x}$ in spherical coordinates (r, ψ, γ) as

$$\hat{x} = \begin{pmatrix} r \sin \psi \ \cos \gamma \\ r \sin \psi \ \sin \gamma \\ r \cos \psi \end{pmatrix}, \qquad r \geq 0, \ \psi \in [0, \pi], \ \gamma \in [0, 2\pi]. \tag{4.26}$$

Then the multiplication of G by $\hat{x}$ from the left and from the right fulfills

$$\hat{x}^{\mathsf{T}} G \hat{x} =: r^3 g(\psi, \gamma), \tag{4.27}$$

where g is a scalar function of two variables $\psi \in [0, \pi]$ and $\gamma \in [0, 2\pi]$:

$$g(\psi, \gamma) := G_{11} \sin^2 \psi \ \cos^2 \gamma + 2G_{12} \sin^2 \psi \ \cos \gamma \ \sin \gamma + 2G_{13} \sin \psi \ \cos \psi \ \cos \gamma$$

$$+ G_{22} \sin^2 \psi \ \sin^2 \gamma + 2G_{23} \sin \psi \ \cos \psi \ \sin \gamma + G_{33} \cos^2 \psi. \tag{4.28}$$

In virtue of (4.27), the surface integral of $(\hat{x}^{\mathsf{T}} G \hat{x})^{s_1}$, $s_1 > 0$, over the unit sphere $\partial \hat{\mathscr{E}} = \{\hat{x} \in \mathbb{R}^3 : |\hat{x}| = 1\}$ reads

$$\int_{\partial \hat{\mathscr{E}}} (\hat{x}^{\mathsf{T}} G \hat{x})^{s_1} \, d\hat{S} = \int_0^\pi \left(\int_0^{2\pi} \sin \psi \ (g(\psi, \gamma))^{s_1} \, d\gamma \right) d\psi \tag{4.29}$$

$$=: G(s_1, s_2, \rho_w, \varrho_w, \boldsymbol{\varphi}_w; \sigma_K, \varsigma_K, \boldsymbol{\phi}_K).$$

Moreover, the volume integral of $(\hat{x}^{\mathsf{T}} G \hat{x})^{s_1}$, $s_1 > 0$, over the unit ball $\hat{\mathscr{E}} = \{\hat{x} \in \mathbb{R}^3 : |\hat{x}| \leq 1\}$ fulfills

$$\int_{\hat{\mathscr{E}}} (\hat{x}^{\mathsf{T}} G \hat{x})^{s_1} \, d\hat{x} = \int_0^1 \left(\int_0^\pi \left(\int_0^{2\pi} r^2 \sin \psi \ (r^2 g(\psi, \gamma))^{s_1} \, d\gamma \right) d\psi \right) dr \tag{4.30}$$

$$= \int_0^1 \left(r^{2s_1+2} \int_0^\pi \left(\int_0^{2\pi} \sin \psi \ (g(\psi, \gamma))^{s_1} \, d\gamma \right) d\psi \right) dr$$

$$= \frac{1}{2s_1+3} G(s_1, s_2, \rho_w, \varrho_w, \boldsymbol{\varphi}_w; \sigma_K, \varsigma_K, \boldsymbol{\phi}_K),$$

where G is given by (4.29).

4.3.1   Estimate in the $L^q(K)$-Norm, $1 \le q < \infty$

We derive the estimate in the $L^q(K)$-norm, $1 \le q < \infty$, using the same approach as in Sect. 4.3.1.

Lemma 4.10 *Let* $q \in [1, \infty)$, K *be a tetrahedron with the geometry* $\{\mu_K, \sigma_K, \varsigma_K, \boldsymbol{\phi}_K\}$ *(cf. Definition 4.1)*, $u \in C^\infty(K)$, $p \in \mathbb{N}$, *and* $w^{\mathrm{int}}_{x_K, p}$ *be the corresponding interpolation error function located at the barycenter* x_K *defined by (4.9). Let* $\{A_w, \rho_w, \varrho_w, \boldsymbol{\varphi}_w\}$ *be the anisotropic bound of* $w^{\mathrm{int}}_{x_K, p}$ *given by Definition 4.7. Then*

$$\left\| w^{\mathrm{int}}_{x_K, p} \right\|^q_{L^q(K)} \le \frac{1}{q(p+1)+3} A^q_w \mu_K^{q(p+1)+3} G\left(\tfrac{q(p+1)}{2}, p+1, \rho_w, \varrho_w, \boldsymbol{\varphi}_w; \sigma_K, \varsigma_K, \boldsymbol{\phi}_K \right),$$
(4.31)

where G *is the function from (4.29).*

Proof Let $\hat{K}$ be the reference tetrahedron (2.41) and F_K be the affine function (4.1) mapping the reference tetrahedron $\hat{K}$ onto K with the corresponding Jacobian matrix $\mathbb{F}_K$. Moreover, for a function $f(x) : K \to \mathbb{R}$, we define the corresponding "reference" function $\hat{f} : \hat{K} \to \mathbb{R}$ by $\hat{f}(\hat{x}) := f(F_K \hat{x}) = f(x)$, $\hat{x} \in \hat{K}$. Using the substitution theorem, relation (4.5), we have

$$\int_K f(x)\, dx = \int_{\hat{K}} \hat{f}(\hat{x}) |\det \mathbb{F}_K|\, d\hat{x} = \mu_K^3 \int_{\hat{K}} \hat{f}(\hat{x})\, d\hat{x}.$$
(4.32)

Furthermore, (4.1) – (4.4) imply

$$x - x_K = \mu_K Q_{\phi_K} S_K Q^{\mathsf{T}}_{\psi_K} \hat{x} \quad \text{and} \quad (x - x_K)^{\mathsf{T}} = \mu_K \hat{x}^{\mathsf{T}} Q_{\psi_K} S_K Q^{\mathsf{T}}_{\phi_K}.$$
(4.33)

Using the $L^q(K)$-norm definition, (4.21), (4.32), and (4.33), we obtain

$$\left\| w^{\mathrm{int}}_{x_K, p} \right\|^q_{L^q(K)} = \int_K \left| w^{\mathrm{int}}_{x_K, p}(x) \right|^q dx \tag{4.34}$$

$$\le \int_K A^q_w \left((x - x_K)^{\mathsf{T}} Q_{\varphi_w} D^{[p+1]}_{\rho_w, \varrho_w} Q^{\mathsf{T}}_{\varphi_w} (x - x_K) \right)^{\frac{q(p+1)}{2}} dx$$

$$= A^q_w \mu_K^{q(p+1)+3} \int_{\hat{K}} \left(\hat{x}^{\mathsf{T}} Q_{\psi_K} G Q^{\mathsf{T}}_{\psi_K} \hat{x} \right)^{\frac{q(p+1)}{2}} d\hat{x},$$

where matrix $\mathbb{G}$ is given by (cf. (4.24) with $s_2 = p + 1$)

$$\mathbb{G} := \mathbb{S}_K \mathbb{Q}_{\phi_K}^\mathsf{T} \mathbb{Q}_{\varphi_w} \mathbb{D}_{\varrho_w,\varrho_w}^{[p+1]} \mathbb{Q}_{\varphi_w}^\mathsf{T} \mathbb{Q}_{\phi_K} \mathbb{S}_K. \tag{4.35}$$

Let $\hat{\mathscr{E}} = \{\hat{x} \in \mathbb{R}^3 : |\hat{x}| \le 1\}$ be the unit ball which circumscribes the reference tetrahedron $\hat{K}$. Then relation (4.34) yields

$$\left\| w_{x_K,p}^{\mathrm{int}} \right\|_{L^q(K)}^q \le A_w^q \mu_K^{q(p+1)+3} \int_{\hat{\mathscr{E}}} \left(\hat{x}^\mathsf{T} \mathbb{G} \hat{x} \right)^{\frac{q(p+1)}{2}} \mathrm{d}\hat{x}, \tag{4.36}$$

where we have used the fact that $\mathbb{Q}_{\psi_K}$ is a 3D rotation matrix, and thus the transformation $\hat{x} \to \mathbb{Q}_{\psi_K}^\mathsf{T} \hat{x}$ maps $\hat{\mathscr{E}}$ onto itself. Using (4.30) with $s_1 := \frac{q(p+1)}{2}$, we obtain (4.31). $\qquad\square$

Corollary 4.11 *Let the assumptions from Lemma 4.11 be valid and $q = 2$. Then*

$$\left\| w_{x_K,p}^{\mathrm{int}} \right\|_{L^2(K)}^2 \le \frac{1}{2(p+1)+3} A_w^2 \mu_K^{2(p+1)+3} G\left(p+1, p+1, \rho_w, \varrho_w, \varphi_w; \sigma_K, \varsigma_K, \phi_K\right), \tag{4.37}$$

where G is the function from (4.29).

4.3.2   Estimate in the $L^2(\partial K)$-Norm

Lemma 4.12 *Let K be a tetrahedron with the geometry $\{\mu_K, \sigma_K, \varsigma_K, \phi_K\}$ (cf. Definition 4.1), $u \in C^\infty(K)$, $p \in \mathbb{N}$, and $w_{x_K,p}^{\mathrm{int}}$ be the corresponding interpolation error function located at the barycenter x_K defined by (4.9). Let $\{A_w, \rho_w, \varrho_w, \varphi_w\}$ be the anisotropic bound of $w_{x_K,p}^{\mathrm{int}}$ given by Definition 4.7. Then*

$$\left\| w_{x_K,p}^{\mathrm{int}} \right\|_{\partial K}^2 \le A_w^2 \mu_K^{2p+3} \sigma_K \varsigma_K G\left(p+1, p+1, \rho_w, \varrho_w, \varphi_w; \sigma_K, \varsigma_K, \phi_K\right), \tag{4.38}$$

where G is the function from (4.29).

Proof We proceed in the same way as in the proof of Lemma 3.29. We employ relation (3.85), i.e.,

$$\int_{\partial K} f(x) \, \mathrm{d}S = \int_{\partial \hat{K}} \hat{f}(\hat{x}) |\mathbb{F}_K \cdot t| \, \mathrm{d}\hat{s} \le \int_{\partial \hat{K}} \hat{f}(\hat{x}) \|\mathbb{F}_K\|_2 \, \mathrm{d}\hat{s}, \tag{4.39}$$

and the 3D variant of (3.86) having the form

$$\|\mathbb{F}_K\|_2 = \mathrm{Sp}(\mathbb{F}_K^\mathsf{T} \mathbb{F}_K)^{1/2} = \mathrm{Sp}(\mathbb{L}_K) = \mu_K \sigma_K \varsigma_K, \tag{4.40}$$

which follows from (4.1)–(4.4). Then, similarly as in (3.87), using (4.39), (4.40), (4.21), and (4.33), we have

$$\left\| w_{x_K,p}^{\text{int}} \right\|_{\partial K}^2 = \int_{\partial K} \left| w_{x_K,p}^{\text{int}}(x) \right|^2 dS \tag{4.41}$$

$$\leq A_w^2 \int_{\partial K} \left((x - \bar{x})^{\mathsf{T}} Q_{\varphi_w} D_{\rho_w,\varrho_w}^{[p+1]} Q_{\varphi_w}^{\mathsf{T}} (x - \bar{x}) \right)^{p+1} dS$$

$$\leq A_w^2 \mu_K^{2p+3} \sigma_K \varsigma_K \int_{\partial \hat{K}} \left(\hat{x}^{\mathsf{T}} Q_{\psi_K} \mathbb{G} Q_{\psi_K}^{\mathsf{T}} \hat{x} \right)^{p+1} d\hat{x},$$

where $\mathbb{G}$ is the same as in (4.35).

Using the same arguments as in the proof of Lemma 3.29, we bound the integral over ∂K by the integral over $\partial \hat{\mathcal{E}}$, and hence

$$\left\| w_{x_K,p}^{\text{int}} \right\|_{\partial K}^2 \leq A_w^2 \mu_K^{2p+3} \sigma_K \varsigma_K \int_{\partial \hat{\mathcal{E}}} \left(\hat{x}^{\mathsf{T}} \mathbb{G} \hat{x} \right)^{p+1} d\hat{x}. \tag{4.42}$$

Finally, using (4.29) with $s_1 = p + 1$, we obtain (4.38). $\qquad\square$

4.3.3 Estimate in the $H^1(\partial K)$-Seminorm

Similarly, as in Sect. 3.3.5, we derive the estimate of the interpolation error function $w_{\bar{x},p}^{\text{int}}$ in the $H^1(\partial K)$-seminorm which is employed in Chap. 8. Let $u \in C^\infty(K)$ be a given function and $x_K = (x_{K,1}, x_{K,2}, x_{K,3})$ be the barycenter of K. A direct computation (similarly as Lemma 3.25) allows to express the square of the magnitude of gradient $|\nabla w_{\bar{x},p}^{\text{int}}(x)|^2$ in the form

$$|\nabla w_{x_K,p}^{\text{int}}(x)|^2 = \sum_{l,j,i=0}^{l+j+i\leq 2p} \gamma_{l,i,j}(x_1 - x_{K,1})^l (x_2 - x_{K,2})^i (x_3 - x_{K,3})^j, \tag{4.43}$$

where $\gamma_{l,i,j} \in \mathbb{R}$ are coefficients depending on the derivatives of u evaluated at x_K for all non-negative integers l, j, i such that $l+j+i \leq 2p$. In virtue of Theorem 4.5, there exists the anisotropic bound

$$|\nabla w_{x_K,p}^{\text{int}}(x)|^2 \leq A_{w'} \left((x - \bar{x})^{\mathsf{T}} Q_{\varphi_{w'}} D_{\rho_{w'},\varrho_{w'}}^{[2p]} Q_{\varphi_{w'}}^{\mathsf{T}} (x - \bar{x}) \right)^p \quad \forall x \in \Omega, \tag{4.44}$$

where $A_{w'} > 0$, $\rho_{w'} \geq 1$, $\varrho_{w'} \geq 1$, and $\varphi_{w'} \in \mathcal{U}$.

Consequently, we formulate the corresponding error estimate, and the proof is completely analogous to the proofs of previous lemmas.

Lemma 4.13 *Let K be a tetrahedron with the geometry $\{\mu_K, \sigma_K, \varsigma_K, \phi_K\}$ (cf. Definition 4.1), $u \in C^\infty(K)$, $p \in \mathbb{N}$, and $w_{x_K,p}^{int}$ be the corresponding interpolation error function located at the barycenter x_K defined by (4.9). Let $\{A_{w'}, \rho_{w'}, \varrho_{w'}, \varphi_{w'}\}$ be the anisotropic bound of $|\nabla w_{x_K,p}^{int}|^2$ given by (4.44). Then*

$$\left\| \nabla w_{x_K,p}^{int} \right\|_{\partial K}^2 \leq A_{w'} \mu_K^{2p+3} \sigma_K \varsigma_K G\left(p, 2p, \varrho_{w'}, \varrho_{w'}, \varphi_{w'}; \sigma_K, \varsigma_K, \phi_K \right),$$

$$(4.45)$$

where G is the function from (4.29).

Chapter 5
Anisotropic Mesh Adaptation Method, h-Variant

We discuss the construction of optimal meshes with respect to the interpolation error introduced in Chaps. 3 and 4. In particular, the goal is to construct a simplicial mesh such that the corresponding interpolation error is minimal, while the number of degrees of freedom is bounded from above. Alternatively, we seek a mesh such that the interpolation error is below a given tolerance, whereas the number of degrees of freedom is minimal. At the core of our approach is the continuous mesh formulation which allows one to use standard tools of variational calculus. Finally, we present an anisotropic mesh adaptation algorithm for the numerical solution of partial differential equations. Its performance is demonstrated by several numerical experiments. Here, we deal with the h-variant only, i.e., the polynomial degree of approximation is arbitrary but fixed. The extension to hp-adaptation is given in the next chapter.

We consider spaces of discontinuous piecewise polynomial functions (cf. (1.22))

$$S_{hp} := \{v_h \in L^2(\Omega);\ v_h|_K \in P^p(K)\ \forall K \in \mathcal{T}_h\}, \tag{5.1}$$

where $\mathcal{T}_h = \{K\}$ is a simplicial mesh of the domain $\Omega \subset \mathbb{R}^d$, $d = 2, 3$, and $p \in \mathbb{N}$ is the (constant) polynomial approximation degree.

For given S_{hp}, we shall consider the interpolation operator

$$\pi_h : C^\infty \to S_{hp}, \qquad \pi_h|_K := \Pi_{x_K, p}|_K \quad \forall K \in \mathcal{T}_h, \tag{5.2}$$

where $C^\infty := C^\infty(\Omega)$ and $\Pi_{x_K, p}$ is the interpolation operator of degree p at the barycenter x_K, as in Definition 3.2. This section revolves around finding optimal, or at least nearly optimal, approximation spaces with respect to this interpolation operator. To that end, consider the following problem:

© The Author(s), under exclusive license to Springer Nature Switzerland AG 2022
V. Dolejší, G. May, *Anisotropic hp-Mesh Adaptation Methods*, Nečas Center Series,
https://doi.org/10.1007/978-3-031-04279-9_5

Problem 5.1 Let $u \in C^\infty$. For $\omega > 0$, find an admissible simplicial mesh $\mathcal{T}_h$, such that, for some suitable norm $\|\cdot\|$,

$$\|u - \pi_h u\| \leq \omega, \tag{5.3}$$

and the number of mesh elements $\#\mathcal{T}_h =: N$ is minimal.

Alternatively, one may consider the problem.

Problem 5.2 Let $u \in C^\infty$. For $N \in \mathbb{N}$, find an admissible simplicial mesh $\mathcal{T}_h$ with $\#\mathcal{T}_h = N$ elements, such that, for some suitable norm $\|\cdot\|$,

$$\|u - \pi_h u\| \to \min. \tag{5.4}$$

In general, Problems 5.1 and 5.2 cannot be solved by analytic means. However, based on the error estimates derived in the previous chapters, one may devise heuristic algorithms to approximate the solution. The general strategy is to apply a two-step approach: First, the anisotropy of the mesh is locally optimized. Then, the mesh element size distribution is fixed. While the first step consistently uses the optimization method put forth in Sects. 5.1 and 5.2 for triangles and tetrahedra, respectively, the second step can be implemented in a variety of ways. This will be discussed in subsequent sections.

5.1 Optimization of the Mesh Element Anisotropy (2D)

In Sect. 3.1, we introduced the geometry of triangle by the triplet $\{\mu_K, \sigma_K, \phi_K\}$, representing the size, the shape, and the orientation of K, respectively. The last two define the anisotropy of the mesh element (cf. Definition 3.1). In Sect. 3.2, we defined, for a given function $u \in C^\infty$, the *interpolation error function* (cf. (3.10))

$$w_{\bar{x},p}^{\text{int}} \approx u - \Pi_{\bar{x},p} u, \tag{5.5}$$

where $\Pi_{\bar{x},p}$ is the polynomial projection of degree p at node $\bar{x}$ defined in Definition 3.2. Furthermore, we described the anisotropy of $w_{\bar{x},p}^{\text{int}}$ by the triplet of quantities $\{A_w, \rho_w \varphi_w\}$, $A_w > 0$, $\rho_w \geq 1$, $\varphi_w \in [0, \pi)$ (cf. Definition 3.18). These quantities are computable by evaluation of the partial derivatives of u at $\bar{x}$. Finally, in Sect. 3.3, we derived estimates of $w_{x_K,p}^{\text{int}}$ in several norms, where x_K is the barycenter of a given triangle K. These estimates depend on $\{\mu_K, \sigma_K, \phi_K\}$ as well as on $\{A_w, \rho_w \varphi_w\}$.

In this section, we address the following task:

Problem 5.3 Let $u \in C^\infty$ be a given function, and let x_K be the barycenter of the given triangle K. Find a new triangle K^{opt} minimizing the norm of the corresponding interpolation error function among all triangles and having the same area and barycenter as K.

Relation (2.28) implies that the area of K is equal (up to a multiplicative constant $3\sqrt{3}/4$) to the square of its size μ_K. Therefore, Problem 5.3 is equivalent to the following minimization problem:

$$\min_{\sigma_K \geq 1,\ \phi_K \in [0,\pi)} \left\| w_{x_K,p}^{\text{int}} \right\|_{X(K)}, \tag{5.6}$$

where $w_{x_K,p}^{\text{int}}$ is the interpolation error given by (3.10) and $\|\cdot\|_{X(K)}$ denotes a Sobolev norm over K. We mentioned at the end of Sect. 3.2.2 that the interpolation error function $w_{x_K,p}^{\text{int}}$ depends on $p + 2$-parameters. Therefore, the minimization problem (5.6) is rather difficult to solve exactly. However, employing the estimates of the interpolation error function derived in Sect. 3.3, we can find analytical formulas for $\sigma_K = \sigma_K(\rho_w)$ and $\phi_K = \phi_K(\varphi_w)$ so as to minimize the upper bound for the interpolation error.

First, recall that the estimates (3.59), (3.69), and (3.80) are a function of (σ_K, ϕ_K) only through the function G given by (3.57), i.e.,

$$G(s_1, s_2, \rho_w, \varphi_w; \sigma_K, \phi_K) := \int_0^{2\pi} \left(\xi^{\mathsf{T}}(t)\, \mathbb{G}\, \xi(t) \right)^{s_1} dt, \tag{5.7}$$

where $\xi(t) = (\cos t, \sin t)^{\mathsf{T}}$,

$$\mathbb{G} = \begin{pmatrix} \sigma_K^2 (\cos^2 \tau_K + \rho_w^{-2/s_2} \sin^2 \tau_K) & -\sin \tau_K \cos \tau_K (1 - \rho_w^{-2/s_2}) \\ -\sin \tau_K \cos \tau_K (1 - \rho_w^{-2/s_2}) & \sigma_K^{-2}(\sin^2 \tau_K + \rho_w^{-2/s_2} \cos^2 \tau_K) \end{pmatrix}, \tag{5.8}$$

and $\tau_K = \phi_K - \varphi_w$. Therefore, instead of (5.6), we consider the problem

$$\min_{\sigma_K \geq 1,\ \phi_K \in [0,\pi)} G(s_1, s_2, \rho_w, \varphi_w; \sigma_K, \phi_K) \tag{5.9}$$

for the given set of parameters s_1, s_2, ρ_w, and φ_w.

Since matrix $\mathbb{G}$ is symmetric, it may be diagonalized in the form

$$\mathbb{G} = \mathbb{Q}_\psi^{\mathsf{T}} \mathbb{D} \mathbb{Q}_\psi, \tag{5.10}$$

where $\mathbb{Q}_\psi$ is a rotation through angle ψ and $\mathbb{D} = \text{diag}(d_1, d_2)$ with $d_i > 0$, $i = 1, 2$, which are the eigenvalues of $\mathbb{G}$. Furthermore, since $\mathbb{Q}_\psi$ is a rotation, we have

$$\mathbb{Q}_\psi \xi(t) = \begin{pmatrix} \cos \psi & -\sin \psi \\ \sin \psi & \cos \psi \end{pmatrix} \begin{pmatrix} \cos t \\ \sin t \end{pmatrix} = \begin{pmatrix} \cos(\psi + t) \\ \sin(\psi + t) \end{pmatrix}. \tag{5.11}$$

Hence, using (5.10) and (5.11), the transformation of variables $t \leftarrow \psi + t$, and employing 2π-periodicity of the argument, the right-hand side of (5.7) reads

$$\int_0^{2\pi} \left(\xi^T(t)\mathbb{G}\xi(t)\right)^{s_1} dt = \int_0^{2\pi} \left(\xi^T(t)\mathbb{Q}_\psi^T\mathbb{D}\mathbb{Q}_\psi\xi(t)\right)^{s_1} dt \qquad (5.12)$$

$$= \int_0^{2\pi} \left(\cos^2(t)d_1 + \sin^2(t)d_2\right)^{s_1} dt.$$

The eigenvalues d_i, $i = 1, 2$, of matrix of $\mathbb{G}$ are the roots of the polynomial

$$(\mathbb{G}_{11} - d_i)(\mathbb{G}_{22} - d_i) - \mathbb{G}_{12}^2 = d_i^2 - (\mathbb{G}_{11} + \mathbb{G}_{22})d_i + \mathbb{G}_{11}\mathbb{G}_{22} - \mathbb{G}_{12}^2 = 0, \qquad (5.13)$$

where $\mathbb{G}_{ij}$, $i, j = 1, 2$, are entries of $\mathbb{G}$. Hence,

$$d_{1,2} = \frac{\mathbb{G}_{11} + \mathbb{G}_{22}}{2} \pm \frac{1}{2}\sqrt{(\mathbb{G}_{11} + \mathbb{G}_{22})^2 - 4(\mathbb{G}_{11}\mathbb{G}_{22} - \mathbb{G}_{12}^2)} \qquad (5.14)$$

$$= \frac{\mathbb{G}_{11} + \mathbb{G}_{22}}{2} \pm \frac{1}{2}\sqrt{(\mathbb{G}_{11} - \mathbb{G}_{22})^2 + 4\mathbb{G}_{12}^2}.$$

Moreover, since $\det \mathbb{G} = \det \mathbb{D} = d_1 d_2$, we have due to (3.54) the equality $d_1 d_2 = \rho_w^{-2/s_2}$. Therefore, we set

$$d_1 = a\delta, \quad d_2 = \frac{a}{\delta}, \qquad (5.15)$$

where $a := \rho_w^{-1/s_2}$ is given and $\delta \geq 1$ is a parameter. Then, the last integral in (5.12), together with (5.15), yields

$$\int_0^{2\pi} \left(d_1 \cos^2 t + d_2 \sin^2 t\right)^{s_1} dt = \int_0^{2\pi} a^{s_1}\left(\delta \cos^2 t + \frac{1}{\delta} \sin^2 t\right)^{s_1} dt. \qquad (5.16)$$

In order to solve the minimization problem (5.9), we have to seek the value $\delta \geq 1$, which minimizes the right-hand side of (5.16). We introduce a technical lemma.

Lemma 5.4 *Let $s \geq 1$. We set*

$$S(\delta) := \int_0^{2\pi} \left(\delta \cos^2 t + \frac{1}{\delta} \sin^2 t\right)^s dt, \qquad \delta \geq 1. \qquad (5.17)$$

Then

$$S(\delta) > S(1) = 2\pi \quad \forall \delta > 1. \qquad (5.18)$$

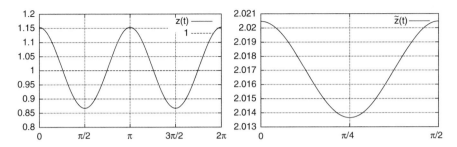

Fig. 5.1 Illustration for the proof of Lemma 5.4, the function $z(t)$ (left) and function $\tilde{z}(t)$ for $\delta = 1.1$ and $s = 1.5$

Proof In (5.17), we set

$$z(t) := \delta \cos^2 t + 1/\delta \sin^2 t. \tag{5.19}$$

For $\delta = 1$, we have $z(t) = 1$ on $[0, 2\pi]$ and thus $S(1) = 2\pi$. So, it is enough to show that $S(\delta) > 2\pi$ for $\delta > 1$.

Let now $\delta > 1$ and $s \geq 1$. An example of function $z(t)$ is shown in Fig. 5.1, left. In the following, we use the inequalities

$$\left(\sqrt{\delta} - \frac{1}{\sqrt{\delta}}\right)^2 > 0 \Rightarrow \delta + \frac{1}{\delta} > 2 \Rightarrow \left(\delta + \frac{1}{\delta}\right)^s > 2^s \Rightarrow \frac{1}{2^s}\left(\delta + \frac{1}{\delta}\right)^s > 1. \tag{5.20}$$

First, we consider the case $s = 1$. Using identity $\int_0^{2\pi} \cos^2(t)\,dt = \int_0^{2\pi} \sin^2(t)\,dt = \pi$, we have

$$S(\delta) = \int_0^{2\pi} \left(\delta \cos^2 t + 1/\delta \sin^2\right) dt = \pi(\delta + 1/\delta) > 2\pi \quad \forall \delta > 1, \tag{5.21}$$

where the last inequality follows from (5.20).

Let $s > 1$. The function $z(t)$ from (5.19) is periodic, and hence we consider the integral of $z(t)$ over $[0, \pi/2]$. Due to the identities $\cos^2(\pi/2 - t) = \sin^2 t$ and $\sin^2(\pi/2 - t) = \cos^2 t$, we have

$$\int_0^{\frac{\pi}{2}} z(t)^s\,dt = \int_0^{\frac{\pi}{4}} \left(z(t)^s + z(\pi/2 - t)^s\right) dt = \int_0^{\frac{\pi}{4}} \tilde{z}(t)\,dt, \tag{5.22}$$

where

$$\tilde{z}(t) := \left(\delta \cos^2 t + \frac{1}{\delta} \sin^2 t\right)^s + \left(\delta \sin^2 t + \frac{1}{\delta} \cos^2 t\right)^s. \tag{5.23}$$

Figure 5.1, right, shows an example of $\tilde{z}$. We show that $\tilde{z}(t) > 2$ $\forall t \in [0, \pi/4]$. Obviously, a direct computation and relation (5.20) imply

$$\tilde{z}(0) = \delta^s + \frac{1}{\delta^s} = \left(\delta^s - 2 + \frac{1}{\delta^s}\right) + 2 = \left(\sqrt{\delta^s} - \frac{1}{\sqrt{\delta^s}}\right)^2 + 2 > 2 \quad \forall \delta > 1 \, \forall s > 1,$$

$$\tilde{z}(\pi/4) = \left(\frac{\delta}{2} + \frac{1}{2\delta}\right)^s + \left(\frac{\delta}{2} + \frac{1}{2\delta}\right)^s = 2\frac{1}{2^s}\left(\delta + \frac{1}{\delta}\right)^s > 2 \quad \forall \delta > 1. \tag{5.24}$$

Furthermore, we show that $\tilde{z}(t)$ is non-increasing on $[0, \pi/4]$. Hence, we have to verify the inequality $\frac{d}{dt}\tilde{z}(t) \leq 0$ for $t \in [0, \pi/4]$, particularly

$$\frac{d}{dt}\tilde{z}(t) = s\left(\delta \cos^2 t + \frac{1}{\delta}\sin^2 t\right)^{s-1}\left(-2\delta \cos t \sin t + \frac{2}{\delta}\sin t \cos t\right) \tag{5.25}$$

$$+ s\left(\delta \sin^2 t + \frac{1}{\delta}\cos^2 t\right)^{s-1}\left(2\delta \sin t \cos t - \frac{2}{\delta}\sin t \cos t\right)$$

$$= s \sin(2t)\left(\frac{1}{\delta} - \delta\right)\left[\left(\delta \cos^2 t + \frac{1}{\delta}\sin^2 t\right)^{s-1} - \left(\delta \sin^2 t + \frac{1}{\delta}\cos^2 t\right)^{s-1}\right] \leq 0.$$

This inequality is satisfied trivially for $t = 0$. Let $t > 0$. Dividing (5.25) by $s \sin(2t)(1/\delta - \delta) < 0$ for $t \in (0, \pi/4)$ and $\delta > 1$, we obtain

$$\left(\delta \cos^2 t + \frac{1}{\delta}\sin^2 t\right)^{s-1} \geq \left(\delta \sin^2 t + \frac{1}{\delta}\cos^2 t\right)^{s-1}$$

$$\Longleftrightarrow \qquad \delta \cos^2 t + \frac{1}{\delta}\sin^2 t \geq \delta \sin^2 t + \frac{1}{\delta}\cos^2 t$$

$$\Longleftrightarrow \qquad \left(\delta - \frac{1}{\delta}\right)\left(\cos^2 t - \sin^2 t\right) \geq 0,$$

which is true for $t \in (0, \pi/4)$ and $\delta > 1$. Moreover, we have $\frac{d}{dt}\tilde{z}(0) = \frac{d}{dt}\tilde{z}(\pi/4) = 0$. Hence, the function $\tilde{z}(t)$ is non-increasing on $(0, \pi/4)$ and attains its minimum for $t = \pi/4$. Using (5.24), we conclude that $\tilde{z}(t) > 2$ on $[0, \pi/4]$, which together with (5.22) implies

$$\int_0^{\frac{\pi}{2}} z(t)^s \, dt = \int_0^{\frac{\pi}{4}} \tilde{z}(t) \, dt > \int_0^{\frac{\pi}{4}} 2 \, dt = \frac{\pi}{2}.$$

Similarly, we can prove that $\int_{(k-1)\pi/2}^{k\pi/2} z(t) \, dt > \pi/2$, $k = 2, 3, 4$, which gives (5.18). $\qquad\square$

Lemma 5.4 implies that the right-hand side of (5.16) is minimal for $\delta = 1$. Therefore, (5.15) gives $d_1 = d_2$, and consequently (5.14) implies that $\mathbb{G}_{11} = \mathbb{G}_{22}$ and $\mathbb{G}_{12} = 0$. From (5.8), we find that $\mathbb{G}_{12} = 0$ if either $\sin \tau_K = 0$ or $\cos \tau_K = 0$. We consider both cases separately:

(i) Let $\cos \tau_K = 0$, then

$$\cos \tau_K = 0 \Rightarrow \tau_K = \phi_K - \varphi_w = \pi/2 \qquad (5.26)$$

and then $\quad \mathbb{G}_{11} = \mathbb{G}_{22} \Rightarrow \sigma_K^2 \rho_w^{-2/s_2} = \sigma_K^{-2} \Rightarrow \sigma_K = \rho_w^{1/(2s_2)}.$

(ii) Let $\sin \tau_K = 0$, then

$$\sin \tau_K = 0 \Rightarrow \tau_K = \phi_K - \varphi_w = 0 \qquad (5.27)$$

and then $\quad \mathbb{G}_{11} = \mathbb{G}_{22} \Rightarrow \sigma_K^2 = \sigma_K^{-2} \rho_w^{-2/s_2} \Rightarrow \sigma_K = \rho_w^{-1/(2s_2)},$

which is unacceptable because $\rho_w \geq 1$ and we require $\sigma_K \geq 1$.

Therefore, from (5.26), we have $\mathbb{G}_{11} = \mathbb{G}_{22} = \rho_w^{-1/s_2}$ and thus

$$\mathbb{G} := \begin{pmatrix} \rho_w^{-1/s_2} & 0 \\ 0 & \rho_w^{-1/s_2} \end{pmatrix}, \qquad (5.28)$$

and consequently, using (5.12) and Lemma 5.4,

$$\int_0^{2\pi} \left(\xi^{\mathsf{T}}(t) \mathbb{G} \xi(t) \right)^{s_1} \mathrm{d}t = 2\pi \rho_w^{-s_1/s_2}. \qquad (5.29)$$

This derivation leads to the following result, which is the solution of Problem (5.9):

Lemma 5.5 *Let $s_1 \geq 1$, $s_2 \geq 1$, $\rho_w \geq 1$, and $\varphi_w \in [0, \pi)$ be given. The function $G = G(s_1, s_2, \rho_w, \varphi_w; \sigma_K, \phi_K)$ given by (5.7) attains its minimum for*

$$\sigma_K = \sigma_K^\star := \rho_w^{1/(2s_2)}, \qquad \phi_K = \phi_K^\star := \pi/2 + \varphi_w. \qquad (5.30)$$

The corresponding minimum is

$$G(s_1, s_2, \rho_w, \varphi_w; \sigma_K^\star, \phi_K^\star) = 2\pi \rho_w^{-s_1/s_2}. \qquad (5.31)$$

We observe that the orientation $\phi_K^\star$ is perpendicular to the orientation of the interpolation error function φ_w, i.e., the "element is small" along the direction of the highest directional derivative $\mathrm{d}^{p+1} u(\bar{x}; \cdot)$, which is in agreement with the general expectation.

Moreover, we need a similar result also for the function $\overline{G}$ given by (3.58), which bounds the interpolation error in the $L^\infty(K)$-norm (cf. Lemma 3.23, estimate (3.70)):

$$\overline{G}(s_1, s_2, \rho_w, \varphi_w; \sigma_K, \phi_K) := \max_{t \in [0, 2\pi]} \left(\xi^{\mathsf{T}}(t) \mathbb{G} \xi(t) \right)^{s_1}, \qquad (5.32)$$

where $\xi(t) = (\cos t, \sin t)^{\mathsf{T}}$ and $\mathbb{G}$ is given by (5.8). Therefore, for the L^∞ case, we consider the minimization problem

$$\min_{\sigma_K \geq 1, \; \phi_K \in [0,\pi)} \overline{G}(s_1, s_2, \rho_w, \varphi_w; \sigma_K, \phi_K) \tag{5.33}$$

for the given set of parameters s_1, s_2, ρ_w, and φ_w. Using the same derivation as in (5.10) – (5.14), we have

$$\max_{t \in [0,2\pi]} \left(\xi^{\mathsf{T}}(t) \mathbb{G} \xi(t)\right)^{s_1} = \max_{t \in [0,2\pi]} \left(\cos^2(t) d_1 + \sin^2(t) d_2\right)^{s_1}, \tag{5.34}$$

where $d_1 \geq 0$ and $d_2 \geq 0$ are the eigenvalues of $\mathbb{G}$ satisfying $\det \mathbb{G} = d_1 d_2 = \rho_w^{-2/s_2}$ due to (3.54). Obviously,

$$\max_{t \in [0,2\pi]} \left(\cos^2(t) d_1 + \sin^2(t) d_2\right)^{s_1} = (\max(d_1, d_2))^{s_1}. \tag{5.35}$$

Hence, the right-hand side of (5.34) is minimal for $d_1 = d_2 = \rho_w^{-1/s_2}$. Then (5.26)–(5.28) are valid also for the L^∞ case and moreover

$$\max_{t \in [0,2\pi]} \left(\xi^{\mathsf{T}}(t) \mathbb{G} \xi(t)\right)^{s_1} = \rho_w^{-s_1/s_2}. \tag{5.36}$$

This leads to the following result which is the solution of problem (5.33):

Lemma 5.6 *Let $s_1 \geq 1$, $s_2 \geq 1$, $\rho_w \geq 1$, and $\varphi_w \in [0, \pi)$ be given. The function* $\overline{G} = \overline{G}(s_1, s_2, \rho_w, \varphi_w; \sigma_K, \phi_K)$ *given by (5.32) attains its minimum for*

$$\sigma_K = \sigma_K^\star := \rho_w^{1/(2s_2)}, \qquad \phi_K = \phi_K^\star := \pi/2 + \varphi_w. \tag{5.37}$$

The corresponding minimum is

$$\overline{G}(s_1, s_2, \rho_w, \varphi_w; \sigma_K^\star, \phi_K^\star) = \rho_w^{-s_1/s_2}. \tag{5.38}$$

Finally, employing Lemmas 5.5 and 5.6, we can formulate the approximate solution of Problem 5.3 for several norms. We start with the L^q-norm for $1 \leq q < \infty$.

Theorem 5.7 *Let $p \in \mathbb{N}$, $u \in C^\infty$ be a given function, and let x_K be the barycenter of the given triangle K with size $\mu_K \sim \sqrt{|K|}$. Let $\{A_w, \rho_w \varphi_w, \}$, $A_w > 0$, $\rho_w \geq 1$, $\varphi_w \in [0, \pi)$ be the quantities defining the anisotropic bound of the interpolation error function $w_{x_K,p}^{\text{int}}$ at x_K. Then, triangle K^{opt}, having the size μ_K and minimizing the upper bound of the interpolation error from Lemma 3.21, has the anisotropy*

$$\sigma_K^\star = \rho_w^{1/(2p+2)}, \qquad \phi_K^\star = \pi/2 + \varphi_w. \tag{5.39}$$

Moreover, the corresponding interpolation error is bounded by

$$\left\| w^{\text{int}}_{x_K,p} \right\|^q_{L^q(K^{\text{opt}})} \leq \frac{2\pi}{q(p+1)+2} A^q_w \rho_w^{-q/2} \mu_K^{q(p+1)+2}, \qquad 1 \leq q < \infty.$$

$$(5.40)$$

Proof The estimate (3.59) from Lemma 3.21 bounds the interpolation error by G with the arguments $s_1 = q(p+1)/2$ and $s_2 = p+1$. Inserting these values in (5.30) gives (5.39). Furthermore, substituting (5.31) into (3.59) gives (5.40).   □

We proceed with the corresponding estimates in the H^1-seminorm, which can be formulated in the same manner.

Theorem 5.8 *Let $p \in \mathbb{N}$, $u \in C^\infty$ be a given function, and let x_K be the barycenter of the given triangle K with size $\mu_K \sim \sqrt{|K|}$. Let $\{A_{w'}, \rho_{w'} \varphi_{w'},\}$, $A_{w'} > 0$, $\rho_{w'} \geq 1$, $\varphi_{w'} \in [0, \pi)$ be the quantities defining the anisotropic bound of the square of the magnitude of the interpolation error function $|\nabla w^{\text{int}}_{x_K,p}|^2$ at x_K. Then, triangle K^{opt}, having the size μ_K and minimizing the upper bound of the interpolation error from Lemma 3.27, has the anisotropy*

$$\sigma^\star_K = \rho_{w'}^{1/(4p)}, \qquad \phi^\star_K = \pi/2 + \varphi_{w'}. \tag{5.41}$$

Moreover, the corresponding interpolation error is bounded by

$$\left\| \nabla w^{\text{int}}_{x_K,p} \right\|^2_{L^2(K^{\text{opt}})} \leq \frac{2\pi}{2p+2} A_{w'} \rho_{w'}^{-1/2} \mu_K^{2p+2}. \tag{5.42}$$

Proof The estimate (3.80) from Lemma 3.27 bounds the interpolation error by G with the arguments $s_1 = p$ and $s_2 = 2p$. Inserting these values in (5.30) gives (5.41). Furthermore, substituting (5.31) into (3.80) gives (5.42).   □

Finally, the optimization of the element shape in the L^∞-norm is given by the following result.

Theorem 5.9 *Let $p \in \mathbb{N}$, $u \in C^\infty$ be a given function, and let x_K be the barycenter of the given triangle K with size $\mu_K \sim \sqrt{|K|}$. Let $\{A_w, \rho_w \varphi_w,\}$, $A_w > 0$, $\rho_w \geq 1$, $\varphi_w \in [0, \pi)$ be the quantities defining the anisotropic bound of the interpolation error function $w^{\text{int}}_{x_K,p}$ at x_K. Then, triangle K^{opt}, having the size μ_K and minimizing the upper bound of the interpolation error from Lemma 3.23, has the anisotropy*

$$\sigma^\star_K = \rho_w^{1/(2p+2)}, \qquad \phi^\star_K = \pi/2 + \varphi_w. \tag{5.43}$$

Moreover, the corresponding interpolation error is bounded by

$$\left\| w^{\text{int}}_{x_K,p} \right\|_{L^\infty(K^{\text{opt}})} \leq A_w \rho_w^{-1/2} \mu_K^{p+1}. \tag{5.44}$$

Proof The estimate (3.70) from Lemma 3.23 bounds the interpolation error by $\overline{G}$ with the arguments $s_1 = (p + 1)/2$ and $s_2 = p + 1$. Inserting these values in (5.37) gives (5.43). Furthermore, substituting (5.38) into (3.70) gives (5.44).   □

5.2   Optimization of the Mesh Element Anisotropy (3D)

In Sect. 4.1, we defined the geometry of a tetrahedron K by $\{\mu_K, \sigma_K, \varsigma_K, \phi_K\}$, representing the size, the shape, and the orientation of K, respectively. The last two define the anisotropy of the mesh element (cf. Definition 4.1). In Sect. 4.2, we defined, for a given function $u \in C^\infty$, the interpolation error function $w^{\text{int}}_{\bar{x},p} \approx u(x) - \Pi_{\bar{x},p}u(x)$, where $\Pi_{\bar{x},p}$ is a polynomial projection associated with the given node $\bar{x}$ (cf. Definition 4.2). We described the anisotropic bound of $w^{\text{int}}_{\bar{x},p}$ by the quantities $\{A_w, \rho_w \varrho_w, \boldsymbol{\varphi}_w\}$, $A_w > 0$, $\rho_w, \varrho_w \geq 1$, $\boldsymbol{\varphi}_w \in \mathscr{U}$ (cf. Definition 4.7). These quantities are computable by evaluation of the partial derivatives at $\bar{x}$. Moreover, in Sect. 4.3, we derived estimates of $w^{\text{int}}_{x_K,p}$ in several norms where x_K is the barycenter of the tetrahedron K. These estimates depend on $\{\mu_K, \sigma_K, \varsigma_K, \phi_K\}$ as well as on $\{A_w, \rho_w \varrho_w, \boldsymbol{\varphi}_w\}$.

In this section, we consider the three-dimensional analog of Problem 5.3.

Problem 5.10 Let $u \in C^\infty$ be a given function, and let x_K be the barycenter of the given tetrahedron K. Find a new tetrahedron K^{opt}, minimizing the norm of the corresponding interpolation error function among all tetrahedra having the same volume and barycenter as K.

Using the same arguments as in Sect. 5.1, we employ the interpolation error estimate (4.31). For fixed size μ_K, the error estimate is directly proportional to the function G, given by (4.29) as

$$G(s_1, s_2, \rho_w, \varrho_w, \boldsymbol{\varphi}_w; \sigma_K, \varsigma_K, \phi_K) = \int_{\partial\hat{\mathscr{C}}} (\hat{x}^\mathsf{T}\mathbb{G}\hat{x})^{s_1} \, \mathrm{d}\hat{S}, \tag{5.45}$$

where the matrix $\mathbb{G}$ is given by (4.24). Therefore, we replace Problem 5.10 by the minimization problem (compare with (5.9))

$$\min_{\sigma_K, \varsigma_K \geq 1, \, \phi_K \in \mathscr{U}} G(s_1, s_2, \rho_w, \varrho_w, \boldsymbol{\varphi}_w; \sigma_K, \varsigma_K, \phi_K) \tag{5.46}$$

for the given set of parameters $s_1 \geq 1$, $s_2 \geq 1$, $\rho_w \geq 1$, $\varrho_w \geq 1$, $\boldsymbol{\varphi}_w \in \mathscr{U}$. To solve this problem, first recall that matrix $\mathbb{G}$ in (5.45) is symmetric. Hence it can be diagonalized as

$$\mathbb{G} = \mathbb{Q}^\mathsf{T}\mathbb{L}\mathbb{Q}, \tag{5.47}$$

where $\mathbb{Q}$ is a unitary (rotation) matrix and $\mathbb{L}$ is diagonal. We write $\mathbb{L}$ in the form

$$\mathbb{L} = \begin{pmatrix} a\delta_1\delta_2 & 0 & 0 \\ 0 & a\delta_2 & 0 \\ 0 & 0 & a/\delta_2 \end{pmatrix}, \tag{5.48}$$

where $a > 0$, $\delta_1 \geq 1$, and $\delta_2 \geq 1$. As the transformation $\hat{y} = \mathbb{Q}\hat{x}$ maps the unit sphere onto itself, we may write

$$\int_{\partial\hat{\mathscr{E}}} (\hat{x}^{\mathsf{T}} \mathbb{G} \hat{x})^{s_1} \, d\hat{S} = \int_{\partial\hat{\mathscr{E}}} (\hat{y}^{\mathsf{T}} \mathbb{L} \hat{y})^{s_1} \, d\hat{S}. \tag{5.49}$$

Moreover, in spherical coordinates, $\hat{y} \in \partial\hat{\mathscr{E}}$ is written

$$\hat{y} = (\cos\theta \sin\phi, \sin\theta \sin\phi, \cos\phi)^T, \tag{5.50}$$

and the integral (5.49) becomes

$$\int_{\partial\hat{\mathscr{E}}} (\hat{y}^{\mathsf{T}} \mathbb{L} \hat{y})^{s_1} \, d\hat{S} = a^{s_1} \int_{\theta=0}^{2\pi} \int_{\phi=0}^{\pi} f(\phi, \theta; \delta_1, \delta_2)^{s_1} \sin\phi \, d\phi \, d\theta =: I_{s_1}(\delta_1, \delta_2), \tag{5.51}$$

where

$$f(\phi, \theta; \delta_1, \delta_2) = \delta_1\delta_2 \sin^2\phi \cos^2\theta + \delta_2 \sin^2\phi \sin^2\theta + \frac{\cos^2\phi}{\delta_2} \geq 0. \tag{5.52}$$

The minimization problem (5.46) is thus equivalent to minimizing the integral (5.51) in the parameters δ_1, δ_2. It turns out that the latter problem is relatively straightforward to solve. The solution can then be expressed in terms of the optimal anisotropy parameters $\sigma_K^\star$, $\varsigma_K^\star$, $\phi_K^\star$.

Lemma 5.11 *The integral (5.51) is minimized for $\delta_1 = \delta_2 = 1$.*

Proof For arbitrary $\bar{\delta}_1, \bar{\delta}_2 \geq 1$, we write the integral (5.51) as

$$I_{s_1}(\bar{\delta}_1, \bar{\delta}_2) = I_{s_1}(1, 1) + \underbrace{\int_{\delta_2=1}^{\delta_2=\bar{\delta}_2} \frac{\partial I_{s_1}}{\partial \delta_2}(1, \delta_2) d\delta_2}_{\mathbf{I}} + \underbrace{\int_{\delta_1=1}^{\delta_1=\bar{\delta}_1} \frac{\partial I_{s_1}}{\partial \delta_1}(\delta_1, \bar{\delta}_2) d\delta_1}_{\mathbf{II}} . \tag{5.53}$$

It is a matter of straightforward calculus to show that terms **I** and **II** are non-negative.

First note that the integral is monotonic in δ_2. Direct differentiation gives

$$\frac{\partial f^{s_1}}{\partial \delta_2} = s_1 f^{s_1-1}\left(\delta_1 \sin^2 \phi \, \cos^2 \theta + \sin^2 \phi \, \sin^2 \theta - \frac{\cos^2 \phi}{\delta_2^2}\right) \qquad (5.54)$$

and

$$\frac{\partial^2 f^{s_1}}{\partial \delta_2{}^2} = s_1(s_1 - 1)f^{s_1-2}\left(\delta_1 \sin^2 \phi \, \cos^2 \theta + \sin^2 \phi \, \sin^2\theta - \frac{\cos^2 \phi}{\delta_2^2}\right)^2 \qquad (5.55)$$

$$+ s_1 f^{s_1-1} 2\frac{\cos^2 \phi}{\delta_2^3} \geq 0,$$

which immediately implies

$$\frac{\partial^2 I_{s_1}}{\partial \delta_2{}^2} = a^{s_1} \int_{\theta=0}^{2\pi} \int_{\phi=0}^{\pi} \frac{\partial^2 f^{s_1}}{\partial \delta_2{}^2} \sin \phi \, d\phi \, d\theta \geq 0. \qquad (5.56)$$

Now, since $f(\cdot, \cdot; 1, 1) = 1$ and

$$\frac{\partial f^{s_1}}{\partial \delta_2}\Big|_{\delta_1=\delta_s=1} = s_1 f^{s_1-1}\left(\sin^2 \phi \, \cos^2 \theta + \sin^2 \phi \, \sin^2 \theta - \cos^2 \phi\right) = -s_1 \cos(2\phi),$$

we obtain

$$\frac{\partial I_{s_1}}{\partial \delta_2}(1, 1) = s_1 a^{s_1} \int_{\theta=0}^{2\pi} \int_{\phi=0}^{\pi} -\sin \phi \, \cos(2\phi) \, d\phi \, d\theta = \frac{4\pi}{3} s_1 a^{s_1} > 0. \qquad (5.57)$$

Therefore, the term **I** in (5.53) is positive. Similarly, for term **II**, we have

$$\frac{\partial I_{s_1}}{\partial \delta_1} = a^{s_1} \int_{\theta=0}^{2\pi} \int_{\phi=0}^{\pi} s_1 (f(\delta_1, \delta_2))^{s_1-1} \delta_2 \sin^2 \phi \cos^2 \theta \, d\phi \, d\theta \geq 0, \qquad (5.58)$$

and thus $I_{s_1}(\bar{\delta}_1, \bar{\delta}_2) \geq I_{s_1}(1, 1)$. $\qquad\qquad\qquad\qquad\qquad\qquad\qquad\qquad\qquad\square$

Inserting the optimal values $\delta_1 = \delta_2 = 1$, into (5.48), one has $\mathbb{L} = a\mathbb{I}_3$, where $\mathbb{I}_3$ is the identity matrix in $\mathbb{R}^3$. But this implies

$$\mathbb{G} = \mathbb{Q}^{\mathsf{T}} \mathbb{L} \mathbb{Q} = a\mathbb{I}_3. \qquad (5.59)$$

Using (4.24), we write

$$\mathbb{S}_K^{-1} \mathbb{G} \mathbb{S}_K^{-1} = \mathbb{Q}_{\phi_K}^{\mathsf{T}} \mathbb{Q}_{\varphi_w} \mathbb{D}_{\varrho_w, \varrho_w}^{[s_2]} \mathbb{Q}_{\varphi_w}^{\mathsf{T}} \mathbb{Q}_{\phi_K}, \qquad (5.60)$$

where

$$\mathbb{S}_K = \text{diag}\left(\sigma_K \varsigma_K, \frac{\varsigma_K}{\sigma_K^2}, \frac{\sigma_K}{\varsigma_K^2}\right) \quad \text{and} \quad \mathbb{D}_{\rho_w,\varrho_w}^{[s_2]} = \text{diag}\left(1, \rho_w^{-2/s_2}, \varrho_w^{-2/s_2}\right)$$

(5.61)

are the diagonal matrices from (4.23). Direct computation gives

$$\mathbb{S}_K^{-1}\mathbb{G}\mathbb{S}_K^{-1} = \begin{pmatrix} \frac{a}{\sigma_K^2 \varsigma_K^2} & 0 & 0 \\ 0 & a\frac{\sigma_K^4}{\varsigma_K^2} & 0 \\ 0 & 0 & a\frac{\varsigma_K^4}{\sigma_K^2} \end{pmatrix}.$$

(5.62)

Note that the entries increase along the diagonal. Consequently, since $\mathbb{S}_K^{-1}\mathbb{G}\mathbb{S}_K^{-1}$ and $\mathbb{D}_{\rho_w,\varrho_w}^{[s_2]}$ are diagonal, the orthogonal matrix $\mathbb{Q}_{\phi_K}^{\mathsf{T}}\mathbb{Q}_{\varphi_w}$ is a permutation matrix, which maps the entries of $\mathbb{D}_{\rho_w,\varrho_w}^{[s_2]}$ in ascending order onto the diagonal of $\mathbb{S}_K^{-1}\mathbb{G}\mathbb{S}_K^{-1}$. In virtue of Remark 4.6, the entries of $\mathbb{D}_{\rho_w,\varrho_w}^{[s_2]}$ decrease since $1 \le \rho_w \le \varrho_w$. Then, we have

$$\mathbb{Q}_{\phi_K}^{\mathsf{T}}\mathbb{Q}_{\varphi_w} = \begin{pmatrix} 0 & 0 & 1 \\ 0 & 1 & 0 \\ 1 & 0 & 0 \end{pmatrix}.$$

(5.63)

Setting

$$\mathbb{Q}_{\varphi_w} := (\xi_1 \ \xi_2 \ \xi_3) \in \mathbb{R}^{3\times3}, \qquad \mathbb{Q}_{\phi_K} := (\zeta_1 \ \zeta_2 \ \zeta_3) \in \mathbb{R}^{3\times3},$$

(5.64)

the eigenvectors are thus aligned as

$$\xi_1 \| \zeta_3 \qquad \xi_2 \| \zeta_2 \qquad \xi_3 \| \zeta_1.$$

(5.65)

Moreover, substituting (5.63) into (4.24) and comparing with (5.59) yield

$$\mathbb{G} = \begin{pmatrix} \varrho_w^{-2/s_2}\sigma_K^2\varsigma_K^2 & 0 & 1 \\ 0 & \rho_w^{-2/s_2}\frac{\varsigma_K^2}{\sigma_K^4} & 0 \\ 0 & 0 & \frac{\sigma_K^2}{\varsigma_K^4} \end{pmatrix} = \begin{pmatrix} a & 0 & 0 \\ 0 & a & 0 \\ 0 & 0 & a \end{pmatrix},$$

(5.66)

and thus

$$a = (\varrho_w \rho_w)^{-\frac{2}{3s_2}}, \qquad \sigma_K = \left(\frac{\varrho_w}{\rho_w}\right)^{\frac{1}{3s_2}}, \qquad \varsigma_K = \varrho_w^{\frac{1}{3s_2}}.$$

(5.67)

In summary, we have the following result:

Lemma 5.12 *Let $s_1 \geq 1$, $s_2 \geq 1$, $\varrho_w \geq \rho_w \geq 1$, and $\boldsymbol{\varphi}_w \in \mathscr{U}$ be given. The function $G(s_1, s_2, \rho_w, \varrho_w, \boldsymbol{\varphi}_w; \sigma_K, \varsigma_K, \boldsymbol{\phi}_K)$ given by (5.45) attains its minimum for*

$$\sigma_K = \sigma_K^\star := \left(\frac{\varrho_w}{\rho_w}\right)^{1/(3s_2)}, \qquad \varsigma_K = \varsigma_K^\star := \varrho_w^{\,1/(3s_2)}, \tag{5.68}$$

and $\boldsymbol{\phi}_K = \boldsymbol{\phi}_K^\star$ such that

$$Q_{\boldsymbol{\phi}_K^\star}^{\mathsf{T}} Q_{\boldsymbol{\varphi}_w} = \begin{pmatrix} 0 & 0 & 1 \\ 0 & 1 & 0 \\ 1 & 0 & 0 \end{pmatrix}. \tag{5.69}$$

The corresponding minimum is

$$G(s_1, s_2, \rho_w, \varrho_w, \boldsymbol{\varphi}_w; \sigma_K^\star, \varsigma_K^\star, \boldsymbol{\phi}_K^\star) = 4\pi \, (\varrho_w \rho_w)^{-2s_1/(3s_2)}. \tag{5.70}$$

Proof The optimal anisotropy (5.68) is directly taken from (5.67). Moreover, by inserting the value for a from (5.67) into (5.59) and (5.45), we obtain (5.70). □

We can now state the final result regarding the solution of Problem 5.10.

Theorem 5.13 *Let $p \in \mathbb{N}$, $u \in C^\infty$ be a given function, and let x_K be the barycenter of the given triangle K having the size $\mu_K \sim |K|^{\frac{1}{3}}$. Let $A_w > 0$, $\varrho_w \geq \rho_w \geq 1$, and $\boldsymbol{\varphi}_w \in \mathscr{U}$ be the quantities defining the anisotropy of the interpolation error function $w_{x_K,p}^{\mathrm{int}}$ at x_K. Then, triangle K^{opt}, having the size μ_K and minimizing the interpolation error in the L^q-norm, $1 \leq q < \infty$, has the anisotropy*

$$\sigma_K^\star = (\varrho_w/\rho_w)^{1/(3(p+1))}, \qquad \varsigma_K^\star = \varrho_w^{\,1/(3(p+1))}, \tag{5.71}$$

and $\boldsymbol{\phi}_K^\star \in \mathscr{U}$ such that

$$Q_{\boldsymbol{\phi}_K^\star}^{\mathsf{T}} Q_{\boldsymbol{\varphi}_w} = \begin{pmatrix} 0 & 0 & 1 \\ 0 & 1 & 0 \\ 1 & 0 & 0 \end{pmatrix}. \tag{5.72}$$

Moreover, the corresponding interpolation error is bounded by

$$\left\| w_{x_K,p}^{\mathrm{int}} \right\|_{L^q(K^{\mathrm{opt}})}^q \leq \frac{4\pi}{q(p+1)+3} A_w^q \, (\varrho_w \rho_w)^{-q/3} \, \mu_K^{q(p+1)+3}, \qquad 1 \leq q < \infty. \tag{5.73}$$

Proof The estimate (4.31) from Lemma 4.10 bounds the interpolation error by G with the arguments $s_1 = q(p+1)/2$ and $s_2 = p+1$. Inserting these values in (5.68) gives (5.71). Furthermore, substituting (5.70) into (4.31) gives (5.73). $\square$

Remark 5.14 In a similar way, it is possible to derive the optimal element shape and the corresponding interpolation error bound also for the L^∞-norm and H^1-seminorm.

5.3   Mesh Adaptation Based on the Equidistribution Principle

In the previous sections, we have derived the solution to Problems 5.3 and 5.10. That is, for a given triangle or tetrahedron K, and given $u \in C^\infty$, we can compute optimal anisotropy parameters, minimizing the bound for the interpolation error. It remains to determine the element size distribution for the new triangulation. For simplicity, we restrict ourselves to the two-dimensional case and optimization with respect to the $L^q(\Omega)$-norm, $1 \le q < \infty$. For this case, Theorem 5.7 will be used. Optimization with respect to the $L^\infty(\Omega)$-norm or the $H^1(\Omega)$-seminorm, using Theorems 5.8 and 5.9, as well as the three-dimensional case, using Theorem 5.13, can be formulated analogously.

First, we note that it is easy, starting from (5.40) in Theorem 5.7, to find the area of the locally optimal triangle, such that the interpolation error is below a given threshold. Indeed, let $\omega_K > 0$ be given for each $K \in \mathscr{T}_h$. Then, in (5.40), we set

$$\left\| u - \Pi_{hp} u \right\|_{L^q(K)} \approx \left\| w^{\text{int}}_{x_K, p} \right\|_{L^q(K)} \le \frac{2\pi}{q(p+1)+2} A_w^q \rho_w^{-q/2} \mu_K^{q(p+1)+2} = \omega_K^q,$$

(5.74)

which can be solved for μ_K to give

$$\mu_K = \left(\frac{q(p+1)+2}{2\pi A_w^q \rho_w^{-q/2}} (\omega_K)^q \right)^{\frac{1}{q(p+1)+2}}. \qquad K \in \mathscr{T}_h.$$

(5.75)

It then follows immediately from (5.40) that, for a triangle characterized by $\{\mu_K, \sigma_K^\star, \phi_K^\star\}$, where $\{\sigma_K^\star, \phi_K^\star\}$ are optimal anisotropy parameters in the sense of Theorem 5.7,

$$\left\| w^{\text{int}}_{x_K, p} \right\|_{L^q(K)} \le \omega_K, \qquad K \in \mathscr{T}_h.$$

(5.76)

This can be used to establish a global size distribution according to an equidistribution principle. Let $\omega > 0$ be a given *global* error bound. We consider two cases:

(E1) *Equal error per element*: we require

$$\left\| u - \Pi_{hp} u \right\|_{L^q(K)} \leq \omega \left(\frac{1}{|\#\mathcal{T}_h|} \right)^{1/q} =: \omega_K \qquad \forall K \in \mathcal{T}_h, \qquad (5.77)$$

where $\#\mathcal{T}_h$ denotes the number of elements of $\mathcal{T}_h$. We note that ω_K is the same for each $K \in \mathcal{T}_h$.

(E2) *Equal error per unit area*: we require

$$\left\| u - \Pi_{hp} u \right\|_{L^q(K)} \leq \omega \left(\frac{|K|}{|\Omega|} \right)^{1/q} =: \omega_K \quad \forall K \in \mathcal{T}_h. \qquad (5.78)$$

Summing over all $K \in \mathcal{T}_h$, it is imminent that if either (5.77) or (5.78) is valid, then the global bound

$$\left\| u - \Pi_{hp} u \right\|_{L^q(\Omega)} \leq \omega \qquad (5.79)$$

is valid as well.

Now, we are able to set the sizes of mesh elements. For (E1), we set

$$\omega_K^q = \frac{\omega^q}{|\#\mathcal{T}_h|} \qquad (5.80)$$

in (5.75), i.e.,

$$\mu_K = \left(\frac{q(p+1)+2}{2\pi A_w^q \rho_w^{-q/2}} \frac{\omega^q}{|\#\mathcal{T}_h|} \right)^{1/(q(p+1)+2)}. \qquad (5.81)$$

Moreover, for case (E2), we take into account that $|K| = \mu_K^2 3\sqrt{3}/4$ (cf. (2.28)). Note that in this case, ω_K in (5.78) depends on μ_K:

$$\omega_K^q = \omega^q \frac{3\sqrt{3}}{4} \left(\frac{\mu_K^2}{|\Omega|} \right). \qquad (5.82)$$

Nevertheless, substituting this expression into (5.75), we can easily solve for μ_K to obtain

$$\mu_K = \left(\frac{q(p+1)+2}{2\pi A_w^q \rho_w^{-q/2}} \omega^q \frac{3\sqrt{3}}{4|\Omega|} \right)^{1/(q(p+1))}. \qquad (5.83)$$

Both size distributions, either (5.81) or (5.83), guarantee the bound (5.76) with ω_K as in either (5.77) or (5.78). Consequently, in both cases, the global bound holds by the implication (5.79). However, we show in the following that (5.81), corresponding to approach (E1) requiring equal error per element, is supported by theoretical results. See the forthcoming Lemma 5.28. Moreover, numerical experiments given in Sect. 5.7 also suggest that (5.81) should be preferred over (5.83) (i.e., approach (E2) requiring equal error per unit area).

5.4 Continuous Mesh Model

A continuous mesh is a mapping $\mathcal{M} : \Omega \to Sym$, where Sym is defined in (2.4) as the space of $d \times d$ symmetric positive definite matrices [87, 88]. Such a mapping defines a Riemannian metric and we refer to $(\mathcal{M}(x))_{x \in \Omega}$ as the metric field. A triangular mesh can be thought of as being encoded by a metric $\mathcal{M}$ if it solves Problem 2.5, i.e., if it (approximately) minimizes

$$Q(\mathcal{T}_h) := \frac{1}{\#\mathscr{F}_h} \sum_{\mathbf{e} \in \mathscr{F}_h} \left(\|\mathbf{e}\|_{\mathcal{M}} - \bar{h} \right)^2, \tag{5.84}$$

over all admissible triangulations $\mathcal{T}_h$. Here, $\mathscr{F}_h$ is the set of all edges and $\|\mathbf{e}\|_{\mathcal{M}}$ is the edge length under the Riemannian metric $\mathcal{M}$ (cf. (2.5)). We fix $\bar{h} := \sqrt{3}$ for $d = 2$ and $\bar{h} := \sqrt{8/3}$ for $d = 3$. Recall from Remark 2.7 that these values correspond to the length of the edges of the maximal simplex inscribed into the unit ball. We formalize the continuous mesh idea in the following definition:

Definition 5.15 Let $\mathcal{M}$ be a Riemannian metric on Ω. We say a triangulation $\mathcal{T}_h$ of Ω is *generated* by the metric $\mathcal{M}$, if it satisfies

$$Q(\mathcal{T}_h) \leq Q(\mathcal{T}_h') \qquad \forall \mathcal{T}_h' \in \{\text{simplicial meshes of } \Omega\}, \tag{5.85}$$

where $Q(\mathcal{T}_h)$ is given by (5.84). The metric $\mathcal{M}$ is called a *continuous mesh*.

Remark 5.16 Various algorithms and software implementations have been proposed, e.g., [41, 86], which construct a mesh $\mathcal{T}_h$ from a given metric $\mathcal{M}$ in the sense of Definition 5.15.

Let $\mathcal{M} : \Omega \to Sym$ be given. For $x \in \Omega$, set $\mathbb{M} := \mathcal{M}(x)$. Let $\lambda_1, \ldots, \lambda_d$ be the eigenvalues of $\mathbb{M}$ given by the spectral decomposition (2.14) ($d = 2$) or (2.38) ($d = 3$). We define the mapping

$$\tau : \Omega \to \mathbb{R}^+, \qquad \tau(x) := \begin{cases} \sqrt{\lambda_1 \lambda_2} & \text{for } d = 2, \\ \sqrt{\lambda_1 \lambda_2 \lambda_3} & \text{for } d = 3. \end{cases} \tag{5.86}$$

Remark 5.17 If K is the element of maximum area inscribed into the ellipse ($d = 2$) or ellipsoid ($d = 3$) implied by $\mathbb{M} := \mathcal{M}(x)$, then, from (2.25) and (2.31) ($d = 2$)

or (2.45) and (2.51) ($d = 3$), we have

$$\tau(x) = \mu_K^{-d}, \qquad d = 2, 3. \tag{5.87}$$

Furthermore, the d-dimensional measure of this element fulfills

$$|K| = \frac{c_d}{\sqrt{\det \mathbb{M}}} = \frac{c_d}{\tau(x)}, \tag{5.88}$$

where $c_d = 3\sqrt{3}/4$ for $d = 2$ and $c_d = 8/(9\sqrt{3})$ for $d = 3$, cf. (2.28) and (2.47).

Definition 5.18 We call the mapping $\tau : \Omega \to \mathbb{R}^+$, defined in (5.86), the *mesh density* distribution.

Consider now a given mesh $\mathscr{T}_h$ and a piecewise constant metric $\mathcal{M}$, such that $\mathcal{M}|_K = \mathbb{M}_K$, where $\mathbb{M}_K$ generates $K \in \mathscr{T}_h$. (Consult Sect. 2.3.5 for a brief summary of the main relations between metric and mesh element.) Then, using (5.88), and setting $\tau|_K = \tau_K$, for $K \in \mathscr{T}_h$,

$$\int_\Omega \tau \, dx = \sum_{K \in \mathscr{T}_h} \tau_K \int_K dx = c_d N, \tag{5.89}$$

where $N = \#\mathscr{T}_h$ is the number of mesh elements. For a more general sufficiently regular mesh density distribution, local Taylor Series approximation of τ, centered at x_K, together with (5.88), yields $\int_\Omega \tau \, dx = c_d N + O(h)$, where $h = \max_K h_K$.

These considerations motivate the following definition:

Definition 5.19 The *complexity* N of the continuous mesh $\mathcal{M}$ is defined as

$$N \equiv N(\mathcal{M}) := \int_\Omega \tau \, dx. \tag{5.90}$$

Remark 5.20 The local variant of (5.89) gives, for any K generated by the piecewise constant metric,

$$\int_K \tau \, dx = \tau_K \int_K dx = c_d \qquad \forall K \in \mathscr{T}_h, \tag{5.91}$$

while for the general case, we have

$$\int_K \tau \, dx = c_d + O(h_K^{1+d}). \tag{5.92}$$

5.5 Continuous Mesh Optimization

In this section, we introduce a method to determine the size distribution of simplex mesh elements using analytic optimization of the continuous mesh model. We focus on error estimates in the L^q-norm, see (3.59) and (4.31). Estimates in the H^1-seminorm and the L^∞-norm can be used in an analogous way.

Let us first recall the local optimization of anisotropy introduced in Sects. 5.1 and 5.2 for the two- and three-dimensional cases, respectively. In particular, let K be a simplex element, centered at $x_K \in \Omega$. Assume that K has locally optimal anisotropy, in the sense of Theorem 5.7 or Theorem 5.13. In other words, for the two-dimensional case, $\{\sigma_K^\star, \phi_K^\star\}$ is given by (5.39), while for the three-dimensional case $\{\sigma_K^\star, \varsigma_K^\star, \phi_K^\star\}$ is given by (5.71) and (5.72). Starting from the error bounds (5.40) and (5.73), corresponding to the locally optimal anisotropy, we now define a continuous interpolation error. To that end, first note the following definition:

Definition 5.21 For $x \in \Omega$, we define the *density of interpolation error* as the mapping $B : \Omega \to \mathbb{R}$, such that

$$x \mapsto B(x) \equiv B(x; p, d) := \begin{cases} A_w \rho_w^{-1/2} & \text{for } d = 2, \\ A_w (\varrho_w \rho_w)^{-1/3} & \text{for } d = 3, \end{cases} \tag{5.93}$$

where A_w and ρ_w are defined in (3.48) for $d = 2$, while A_w, ρ_w, and ϱ_w are defined as in (4.20) for $d = 3$.

Remark 5.22 For a simplex of locally optimal anisotropy and fixed size, the error bounds (5.40) and (5.73) are directly proportional to the error density B. The values used in the definition of B correspond to the guaranteed error bounds of Theorem 3.17 ($d=2$) and Theorem 4.5 ($d=3$). If we use instead the non-guaranteed analogs, i.e., $\tilde{A}_p$, $\tilde{A}_p^\perp$, $\tilde{\rho}_p$ defined in (3.37), or $\tilde{A}_p$ $\tilde{A}_p^{(2)}$, $\tilde{A}_p^{(3)}$, $\tilde{\rho}_p$, $\tilde{\varrho}_p$ defined in (4.16)), we have

$$B = \begin{cases} \tilde{A}_p (\tilde{\rho}_p)^{-1/2} = \left(\tilde{A}_p \tilde{A}_p^\perp\right)^{1/2} & \text{for } d = 2, \\ \tilde{A}_p \left(\tilde{\rho}_p \tilde{\varrho}_p\right)^{-1/3} = \left(\tilde{A}_p \tilde{A}_p^{(2)} \tilde{A}_p^{(3)}\right)^{1/3} & \text{for } d = 3. \end{cases} \tag{5.94}$$

In this case, the density of the interpolation error is simply the geometric average of the maximum directional derivative of order p and the respective derivatives in orthogonal directions.

We are now ready to give the definition of the continuous interpolation error.

Definition 5.23 For $x \in \Omega$, we define the *continuous interpolation error* as the mapping $e_\tau : \Omega \to \mathbb{R}$, such that

$$e_\tau(x) := b^{1/q} B(x) \tau(x)^{-(p+1)/d}, \tag{5.95}$$

where

$$b \equiv b(p, q, d) := \frac{|\partial \hat{\mathcal{E}}|}{q(p+1)+d} = \begin{cases} \frac{2\pi}{q(p+1)+2} & \text{for } d = 2, \\ \frac{4\pi}{q(p+1)+3} & \text{for } d = 3. \end{cases} \tag{5.96}$$

The significance of Definition 5.23 is the following: for any simplex element of locally optimal anisotropy, centered at $x_K \in \Omega$, the estimates (5.40) and (5.73), together with (5.95), (5.96), (5.87), and (5.93), imply the bound

$$\left\| w_{x_K,p}^{\text{int}} \right\|_{L^q(K)}^q \leq b B(x_K)^q \mu_K^{q(p+1)+d} = b B(x_K)^q \tau(x_K)^{-\frac{q(p+1)+d}{d}} \tag{5.97}$$

$$= b B(x_K)^q \tau(x_K)^{-\frac{q(p+1)}{d}} \tau(x_K)^{-1} = e_\tau(x_K)^q \tau(x_K)^{-1}.$$

Consider now a triangulation $\mathcal{T}_h = \{K\}$, composed of simplex elements with arbitrary size distribution, and a corresponding piecewise polynomial approximation space as in (5.1). Note, in particular, that the polynomial degree of approximation is constant. Assume that each simplex has locally optimal anisotropy in the sense of Theorem 5.7 or Theorem 5.13. Then, for the interpolation operator π_h, given by (5.2), we have (due to (3.9) and (5.97))

$$\| u - \pi_h u \|_{L^q(\Omega)}^q = \sum_{K \in \mathcal{T}_h} \left\| u - \Pi_{x_K,p} u \right\|_{L^q(K)}^q \tag{5.98}$$

$$\approx \sum_{K \in \mathcal{T}_h} \left\| w_{x_K,p}^{\text{int}} \right\|_{L^q(K)}^q \leq \sum_{K \in \mathcal{T}_h} \frac{e_\tau(x_K)^q}{\tau(x_K)}.$$

Consequently, noting (5.88),

$$\sum_{K \in \mathcal{T}_h} \frac{e_\tau(x_K)^q}{\tau(x_K)} = c_d^{-1} \sum_{K \in \mathcal{T}_h} e_\tau(x_K)^q |K| \to c_d^{-1} \int_\Omega e_\tau(x)^q \, dx, \quad (h \to 0). \tag{5.99}$$

Thus, if we think of the mesh as being generated by the metric $\mathcal{M}$, and using (5.95), we obtain the global *continuous mesh error estimate*

$$E(\mathcal{M}) = E^\star(\tau) := \int_\Omega e_\tau^q \, dx = b \int_\Omega B^q \tau^{-(p+1)q/d} \, dx, \tag{5.100}$$

where it is understood that $\mathcal{M} = \mathcal{M}(\tau)$ is of locally optimal anisotropy. In order to shorten the notation, we define

$$s = (p + 1)q/d. \tag{5.101}$$

Then, (5.100) reads

$$E^\star(\tau) = b \int_\Omega B(x)^q \tau(x)^{-s} \, dx. \tag{5.102}$$

As the local anisotropy is fixed to optimal values, the continuous mesh $\mathcal{M}$ is a function only of the mesh density distribution τ. The latter can be subjected to a constraint based on a given mesh complexity. Thus, one finally arrives at the continuous mesh analog of Problem 5.2:

Problem 5.24 Let $N \in \mathbb{N}$ be given, and let $E^\star(\tau)$ be defined as in (5.100). Find the mesh density distribution $\tau^\star$ that satisfies

$$\tau^\star = \arg\min_\tau E^\star(\tau), \qquad \text{such that} \quad N = \int_\Omega \tau \, dx. \tag{5.103}$$

Similarly, the continuous mesh analog of Problem 5.1 reads:

Problem 5.25 Let $\omega > 0$ be given tolerance, and let $E^\star(\tau)$ be defined as in (5.100). Find the mesh density distribution $\tau^\star$ that satisfies

$$\tau^\star = \arg\min_\tau \int_\Omega \tau \, dx, \qquad \text{such that} \quad \int_\Omega E^\star(\tau) \, dx = \omega^q. \tag{5.104}$$

5.5.1 Solution of Problem 5.24

Straightforward application of calculus of variations to (5.102) leads to the first-order optimality condition

$$\delta E = -bs \int_\Omega B^q \tau^{-s-1} \delta\tau \, dx = 0. \tag{5.105}$$

From the constraint equation in (5.103), we obtain

$$\delta N = \int_\Omega \delta\tau \, dx = 0. \tag{5.106}$$

Hence, an admissible solution to (5.105) is given by

$$B^q \tau^{-(s+1)} = \text{const.}$$

Solving for the density distribution τ, and substituting the result into the constraint equation finally, yields

$$\tau^\star(x) = C B(x)^{\frac{q}{s+1}}, \tag{5.107}$$

where

$$C = N \left(\int_\Omega B^{\frac{q}{s+1}} \, dx \right)^{-1}. \tag{5.108}$$

The only parameter to be specified by the user is the complexity N. Recall from (5.89) that the complexity is directly proportional to N, the number of elements in $\mathcal{T}_h$.

Finally, we evaluate the error $E^\star(\tau^\star)$, which is attained for the optimal mesh density distribution. Inserting (5.107)–(5.108) into (5.102), we obtain

$$E^\star(\tau^\star) = b \int_\Omega B(x)^q (\tau^\star(x))^{-s} \, dx = C^{-s} b \int_\Omega B^q \left(B^{\frac{q}{s+1}} \right)^{-s} dx \tag{5.109}$$

$$= b N^{-s} \left(\int_\Omega B^{\frac{q}{s+1}} \, dx \right)^s \int_\Omega B^{\frac{q}{1+s}} \, dx = b N^{-s} \left(\int_\Omega B^{\frac{q}{s+1}} \, dx \right)^{s+1}.$$

5.5.2 Solution of Problem 5.25

Employing the notation (5.101), the constraint in (5.104) reads

$$b \int_\Omega B(x)^q \tau(x)^{-s} \, dx = \omega^q. \tag{5.110}$$

The constrained minimization Problem 5.25 is solved by the Lagrangian multiplier technique. Then we define

$$\mathcal{K}(\tau, L) := \int_\Omega \tau \, dx + L \left(b \int_\Omega B^q \tau^{-s} \, dx - \omega^q \right), L \neq 0. \tag{5.111}$$

The solution of Problem 5.25 fulfills

$$\frac{d}{dt} \mathcal{K}(\tau + t\tilde{\tau}, L)|_{t=0} = 0 \quad \text{for any perturbation } \tilde{\tau} : \Omega \to \mathbb{R}^+. \tag{5.112}$$

Hence, the direct differentiation of (5.111) reads

$$\frac{d}{dt} \mathcal{K}(\tau + t\tilde{\tau}, L)|_{t=0} = \int_\Omega \tilde{\tau} \, dx + L b \int_\Omega (-s) B^q \tau^{-s-1} \tilde{\tau} \, dx. \tag{5.113}$$

Since (5.113) has to vanish for any $\tilde{\tau}$, we obtain the relation

$$1 - L b s B^q \tau^{-s-1} = 0, \qquad (5.114)$$

which leads to the relation for the optimal mesh density distribution

$$\tau^\star(x) = \left(L b s B^q(x)\right)^{1/(s+1)}. \qquad (5.115)$$

In order to eliminate L from (5.115), we insert it in the constraint equation (5.110), which gives

$$\omega^q = b \int_\Omega B^q(\tau^\star)^{-s}\, dx = (Ls)^{-\frac{s}{s+1}} b^{\frac{1}{s+1}} \int_\Omega B^{\frac{q}{s+1}}\, dx. \qquad (5.116)$$

Eliminating the Lagrangian multiplier L (multiplied by s) from (5.116), we obtain

$$Ls = \left(\frac{\omega^q}{b^{\frac{1}{s+1}} \int_\Omega B^{\frac{q}{s+1}}\, dx} \right)^{-\frac{s+1}{s}}. \qquad (5.117)$$

Inserting this into (5.115) gives the optimal mesh density distribution

$$\tau^\star(x) = \left(\frac{\omega^q}{b^{\frac{1}{s+1}} \int_\Omega B^{\frac{q}{s+1}}\, dx} \right)^{-1/s} b^{\frac{1}{s+1}} (B(x))^{\frac{q}{s+1}} = \left(\frac{b}{\omega^q} \int_\Omega B^{\frac{q}{s+1}}\, dx \right)^{1/s} (B(x))^{\frac{q}{s+1}}. \qquad (5.118)$$

Furthermore, we evaluate the complexity N of the optimal continuous mesh given by (5.118), i.e., inserting (5.118) into (5.90), we have

$$N = \int_\Omega \tau^\star\, dx = \left(\frac{b}{\omega^q} \right)^{1/s} \left(\int_\Omega B^{\frac{q}{s+1}}\, dx \right)^{1/s} \int_\Omega (B(x))^{\frac{q}{s+1}}\, dx \qquad (5.119)$$

$$= \left(\frac{b}{\omega^q} \right)^{1/s} \left(\int_\Omega B^{\frac{q}{s+1}}\, dx \right)^{\frac{s+1}{s}}.$$

Relation (5.119) implies also the identity

$$\omega^q = \frac{b}{N^s} \left(\int_\Omega B^{\frac{q}{s+1}}\, dx \right)^{s+1}, \qquad (5.120)$$

which is identical with (5.109) provided that $E^\star(\tau^\star) = \omega^q$.
We summarize the main results of Sect. 5.5.

Theorem 5.26 *Let $q \geq 1$, $p \in \mathbb{N}$ be the polynomial approximation degree, and $s = (p+1)q/d$. Moreover, assume $B : \Omega \to \mathbb{R}$ is the given density of interpolation error, and b is as in (5.96).*

  (i) *Let $N \in \mathbb{N}$, then the mesh density distribution $\tau^{\star}$ solving Problem 5.24 is*

$$\tau^{\star}(x) = N \left(\int_{\Omega} B^{\frac{q}{s+1}} \, \mathrm{d}x \right)^{-1} B(x)^{\frac{q}{s+1}}. \qquad (5.121)$$

 (ii) *Let $\omega > 0$ be the prescribed tolerance, then the mesh density distribution $\tau^{\star}$ solving Problem 5.25 is given by*

$$\tau^{\star}(x) = \left(\frac{b}{\omega^q} \int_{\Omega} B^{\frac{q}{s+1}} \, \mathrm{d}x \right)^{1/s} (B(x))^{\frac{q}{s+1}}. \qquad (5.122)$$

(iii) *Problems 5.24 and 5.25 are equivalent, i.e., if their input data N and ω fulfill the identity*

$$\omega^q N^s = b \left(\int_{\Omega} B^{\frac{q}{s+1}} \, \mathrm{d}x \right)^{s+1}, \qquad (5.123)$$

then the mesh density distributions (5.121) and (5.122) are identical.

Proof The assertion (i) follows from (5.107)–(5.108), and assertion (ii) follows from (5.118). Finally, both relations (5.118) and (5.107) give $\tau^{\star} \sim B^{q/(s+1)}$, and (5.109) together with (5.120) implies the equivalence of Problems 5.24 and 5.25.   □

Remark 5.27 Theorem 5.26 also has a practical aspect. We can prescribe either the tolerance for the interpolation error and solve Problem 5.24 or the number of mesh elements and solve Problem 5.25. Both problems lead to the same continuous meshes.

The consequence of Theorem 5.26 is the following equidistribution of the error.

Lemma 5.28 *Let $\tau^{\star}$ be the mesh density distribution given by (5.122), then, asymptotically, as $h \to 0$,*

$$\|u - \pi_h u\|_{L^q(K)}^q \approx E_K^{\star}(\tau^{\star}) := b \int_K B(x)^q \tau^{\star}(x)^{-s} \, \mathrm{d}x \to C, \quad \forall K \in \mathscr{T}_h, \qquad (5.124)$$

where C is a constant, independent of K.

Proof We write (5.122) as

$$\tau^\star(x) = M\,(B(x))^{\frac{q}{s+1}}, \qquad M := \left(\frac{b}{\omega^q}\int_\Omega B^{\frac{q}{s+1}}\,dx\right)^{1/s}, \tag{5.125}$$

which is equivalent to

$$B(x) = M^{-\frac{s+1}{q}}\left(\tau^\star(x)\right)^{\frac{s+1}{q}}. \tag{5.126}$$

Inserting (5.126) in the definition of $E_K^\star$ and using (5.92), we obtain

$$E_K^\star(\tau^\star) = b\int_K B(x)^q \tau^\star(x)^{-s}\,dx = b\,M^{-(s+1)}\int_K (\tau^\star(x))^{s+1}\tau^\star(x)^{-s}\,dx \tag{5.127}$$

$$= b\,M^{-(s+1)}\int_K \tau^\star(x)\,dx = C + O(h_K^{1+d}),$$

where $C = b\,M^{-(s+1)}c_d$ (see (5.92)). This proves (5.124). $\qquad\square$

Remark 5.29 Lemma 5.28 shows that the optimization of continuous mesh models leads to the choice (E1) (equal error per element) in Sect. 5.3, relation (5.77).

5.6   Adaptive Solution of Partial Differential Equations

In previous sections, we treated the following problem: for a given function $u \in C^\infty$, find a simplicial mesh which is optimal with respect to the interpolation error $\|u - \pi_h u\|$ in the given norm, where $\pi_h u$ is a discontinuous piecewise polynomial approximation of degree $p \geq 1$ of u, cf. (5.2). Determination of the optimal mesh is based on the knowledge of higher-order derivatives of u, namely all partial derivatives of degree $p + 1$. Here we introduce corresponding anisotropic mesh adaptation algorithms for the numerical solution of partial differential equations (PDEs).

5.6.1   Anisotropic Mesh Adaptation Algorithms for PDEs

In practical problems, if a function u represents an unknown exact solution of a PDE, the higher-order derivatives are not available. A way to find an approximation of the optimal mesh for the given PDE is to employ an adaptive algorithm where we start with an initial mesh $\mathcal{T}_h^k$, $k = 0$, compute the approximate solution u_h^k by a suitable numerical method, evaluate approximations of the $p + 1$th-order derivatives using u_h^k, and construct new mesh $\mathcal{T}_h^{k+1}$ (closer to the optimal one). This procedure is repeated for $k = 1, 2, \ldots$ until the prescribed stopping criterion is achieved. We expect that this procedure gives a sequence of approximate solutions u_h^k, $k = 0, 1, \ldots$, which approaches to the exact solution u, and therefore the corresponding approximations of the higher-order derivatives converge to the higher-order derivatives of u. Hence, the resulting mesh is close to the optimal one.

Let $p \geq 1$ be the given polynomial approximation degree. In virtue of (5.1), we define the spaces of discontinuous piecewise polynomial functions

$$S_{hp}^k := \{v_h \in L^2(\Omega);\ v_h|_K \in P^p(K)\ \forall K \in \mathcal{T}_h^k\}, \qquad k = 0, 1, \ldots, \qquad (5.128)$$

where $\mathcal{T}_h^k$, $k = 0, 1, \ldots$, is a sequence of simplicial grids. In order to approximate the high-order derivatives of u, we consider spaces

$$S_{hp+1}^k := \{v_h \in L^2(\Omega);\ v_h|_K \in P^{p+1}(K)\ \forall K \in \mathcal{T}_h^k\}, \qquad k = 0, 1, \ldots$$
$$(5.129)$$

given by the enrichment of S_{hp}^k by polynomials of degree $p + 1$. Then, for the given $u_h \in S_{hp}^k$, we define a higher-order reconstruction

$$u_h^+ = \mathcal{R}(u_h), \qquad (5.130)$$

where $\mathcal{R} : S_{hp}^k \to S_{hp+1}^k$ denotes a reconstruction operator. Several possibilities are discussed in Sect. 9.1.

Having $u_h^+ \in S_{hp+1}^k$, we can employ the projection π_h defined by (5.2) since it is defined elementwise. Then we approximate the interpolation error

$$u - \pi_h u \approx u_h^+ - \pi_h u_h^+ \qquad (5.131)$$

and introduce the following stopping criterion for the mesh adaptation algorithm:

$$\eta := \left\| u_h^+ - \pi_h u_h^+ \right\| \leq \omega, \qquad (5.132)$$

where $\omega > 0$ is the given tolerance and $\|\cdot\|$ is the norm of our interest, e.g., the L^q-norm, $q \in [1, \infty]$, or the H^1-seminorm. Since the reconstructed approximation u_h^+ belongs to S_{hp+1}^k, its $(p + 1)$th-order derivatives are piecewise constant on each mesh element. We employ these derivatives in the interpolation error estimates presented in Chaps. 3 and 4 and the optimization of the anisotropy of elements given in Sects. 5.1–5.2.

Algorithm 5.1 shows an abstract mesh adaptation process for the two-dimensional case, and the 3D case can be written analogously.

Algorithm 5.1: Mesh adaptation based on interpolation error control (2D)

1: set tolerance $\omega > 0$ and initial mesh $\mathscr{T}_h^0$
2: **for** $k = 0, 1, \ldots$ **do**
3:     compute approximate solution $u_h^k \in S_{hp}^k$
4:     compute higher-order reconstruction $u_h^+ = \mathscr{R}(u_h^k) \in S_{hp+1}^k$, cf. Sect. 9.1
5:     **if** $\left\| u_h^+ - \pi_h u_h^+ \right\| \leq \omega$ **then**
6:         STOP the computation
7:     **else**
8:         **for** each element K **do**
9:             determine local anisotropy $\{\sigma_K^\star, \phi_K^\star\}$ as in Theorems 5.7–5.9
10:             compute optimal size parameter $\mu_K^\star$, cf. Sects. 5.3–5.5
11:             construct new local metric $\mathbb{M}_K^\star = \mathcal{M}(x_K)$, cf. Eq. (2.54)
12:         **end for**
13:         construct continuous mesh $\mathcal{M}$ from $(\mathbb{M}_K^\star)_{K \in \mathscr{T}_h}$, cf. Sect. 2.5.2
14:         construct new mesh $\mathscr{T}_h^{k+1}$ from $\mathcal{M}$, cf. Sect. 2.5.3
15:     **end if**
16: **end for**

- In step 9 of Algorithm 5.1, we use the local optimization based on interpolation error, discussed in Sect. 5.1, to determine locally optimal (in the sense of Theorem 5.7) anisotropy parameters $\{\sigma_K^\star, \phi_K^\star\}_{K \in \mathscr{T}_h^k}$.
- In step 10 of Algorithm 5.1, we set the optimal size parameter $\mu_K^\star$ for each $K \in \mathscr{T}_h^k$, which is described in Sect. 5.6.2.
- In step 11 of Algorithm 5.1, the nodal metric is constructed using the technique given in Sect. 2.5.1.

5.6.2   Setting of Optimal Size of Mesh Elements

We describe several variants of the setting of the optimal size parameter $\mu_K^\star$ in step 10 of Algorithm 5.1. The first one exhibits a heuristic extension of the common marking strategy where a fixed fraction of elements is refined. The other ones are based on theoretical results from Sects. 5.3 and 5.5. First, note that the element size can be set directly from the optimal mesh density distribution (5.107) or (5.118). Alternatively, we may use the error equidistributions (5.77) and (5.78). We recall that the former one (5.77) is equivalent to the theoretical results, cf. Lemma 5.28 and Remark 5.29. These techniques employ the local interpolation error estimates given by

$$\eta_K := \left\| u_h^+ - \pi_h u_h^+ \right\|_K, \qquad K \in \mathscr{T}_h, \tag{5.133}$$

where $\|\cdot\|_K$ means either $L^q(K)$-norm, $q \in [1, \infty]$, or $H^1(K)$-seminorm depending on the norm of our interest.

5.6.2.1   A Simple Approach Based on Thresholds

This approach generalizes the *fixed fraction* strategy which is frequently employed in standard isotropic refinement methods, where a fixed ratio (e.g., 10%) of elements with the largest local estimator η_K is split into several subelements. Without much difficulty, one extends this idea to metric-based adaptation, as we shall see below.

We note that for methods based on cell subdivision, the local refinement ratio remains fixed by the chosen subdivision method. Moreover, mesh coarsening is difficult to implement. In contrast, advantages of the metric-based approach, which is based on mesh regeneration, include (a) the local refinement ratio can be freely chosen and (b) mesh coarsening can be done as easily as refinement.

The idea of the presented approach is to define two thresholds η_c and η_r satisfying $0 < \eta_c < \eta_r$. If $\eta_K > \eta_r$, then element K is refined and the new size parameter $\mu_K^\star$ has to be smaller than μ_K. On the other hand, if $\eta_K < \eta_c$, then element K is coarsened and the new size parameter $\mu_K^\star$ has to be larger than μ_K. Hence, we write

$$\mu_K^\star := \alpha_K \mu_K, \tag{5.134}$$

where $\alpha_K > 0$ is a refined/coarsened parameter depending on η_K which has to be determined. Since the mesh optimization process is based on continuous mesh construction, parameter α_K should depend continuously on η_K. Moreover, it makes sense to use a logarithmic scaling of the error estimators. Hence, we define the function

$$\xi_K = \xi_K(\eta_K) := \begin{cases} \dfrac{\log(\eta_K) - \log(\eta_r)}{\log(\eta_{\max}) - \log(\eta_r)} & \text{for } \eta_K \geq \eta_r, \\[2ex] \dfrac{\log(\eta_K) - \log(\eta_c)}{\log(\eta_{\min}) - \log(\eta_c)} & \text{for } \eta_K \leq \eta_c, \end{cases} \tag{5.135}$$

where $\eta_{\max} = \max_{K \in \mathscr{T}_h^k} \eta_K$ and $\eta_{\min} = \min_{K \in \mathscr{T}_h^k} \eta_K$. Obviously, $\xi_K \in [0, 1]$ and

$$\xi_K(\eta_r) = \xi_K(\eta_c) = 0 \qquad \text{and} \qquad \xi_K(\eta_{\max}) = \xi_K(\eta_{\min}) = 1.$$

Furthermore, we have to set the maximum refinement ratio $r_{\max} < 1$ and the maximum coarsening ratio $c_{\max} > 1$ and define *activation functions*,

$$r(\,\cdot\,; r_{\max}) : [0, 1] \to [r_{\max}, 1], \tag{5.136}$$

$$c(\,\cdot\,; c_{\max}) : [0, 1] \to [1, c_{\max}]$$

to translate the error estimate to local refinement or coarsening. We generally assume that both functions are monotonous. Finally, the parameter α_K (cf. (5.134)) is set as

$$\alpha_K := \alpha(\eta_K; r_{\max}, c_{\max}) := \begin{cases} r(\xi_K(\eta_K); r_{\max}) & \eta_K \geq \eta_r \\ c(\xi_K(\eta_K); c_{\max}) & \eta_K < \eta_c \\ 1 & \text{otherwise,} \end{cases} \tag{5.137}$$

where ξ_K is given by (5.135).

We present two choices of the activation functions r and c. Following [13, Section 4.3], we set

$$r(\eta_K; r_{\max}) := 1 + (r_{\max} - 1)\xi_K^2, \tag{5.138}$$

$$c(\eta_K; c_{\max}) := 1 + (c_{\max} - 1)\xi_K^2.$$

According to [16], the efficiency of the whole algorithm can be still slightly improved by a modification of (5.138) in such a way that α_K is close to a constant in vicinity of the limit values $\eta_{\min}$ and $\eta_{\max}$. Namely, we put

$$r(\eta_K; r_{\max}) := \tfrac{1}{2}(1 - r_{\max})(\cos(\pi\xi_K) + 1) + r_{\max}, \tag{5.139}$$

$$c(\eta_K; r_{\max}) := \tfrac{1}{2}(c_{\max} - 1)(\cos(\pi(\xi_K + 1)) + 1) + 1.$$

Figure 5.2 shows the dependence of α_K on η_K (cf. (5.137)) for the choices of the activation functions r and c by (5.138) and (5.139).

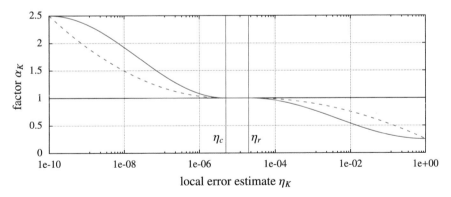

Fig. 5.2 The dependence of the factor α_K on η_K for the activation functions given by (5.139) (blue graph) and (5.138) (red graph) with $\eta_{\max}^l = 1$, $\eta_{\min}^l = 10^{-10}$, $\eta_c = 5 \cdot 10^{-6}$, $\eta_r = 2 \cdot 10^{-5}$, $r_{\max} = 0.25$, and $c_{\max} = 2.5$

In remains to set the thresholds η_r and η_c. A simple possibility is to use the half and the double of the mean value of estimates η_K, $K \in \mathcal{T}_h$, i.e.,

$$\eta_r := \frac{1}{2}\left(\frac{1}{\#\mathcal{T}_h}\sum_{K\in\mathcal{T}_h}\eta_K\right), \qquad \eta_c := 2\left(\frac{1}{\#\mathcal{T}_h}\sum_{K\in\mathcal{T}_h}\eta_K\right). \tag{5.140}$$

Finally, the choice of maximum refinement ratio r_{max} and the maximum coarsening ratio c_{max} is a little delicate. Too small r_{max} (and large c_{max}) allows large changes of the mesh in one adaptation level, but it leads frequently to multiple repetition of the refinement/coarsening. On the other hand, too large r_{max} (and small c_{max}) admits only small changes of the mesh in one adaptation level which increases the number of adaptive loops necessary to achieve the given stopping criterion. Based on a large number of numerical experiments, we use in our code the values $r_{max} = 0.25$ and $c_{max} = 2.5$. We note that the value $r_{max} = 0.25$ admits the maximal decrease of the size of elements by factor $1/r_{max} = 4$ at one adaptive loop, and hence in 2D, the area of the element is decreased 16 times.

5.6.2.2 Approaches Based on Error Equidistribution

The forthcoming techniques are based on the theoretical results derived in Sect. 5.5. Theorem 5.26 gives the optimal element size distribution for both Problems 5.24 and 5.25: we prescribe the number of mesh elements N and minimize the error (Problem 5.24) or we prescribe the error tolerance ω and minimize the number of mesh elements (Problem 5.25). Moreover, if N and ω fulfill (5.123), then the corresponding optimal element size distribution is the same for both problems.

Therefore, in practical computations, we can prescribe either a sequence of increasing number of mesh elements N_k, $k = 0, 1, \ldots$, or a sequence of decreasing error tolerances $\bar{\omega}_k$, $k = 0, 1, \ldots$. Both possibilities are equivalent if N_k and $\bar{\omega}_k$ are related through (5.123) for all adaptation levels $k = 0, 1, \ldots$. Here, we discuss the second approach, the first one can be set analogously. We define the *tolerances for the mesh adaptation level* $\bar{\omega}_k$, $k = 0, 1, \ldots$, by

$$\bar{\omega}_k = \begin{cases} \zeta\eta & \text{for } k = 0 \\ \zeta\omega_{k-1} & \text{for } k > 0, \end{cases} \tag{5.141}$$

where η is the error estimator given by (5.132) computed on the initial mesh $\mathcal{T}_h^0$ and $\zeta \in (0, 1)$ is a suitably chosen factor. Too small ζ leads to too high number of elements for low k, and on the other hand, choosing ζ close to 1 requires too many mesh adaptation levels to achieve (5.132). Typically, $\zeta = \frac{1}{2}$ and $\zeta = \frac{3}{4}$ are good choices.

Lemma 5.28 asserts that the optimal element size distribution leads to the same error for each element. Hence, the goal is to set the sizes of elements such that

$$\eta_K^\star := \bar{\omega}_k \left(\# \mathscr{T}_h^k \right)^{-1/q} \qquad \forall K \in \mathscr{T}_h^k, \ k = 0, 1, 2 \dots, \tag{5.142}$$

where $\eta_K^\star$ defines the desired local error estimate on $K \in \mathscr{T}_h^k$, $\#\mathscr{T}_h^k$ denotes the number of elements of $\mathscr{T}_h^k$ on the kth level of adaptation, and $\bar{\omega}_k$ is prescribed by (5.141). We note that the right-hand side of (5.142) is independent of K.

In order to achieve the error equidistribution (5.142), if $\eta_K > \eta_K^\star$, then the size of element K has to be decreased, and if $\eta_K < \eta_K^\star$, then the size of element K should be increased. Hence, similarly as in (5.134), we set

$$\mu_K^\star := \alpha_K \mu_K, \qquad K \in \mathscr{T}_h^k, \tag{5.143}$$

where α_K depends on η_K and it has to satisfy $\alpha_K < 1$ for $\eta_K > \eta_K^\star$ and $\alpha_K \geq 1$ for $\eta_K \leq \eta_K^\star$. Hereafter, we present several possibilities of how to set α_K in (5.143).

- An approach based on the rate of convergence employs the assumption resulting from the interpolation theory in the form

$$\eta_K = \left\| u_h^+ - \pi_h u_h^+ \right\|_K \approx C(\mu_K)^\beta, \qquad K \in \mathscr{T}_h, \tag{5.144}$$

where μ_K is the size of K (proportional to its diameter since the shape of element has been set by shape optimization) and $\beta > 0$ is the rate of convergence. The standard approximation theory, e.g., [31], gives $\beta = p + 1$ for the L^2-norm and $\beta = p$ for the H^1-seminorm provided that the interpolated function is sufficiently regular. In virtue of (5.144), we assume also

$$\eta_K^\star \approx C(\mu_K^\star)^\beta, \qquad K \in \mathscr{T}_h, \tag{5.145}$$

where $\eta_K^\star$ is the desired local error estimate, cf. (5.142), and $\mu_K^\star$ is the new optimal element size. From (5.143)–(5.145), we set

$$\alpha_K = \left(\frac{\bar{\omega}_k}{\eta_K^\star} \right)^{1/\beta}, \qquad K \in \mathscr{T}_h^k. \tag{5.146}$$

Although this approach is rather naive since the rate of convergence β depends on local regularity of the unknown exact solution, it works efficiently for some problems.
- We adopt the approach based on thresholds from Sect. 5.6.2.1 in the following way. We define the maximal and minimal value of the estimator η_K, $K \in \mathscr{T}_h$, by

$$\eta_{\max} = \max_{K \in \mathscr{T}_h^k} \eta_K, \quad \eta_{\min} = \min_{K \in \mathscr{T}_h^k} \eta_K, \tag{5.147}$$

and two additional parameters, the *maximal refinement factor* $r_{max} \in (0, 1)$ and the *maximal coarsening factor* $c_{max} > 1$. Then, in virtue of (5.135) and (5.138), we set

$$
\alpha_K = \begin{cases} 1 + (r_{max} - 1)\xi_K^2, & \xi_K := \frac{\log(\eta_K) - \log(\bar{\omega}_k)}{\log(\eta_{max}) - \log(\bar{\omega}_k)} & \text{for } \eta_K \geq \bar{\omega}_k, \\[3mm] 1 + (c_{max} - 1)\xi_K^2, & \xi_K := \frac{\log(\eta_K) - \log(\bar{\omega}_k)}{\log(\eta_{min}) - \log(\bar{\omega}_k)} & \text{for } \eta_K < \bar{\omega}_k. \end{cases}
$$
(5.148)

Alternatively, employing (5.135) and (5.139), we set

$$
\alpha_K = \begin{cases} \frac{1}{2}(1 - r_{max})(\cos(\pi \xi_K) + 1) + r_{max}, \\[1mm] \qquad \text{with } \xi_K := \frac{\log(\eta_K) - \log(\bar{\omega}_k)}{\log(\eta_{max}) - \log(\bar{\omega}_k)} & \text{for } \eta_K \geq \bar{\omega}_k, \\[3mm] \frac{1}{2}(c_{max} - 1)(\cos(\pi(\xi_K + 1)) + 1) + 1, \\[1mm] \qquad \text{with } \xi_K := \frac{\log(\eta_K) - \log(\bar{\omega}_k)}{\log(\eta_{min}) - \log(\bar{\omega}_k)} & \text{for } \eta_K < \bar{\omega}_k. \end{cases}
$$
(5.149)

- For the sake of completeness, we mention another possibility, which is based on the heuristic approach given by (5.78) (*equidistribution per unit area*). The setting of the new element size (5.142) is replaced by

$$
\eta_K^\star := \bar{\omega}_k \, (|K|/|\Omega|)^{1/q} \qquad \forall K \in \mathscr{T}_h^k, \ k = 0, 1, 2 \ldots,
$$
(5.150)

and we further we use (5.148) or (5.149).

Figure 5.3 shows the comparison techniques given by (5.146), (5.148), and (5.149). We observe that for (5.146), α_K is changing rapidly for η close to $\bar{\omega}_k$, which usually leads to frequent repetitions of refinement and consequently to the increase of the number of mesh adaptation levels. This effect is illustrated in Fig. 8.4 in the context of the goal-oriented error estimates in Chap. 8 where we observe repetition of refinement and coarsening in the vicinity of the interior angle. On the other hand, α_K given by (5.148) and (5.149) is almost equal to 1 in the vicinity of $\eta_K \approx \bar{\omega}_k$. Consequently, this technique does not suffer from this drawback.

Remark 5.30 Comparing Figs. 5.2 and 5.3, we found that techniques (5.138) and (5.148) give similar graphs, and in the same manner, (5.139) and (5.149) are also close to each other. However, there is a fundamental difference. Whereas the threshold techniques (5.138) and (5.139) require the setting of the user-defined thresholds η_c and η_r, the approaches (5.148) and (5.149) need only the prescribed error tolerance for the given adaptation level. Nevertheless, if $\eta_c \sim \eta_r \approx \bar{\omega}_k$, then these techniques give a similar convergence since the threshold techniques also equidistribute the error over the computational domain.

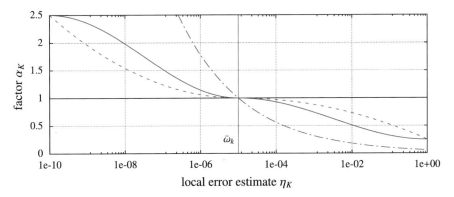

Fig. 5.3 The dependence of the factor α_K on η_K for techniques (5.146) (green), (5.148) (red), and (5.149) (blue) with $\eta^I_{max} = 1$, $\eta^I_{min} = 10^{-10}$, $\bar{\omega}_k = 10^{-5}$, $r_{max} = 0.25$, and $c_{max} = 2.5$

5.7 Numerical Experiments

We demonstrate the performance of the mesh adaptation algorithm for two partial differential equations which are discretized by the *discontinuous Galerkin method.* The corresponding numerical scheme is described in Sect. 7.3, where a general linear convection–diffusion–reaction equation is considered. For the numerical analysis of the method, we refer to, e.g., [45].

We consider the following settings of the size of mesh elements from Sect. 5.6.2:

- **thres**—*thresholds* using (5.134), (5.137), (5.139), and (5.140)
- **rate**—*rate of convergence* using (5.142), (5.143), and (5.146) with $\beta = p + 1$ for the optimization in the $L^2(\Omega)$-norm and $\beta = p$ for the optimization in the $H^1(\Omega)$-seminorm
- **e-elem**—*equidistribution per element* using (5.142), (5.143), and (5.149).
- **e-area**—*equidistribution per unit area* using (5.150), (5.143), and (5.149).

We skip the other possible combinations mentioned in Sect. 5.6.2. We note that approach **e-area** is included for comparison, although it is not supported by theoretical results in the same way as the techniques **thres, rate** and **e-elem**.

5.7.1 Boundary Layer Problem

Similarly as in [32, 49], we consider the standard benchmark for the numerical solution of convective-dominated problems

$$-\varepsilon\Delta u + \boldsymbol{b}\cdot\nabla u = g \qquad \text{in } \Omega := (0, 1)^2, \tag{5.151}$$

$$u = u_D \qquad \text{on } \Gamma := \partial\Omega,$$

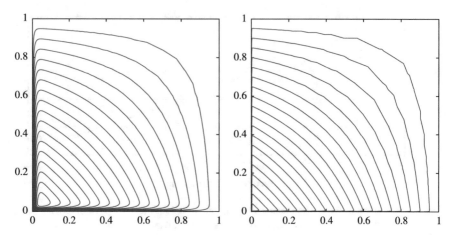

Fig. 5.4 Boundary layer problem (5.151)–(5.152), isolines of the approximate solutions for $\varepsilon = 10^{-2}$ (left) and $\varepsilon = 10^{-3}$ (right)

where $\varepsilon > 0$ is a constant diffusion coefficient and $\boldsymbol{b} = (-1, -1)^{\mathsf{T}}$ is the constant convective field. The boundary function u_D and the source term g are prescribed such that the exact solution of (5.151) is

$$u(x_1, x_2) = \big(c_1 + c_2(1 - x_1) + e^{-x_1/\varepsilon}\big)\big(c_1 + c_2(1 - x_2) + e^{-x_2/\varepsilon}\big) \qquad (5.152)$$

with $c_1 = -e^{-1/\varepsilon}$ and $c_2 = -1 - c_1$. The solution has two boundary layers along $x_1 = 0$ and $x_2 = 0$, whose width is proportional to the diffusion coefficient ε. Figure 5.4 shows the approximate solutions of (5.152) obtained on adaptively refined grids. The isolines are not smooth near the top right corner where the mesh was not sufficiently refined. Nevertheless, the boundary layers are resolved sufficiently. Since the exact solution is known analytically, we can evaluate the computational error $u - u_h$ in the norms of interest.

The first numerical experiments compare the various techniques for setting the sizes of mesh elements **thres**, **rate**, **e-elem**, and **e-area** defined above. We use the fixed polynomial approximation degree $p = 3$ and run the mesh adaptation Algorithm 5.1 which optimizes the mesh either with respect to the $L^2(\Omega)$-norm or with respect to the $H^1(\Omega)$-seminorm.

Figure 5.5 shows the comparison of convergence of the error $u - u_h$ with respect to the square root of the number of degrees of freedom (DoF $= \frac{1}{2}(p+1)(p+2)\#\mathscr{T}_h$) for the case $\varepsilon = 10^{-2}$. We observe that there is no essential difference among all tested approaches, but technique **rate** has slightly faster convergence. On the other hand, the differences with $\varepsilon = 10^{-3}$ are more essential, see Figure 5.6. Technique **rate** produces irregular convergence when the meshes are optimized with respect to the $L^2(\Omega)$-norm. We observed this phenomenon also for other examples. Sometimes, **rate** is superior to the other techniques, but sometimes it is very inefficient. Moreover, comparing the optimization with respect to the

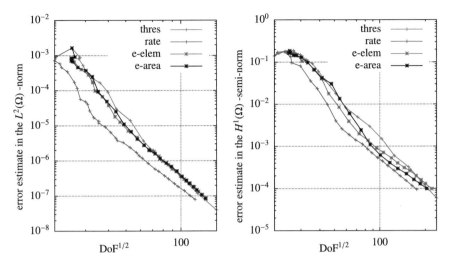

Fig. 5.5 Boundary layer problem (5.151)–(5.152) with $\varepsilon = 10^{-2}$, convergence of the error with respect to $\mathrm{DoF}^{1/2}$ for techniques thres, rate, e-elem, and e-area, mesh optimization with respect to the $L^2(\Omega)$-norm (left) and $H^1(\Omega)$-seminorm (right)

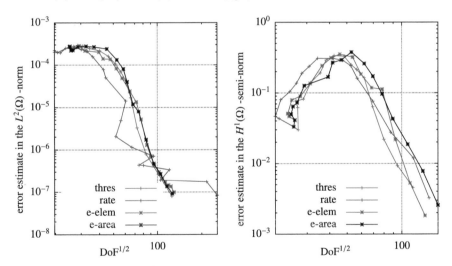

Fig. 5.6 Boundary layer problem (5.151)–(5.152) with $\varepsilon = 10^{-3}$, convergence of the error with respect to $\mathrm{DoF}^{1/2}$ for techniques thres, rate, e-elem, and e-area, mesh optimization with respect to the $L^2(\Omega)$-norm (left) and $H^1(\Omega)$-seminorm (right)

$H^1(\Omega)$-seminorm, e-elem technique is significantly better than e-area. This is in agreement with the theoretical results, cf. Lemma 5.28. Technique thres is comparable with e-elem, see Remark 5.30.

Furthermore, we demonstrate the accuracy of the interpolation error estimator. Fig. 5.7 shows the convergence of the computational error $e_h := u - u_h$ and its

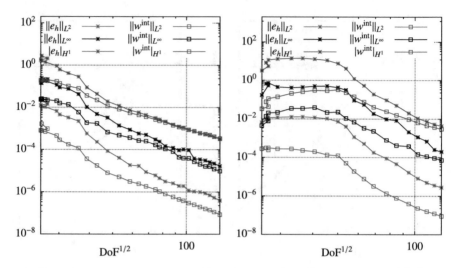

Fig. 5.7 Boundary layer problem (5.151)–(5.152) with $\varepsilon = 10^{-2}$ (left) and $\varepsilon = 10^{-3}$ (right), convergence of the error e_h and the interpolation error estimator w^{int} in the $L^2(\Omega)$-norm, the $L^\infty(\Omega)$-norm, and the $H^1(\Omega)$-seminorm, mesh optimization with respect to the $L^2(\Omega)$-norm, results obtained by e-elem method

interpolation error estimator $w^{\text{int}} := u_h^+ - \pi_h u_h^+$ (cf. (5.132)) in the $L^2(\Omega)$-norm, the $L^\infty(\Omega)$-norm, and the $H^1(\Omega)$-seminorm for both diffusive coefficients. We observe similar rates of convergence of all investigated pairs. Obviously, the interpolation error estimators are not upper bounds, but they give reasonable approximations, cf. (5.131). For the case with $\varepsilon = 10^{-3}$, the error is first increasing since it takes a few iterations for the method to properly sense the boundary layer. Proper convergence sets in after several levels of mesh adaptations. Finally, we note that there is only a negligible difference between the convergences of the mesh optimization with respect to the $L^2(\Omega)$-norm and the $H^1(\Omega)$-seminorm.

Moreover, we present the convergence of the algorithm for different polynomial approximation degrees. Figures 5.8 and 5.9 show the convergence of the error for polynomial approximations P_p, $p = 2, 3, 4$ with the diffusion $\varepsilon = 10^{-2}$ and $\varepsilon = 10^{-3}$, respectively. Obviously, higher polynomial approximation degrees give higher rate of convergence. Even in terms of DoF, higher-order methods are more efficient.

Furthermore, Fig. 5.10 shows the histogram of the distribution of the element error estimator η_K, $K \in \mathscr{T}_h^k$, on the last adaptation level obtained by Algorithm 5.1 using e-elem technique. Omitting the left parts of the graph, where the elements have almost vanishing error estimators, we observe a Gaussian-like distribution of η_K, $K \in \mathscr{T}_h^k$.

The final grids generated by the mesh adaptive algorithm are shown in Figs. 5.11 and 5.12. We observe a strong anisotropic capturing of both boundary layers. Moreover, the grids generated using higher polynomial degrees have elements with

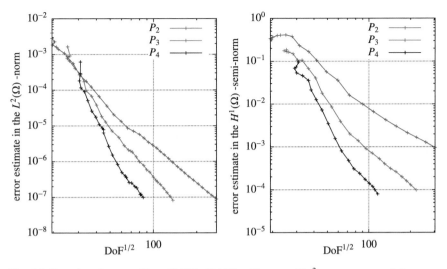

Fig. 5.8 Boundary layer problem (5.151)–(5.152) with $\varepsilon = 10^{-2}$, convergence of the error with respect to $\text{DoF}^{1/2}$ for P_2, P_3, and P_4 polynomial approximation, e-elem method, mesh optimization with respect to the $L^2(\Omega)$-norm (left) and $H^1(\Omega)$-seminorm (right)

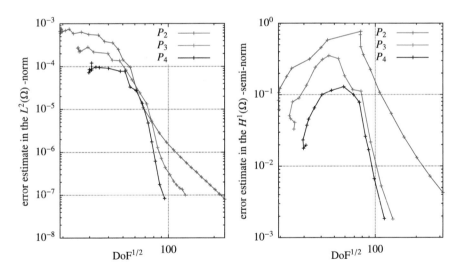

Fig. 5.9 Boundary layer problem (5.151)–(5.152) with $\varepsilon = 10^{-3}$, convergence of the error with respect to $\text{DoF}^{1/2}$ for P_2, P_3, and P_4 polynomial approximation, e-elem method, mesh optimization with respect to the $L^2(\Omega)$-norm (left) and $H^1(\Omega)$-seminorm (right)

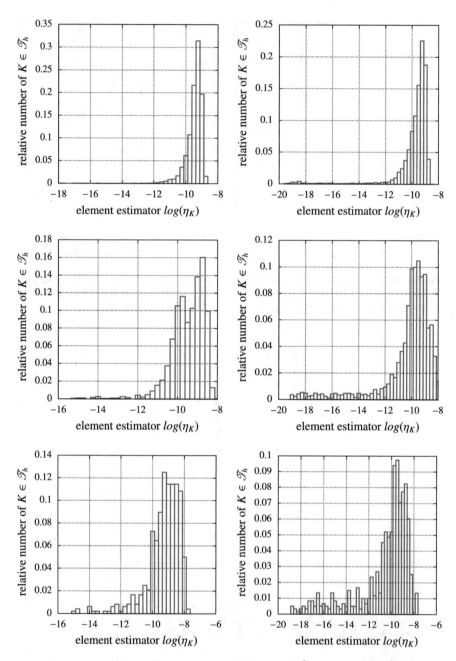

Fig. 5.10 Boundary layer problem (5.151)–(5.152) with $\varepsilon = 10^{-2}$ (left) and $\varepsilon = 10^{-3}$ (right), the histogram of the distribution of η_K, $K \in \mathscr{T}_h^k$ on the last adaptation level obtained by Algorithm 5.1 using e-elem technique, P_2 (first row), P_3 (second row), and P_4 approximations (third row)

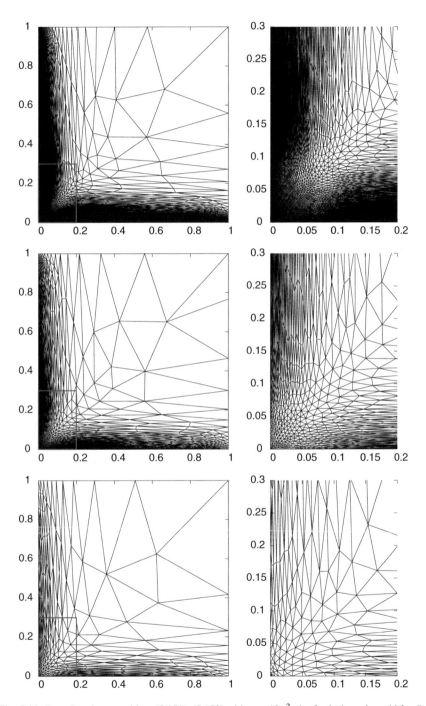

Fig. 5.11 Boundary layer problem (5.151)–(5.152) with $\varepsilon = 10^{-2}$, the final triangular grid for P_2 (first row), P_3 (second row), and P_4 (third row) polynomial approximations using e-elem method, mesh optimization with respect to the $L^2(\Omega)$-norm, total view (left) and the detail close to the left-bottom corner (right). Red box in the left-hand side figure is the zoomed region

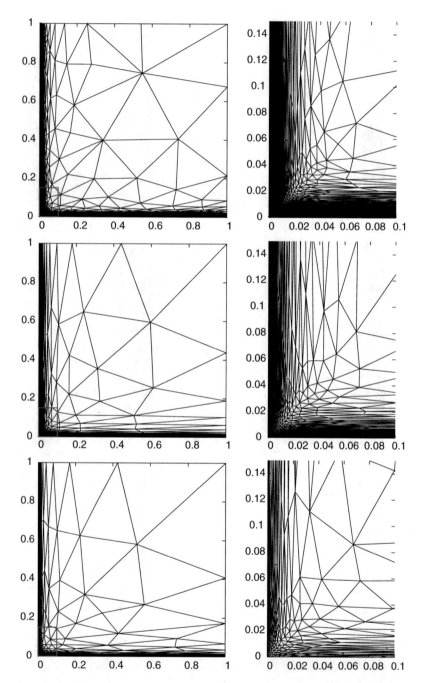

Fig. 5.12 Boundary layer problem (5.151)–(5.152) with $\varepsilon = 10^{-3}$, the final triangular grid for P_2 (first row), P_3 (second row), and P_4 (third row) polynomial approximations using e-elem method, mesh optimization with respect to the $L^2(\Omega)$-norm, total view (left) and the detail close to the left-bottom corner (right). Red box in the left-hand side figure is the zoomed region

lower aspect ratio (parameter σ_K). Furthermore, comparing both cases $\varepsilon = 10^{-2}$ and $\varepsilon = 10^{-3}$, the refined regions are significantly smaller for the latter case since the boundary layers are sharper. From the same reason, the aspect ratio of the generated elements is much larger for the case $\varepsilon = 10^{-3}$.

5.7.2 Multiple Difficulties Problem

This test case comes from [95] and it combines several different features, namely a point singularity due to a re-entrant corner, a circular wave front, a sharp peak, and a boundary layer. We solve the Poisson problem

$$-\Delta u = g \qquad \text{in } \Omega = (0, 1)^2 \setminus [0, 1] \times [-1, 1], \qquad (5.153)$$

$$u = u_D \qquad \text{on } \Gamma := \partial \Omega,$$

where g and u_D are chosen such that the exact solution of (5.153) is

$$u(x_1, x_2) = r^{\pi/\alpha_s} \sin(\varphi \pi / \alpha_s) \qquad (5.154)$$
$$+ \tan^{-1}\{\alpha_w[((x_1 - x_2^w)^2 + (x_2 - x_2^w)^2)^{1/2} - r_w]\}$$
$$+ \exp[-\alpha_p((x_1 - x_2^p)^2 + (x_2 - x_2^p)^2)]$$
$$+ \exp[-\alpha_e(1 + x_2)],$$

where the re-entrant angle is $\alpha_s = 3\pi/2$ and r and φ denote the polar coordinates. The interior wave is defined by $x_1^w = 0$, $x_2^w = -3/4$, $r_w = 3/4$, and $\alpha_w = 200$. The peak is centered at $x_1^p = -\sqrt{5}/4$, $x_2^p = -1/4$ with strength $\alpha_p = 1000$. The boundary layer is given by $\alpha_e = 100$. Figure 5.13 shows the isolines of the solution together with a description of its main features. The salient difficulties are the steep interior wave and the exponential boundary layer, which behave like singularities on a coarse grid.

Again, we compare the setting of sizes of mesh elements thres, rate, e-elem, and e-area. We use the polynomial approximation degree $p = 3$ and run Algorithm 5.1 which optimizes the mesh either with respect to the $L^2(\Omega)$-norm or with respect to the $H^1(\Omega)$-seminorm. Figure 5.14 shows the comparison of convergences of the error $u - u_h$ with respect to the square root of the number of degrees of freedom. Although rate gives the fastest convergence for the optimization with respect to the $L^2(\Omega)$-norm, the previous example (Fig. 5.6) prevents us to consider it as a suitable approach. The other convergence rates from Fig. 5.14 support the expectation: e-elem is the most efficient.

Figure 5.15 shows the convergence of the computational error $e_h := u - u_h$ and its interpolation error estimator $w^{\text{int}} := u_h^+ - \pi_h u_h^+$ (cf. (5.132)) in the $L^2(\Omega)$-norm, the $L^\infty(\Omega)$-norm, and the $H^1(\Omega)$-seminorm for the mesh optimization with respect

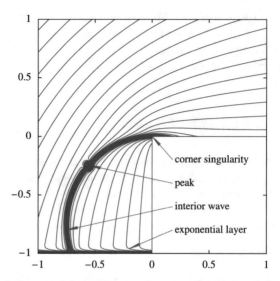

Fig. 5.13 Multiple difficulties problem (5.153)–(5.154): description of the features of the exact solution

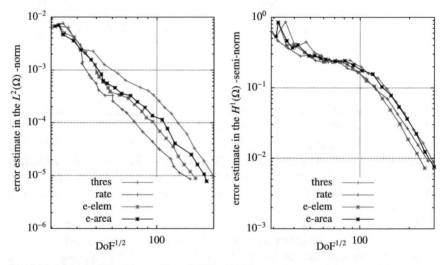

Fig. 5.14 Multiple difficulties problem (5.153)–(5.154), convergence of the error with respect to DoF$^{1/2}$ for techniques thres, rate, e-elem, and e-area, mesh optimization with respect to the $L^2(\Omega)$-norm (left) and $H^1(\Omega)$-seminorm (right)

to the $L^2(\Omega)$-norm and $H^1(\Omega)$-seminorm. We observe that the differences between the error and the corresponding estimate are larger in comparison to the boundary layer problem from Sect. 5.7.1. However, these differences are decreasing for the increasing number of mesh elements, and on the final grid, the ratios between the

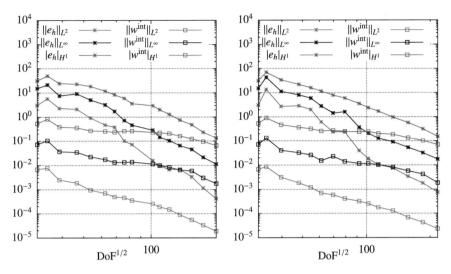

Fig. 5.15 Multiple difficulties problem (5.153)–(5.154), convergence of the error e_h and the interpolation error estimator w^{int} in the $L^2(\Omega)$-norm, the $L^\infty(\Omega)$-norm, and the $H^1(\Omega)$-seminorm, e-elem technique, mesh optimization with respect to the $L^2(\Omega)$-norm (left) and $H^1(\Omega)$-seminorm (right)

error and the corresponding estimate are similar to those ones from the boundary layer problem, cf. Fig. 5.7.

The final grid generated by the mesh adaptive algorithm is shown in Fig. 5.16. We observe a strong anisotropic capturing of the wave and the exponential boundary layer.

5.7.3   Summary of the Numerical Examples

We summarize the observations from the presented numerical experiments. The technique **e-elem**, which *equidistributes the interpolation error per element*, is superior in comparison to the other tested techniques. This justifies theoretical results, namely Theorem 5.26 and its consequence Lemma 5.28. Therefore, this approach is used in next chapters.

The interpolation error estimate $u_h^+ - \pi_h u_h^+$ gives a reasonable approximation of the interpolation error $u - \pi_h u$ in the $L^2(\Omega)$-norm, the $L^\infty(\Omega)$-norm, and the $H^1(\Omega)$-seminorm. Both the errors and their estimates have similar rate of convergence, hence the $u_h^+ = \mathscr{R}(u_h)$ (cf. (5.130)) gives a good approximation of the exact solution u, and therefore the final grid is close to the optimal mesh for the solution u.

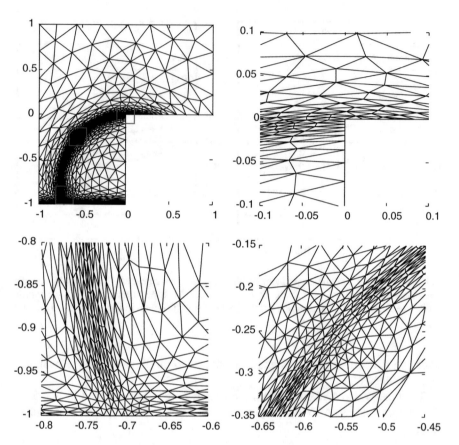

Fig. 5.16 Multiple difficulties problem (5.153)–(5.154): the final triangular grid obtained using P_3 polynomial approximation, mesh optimization with respect to the $L^2(\Omega)$-norm, total view (top left), and the details of the red frames

Last interesting observation is that the optimal meshes consist of mesh elements with smaller aspect ratio for increasing polynomial approximation degree p. Obviously, increasing p leads to a smaller number of degrees of freedom necessary to achieve the same error tolerance.

Chapter 6
Anisotropic Mesh Adaptation Method, hp-Variant

We present an hp-variant of the anisotropic mesh adaptation methods discussed so far. That is, contrary to our discussion in Chap. 5, we now let the polynomial degree of approximation vary from mesh element to mesh element. This is essential in situations where the exact solution contains local singularities but is very smooth in other parts of the computational domain. The main concern here is the extension of the continuous mesh and error models to the hp-variant. However, as the analytical solution of the pertinent optimization problem is not as straightforward as in the h-variant, we derive a semi-analytical iterative approach.

Consider polynomial approximation spaces of the type (1.23), i.e.,

$$S_{hp} := \{v_h \in L^2(\Omega); \ v_h|_K \in P^{p_K}(K) \ \forall K \in \mathscr{T}_h\}, \tag{6.1}$$

where $\mathscr{T}_h$ is a triangulation of the domain $\Omega \subset \mathbb{R}^d$, $d = 2, 3$, and $\boldsymbol{p}$ denotes the polynomial degree vector $\boldsymbol{p} = \{p_K : p_K \in \mathbb{N}, \ K \in \mathscr{T}_h\}$. Here, p_K is the polynomial degree assigned to element K. In the hp-variant of anisotropic mesh adaptation, the polynomial degree vector serves as an additional design variable, which can be used to find optimal, or near-optimal approximation spaces.

Definition 6.1 Let $\mathscr{T}_h$ be an admissible triangulation in the sense of Definition 1.12, and let $\boldsymbol{p} = \{p_K : p_K \in \mathbb{N}, \ K \in \mathscr{T}_h\}$. We call $\{\mathscr{T}_h, \boldsymbol{p}\}$ an hp-mesh.

6.1 Continuous Mesh Model

Recall from Chap. 5 that a continuous mesh is a mapping $\mathcal{M} : \Omega \to Sym$, where Sym is the space of $d \times d$ symmetric positive definite matrices. The continuous mesh analog of the discrete hp-mesh $\{\mathscr{T}_h, \boldsymbol{p}\}$ requires a polynomial degree distribution

V. Dolejší, G. May, *Anisotropic hp-Mesh Adaptation Methods*, Nečas Center Series, https://doi.org/10.1007/978-3-031-04279-9_6

function. For our purposes, any mapping $\mathcal{P} : \Omega \to \mathbb{R}^+$, such that

$$p_K = \text{int}\left[\frac{1}{|K|}\int_K \mathcal{P}dx\right], \tag{6.2}$$

may be used. Here, $\text{int}[a] := \lfloor a + 1/2\rfloor$ denotes the integer part of the number $a + 1/2$, $a \geq 0$.

Definition 6.2 Let $\mathcal{M}$ be a continuous mesh, and let $\mathcal{P} : \Omega \to \mathbb{R}^+$ be given. We call the tuple $(\mathcal{M}, \mathcal{P})$ a *continuous hp-mesh*.

Further, we recall the *mesh density* distribution function $\tau : \Omega \to \mathbb{R}^+$ given by (5.86). In addition, we define the *local* DoF function $\gamma : \Omega \to \mathbb{R}^+$, such that, for $x \in \Omega$,

$$\gamma(x) := \begin{cases} \frac{1}{2}(\mathcal{P}(x) + 1)(\mathcal{P}(x) + 2) & \text{for } d = 2, \\ \frac{1}{6}(\mathcal{P}(x) + 1)(\mathcal{P}(x) + 2)(\mathcal{P}(x) + 3) & \text{for } d = 3. \end{cases} \tag{6.3}$$

In analogy to (5.89), this definition may be motivated by noticing that for a given triangulation $\mathcal{T}_h = \{K\}$ and piecewise constant functions τ and $\mathcal{P}$ such that $\tau|_K = \tau_K = c_d/|K|$, and $\mathcal{P}|_K = p_K$, one obtains (for $d = 2$)

$$\int_\Omega \gamma\tau dx = c_d \sum_{K\in\mathcal{T}_h} \frac{1}{2}(p_K + 1)(p_K + 2) = c_d N_{\text{dof}}, \tag{6.4}$$

where

$$N_{\text{dof}} := \sum_{K\in\mathcal{T}_h} \frac{1}{2}(p_K + 1)(p_K + 2) \tag{6.5}$$

is the total number of degrees of freedom (DoF) of the discrete hp-mesh. Analogous relations are valid also for $d = 3$. Hence the product $\gamma(x)\tau(x)$ can be interpreted as the *density of degrees of freedom* at $x \in \Omega$. In analogy to Definition 5.19 we have

Definition 6.3 The *complexity* N of the continuous hp-mesh $(\mathcal{M}, \mathcal{P})$ is defined as

$$N \equiv N(\mathcal{M}, \mathcal{P}) := \int_\Omega \gamma\tau dx. \tag{6.6}$$

From here, the main arguments of the continuous mesh optimization in Sect. 5.5 remain intact. Again, we restrict to the estimates in the L^q-norm, $1 \leq q < \infty$. First, in virtue of Definition 5.23, we generalize the definition of the continuous interpolation error:

Definition 6.4 For $x \in \Omega$, we define the *continuous interpolation error* as the mapping $e_\tau : \Omega \to \mathbb{R}$, such that

$$e_\tau(x) \equiv e_\tau(x; p_K, q) := b(\mathcal{P}(x))^{1/q} B(x, \mathcal{P}(x)) \tau(x)^{-(\mathcal{P}(x)+1)/d}, \qquad (6.7)$$

where

$$b(\mathcal{P}(x)) = \begin{cases} \dfrac{2\pi}{q(\mathcal{P}(x)+1)+2} & d = 2, \\ \dfrac{4\pi}{q(\mathcal{P}(x)+1)+3} & d = 3, \end{cases} \qquad B(x, \mathcal{P}(x)) = \begin{cases} A_w \rho_w^{-1/2} & d = 2, \\ A_w (\varrho_w \rho_w)^{-1/3} & d = 3, \end{cases}$$

$$(6.8)$$

and A_w, ϱ_w, ρ_w are the parameters defining the anisotropic bound of the interpolation error function at $x \in \Omega$, cf. Definition 3.18 for $d = 2$ and Definition 4.7 for $d = 3$. These parameters depend explicitly on the polynomial approximation degrees.

Similarly as in (5.97), using (5.40) and (6.7) (for $d = 2$) or (5.73) and (6.7) (for $d = 3$), we have, for any simplex, centered at $x_K \in \Omega$, and having locally optimal anisotropy, the bound

$$\left\| w_{x_K, p_K}^{\text{int}} \right\|_{L^q(K)}^q \le e_\tau(x_K)^q \tau^{-1}(x_K). \qquad (6.9)$$

Further, (5.98) remains valid under a variable p-distribution. That is, if we introduce the interpolation operator

$$\pi_{h,p} : C^\infty \to S_{hp} : \qquad \pi_{h,p}|_K := \Pi_{x_K, p_K}|_K, \quad K \in \mathcal{T}_h, \qquad (6.10)$$

cf. Definition 3.2, then

$$\left\| u - \pi_{h,p} u \right\|_{L^q(\Omega)}^q = \sum_{K \in \mathcal{T}_h} \left\| u - \Pi_{x_K, p_K} u \right\|_{L^q(K)}^q \qquad (6.11)$$

$$\approx \sum_{K \in \mathcal{T}_h} \left\| w_{x_K, p_K}^{\text{int}} \right\|_{L^q(K)}^q \le \sum_{K \in \mathcal{T}_h} \frac{e_\tau(x_K)^q}{\tau(x_K)},$$

where $e_\tau(x)$ is the continuous interpolation error (cf. Definition 5.23), and we set $e_\tau(x_K) \equiv e_\tau(x_K; p_K, q)$, which, by (6.2), is consistent with the subsequent use of the midpoint rule to establish

$$\sum_{K \in \mathcal{T}_h} \frac{e_\tau(x_K)^q}{\tau(x_K)} \sim \sum_{K \in \mathcal{T}_h} e_\tau(x_K)^q |K| \to \int_\Omega e_\tau(x, \mathcal{P}(x); q)^q \, dx, \quad (h \to 0).$$

$$(6.12)$$

Then, in analogy to (5.100), we define the following global *continuous mesh error estimate*

$$E(\mathcal{M}, \mathcal{P}) = E^{\star}(\tau, \mathcal{P}) := \int_{\Omega} e_{\tau}^{q} \, dx \qquad (6.13)$$

$$= \int_{\Omega} b(\mathcal{P}(x)) \, B(x, \mathcal{P}(x))^{q} \, \tau(x)^{-(\mathcal{P}(x)+1)q/d} \, dx,$$

where it is understood that $\mathcal{M} = \mathcal{M}(\tau)$ is characterized by locally optimal anisotropy, given by Theorem 5.7 for $d = 2$ or Theorem 5.13 for $d = 3$.

In an abstract setting, the hp-minimization problem can now be cast as the following task:

Problem 6.5 Let $E^{\star}$ be defined in (6.13) and $N \in \mathbb{N}$ be given. Find the continuous hp-mesh $\{\mathcal{M}(\tau^{\star}), \mathcal{P}^{\star}\}$ of locally optimal anisotropy that satisfies

$$(\tau^{\star}, \mathcal{P}^{\star}) = \arg\min_{(\tau, \mathcal{P})} E^{\star}(\tau, \mathcal{P}), \qquad \text{such that} \quad N(\mathcal{M}(\tau), \mathcal{P}) = \int_{\Omega} \gamma \tau \, dx = N,$$
$$(6.14)$$

where γ is the density of degrees of freedom defined in (6.4).

Problem 6.5 is the hp-variant of Problem 5.24. In a similar manner, one introduces the hp-variant of Problem 5.25. However, for the sake of conciseness, we deal only with the first variant here.

6.2   Semi-analytical Optimization

The solution of the abstract minimization problem (6.14) is not straightforward. However, it is possible to solve it iteratively in two steps.

(i) For an initial polynomial degree distribution function, we solve an analogue of Problem 5.24, dealing with the h-adaptation.
(ii) Having the resulting mesh density distribution, we locally modify the polynomial degree distribution function in order to decrease the interpolation error estimate while the complexity of the mesh is kept constant.

Then the whole process can be repeated.

The analogue Problem 5.24 with a *fixed* (but non-constant) polynomial degree distribution function reads:

Problem 6.6 Let $E^{\star}$ be defined in (6.13) and $N \in \mathbb{N}$ be given. Moreover, let polynomial distribution $\mathcal{P}$ be given. Find the mesh density distribution function $\tau^{\star}$

such that

$$\tau^{\star} = \arg\min_{\tau} E^{\star}(\tau, \mathcal{P}), \qquad \text{such that} \quad N = \int_{\Omega} \gamma \tau \, dx, \tag{6.15}$$

where γ is the local density of DoF defined in (6.3).

In order to shorten the notation, we set

$$s(\mathcal{P}(x)) = \tfrac{1}{d}(\mathcal{P}(x) + 1)q, \quad x \in \Omega. \tag{6.16}$$

Then, (6.13) reads

$$E^{\star}(\tau) = \int_{\Omega} b(\mathcal{P}(x)) \, (B(x, \mathcal{P}(x))^q \, \tau(x)^{-s(\mathcal{P}(x))} \, dx. \tag{6.17}$$

It is now again a matter of applying simple calculus of variations to obtain from (6.17)

$$\delta E^{\star} = -\int_{\Omega} s(\mathcal{P}(x)) \, b(\mathcal{P}(x)) \, (B(x, \mathcal{P}(x))^q \tau^{-s(\mathcal{P}(x))-1} \delta\tau \, dx. \tag{6.18}$$

Using the constraint,

$$\int_{\Omega} \gamma \delta\tau \, dx = 0, \tag{6.19}$$

one sees immediately that

$$S := s(\mathcal{P}(x)) \, b(\mathcal{P}(x)) \, (B(x, \mathcal{P}(x))^q \, \gamma(x)^{-1} \tau^{-s(\mathcal{P}(x))-1} = \text{const.} \tag{6.20}$$

is an admissible solution. This can be solved for the mesh density distribution to yield:

$$\tau^{\star}(x) = \left(s(\mathcal{P}(x)) \, b(\mathcal{P}(x)) \, (B(x, \mathcal{P}(x))^q \, \gamma(x)^{-1} S^{-1} \right)^{1/(s\mathcal{P}(x)+1)}. \tag{6.21}$$

Substituting this back into the constraint equation in (6.15) allows one to determine the constant S. However, it is not possible to determine the constant analytically. One must resort instead to a numerical computation. In practice, an application of the midpoint rule may be used to replace the constraint with

$$N = \sum_{K \in \mathcal{T}_h} \gamma(x_K)\tau(x_K)|K|, \tag{6.22}$$

which may be evaluated with $\tau^\star(x_K)$, $\mathcal{P}(x_K) \approx p_K$ and

$$\gamma_K := (p_K + 1) \dots (p_K + d)/d!$$

(cf. (6.3)), which is consistent with the midpoint rule. Hence, using (6.8) and (6.16), the discrete analogue of the constraint in (6.15) reads

$$N = \sum_{K \in \mathcal{T}_h} \gamma_K |K| \left(\frac{q(p_K + 1)}{d} \frac{2(d-1)\pi}{q(p_K+1)+d} \frac{B^q(x_K)}{\gamma_K S} \right)^{\frac{d}{q(p_K+1)+d}}. \tag{6.23}$$

Relation (6.23) is a scalar nonlinear equation for $S \in \mathbb{R}$ which can be solved numerically.

Now, we proceed to step (ii) which locally optimizes the polynomial degree distribution function and returns a new function $\mathcal{P}^\star$. Taking into account that we are in possession of a reference error estimate for the distribution $\tau^\star$, the following heuristic procedure presents itself:

First, for all $K \in \mathcal{T}_h$, one evaluates the continuous error function $e_\tau(x_K) =:$ e_K^{ref} with the optimal mesh density distribution $\tau^\star(x_K) =: \tau_0(x_K)$ given by (6.21). Further, we evaluate the optimal anisotropy parameters given by Theorem 5.7 ($d = 2$) or Theorem 5.13 for ($d = 3$) for nearby values $p_K + i$, $i \in \mathcal{I} \subset \mathbb{Z}\backslash\{0\}$. In practice, we admit an increase/decrease of p_K by one, i.e., $\mathcal{I} = \{-1, 1\}$. Then we evaluate $\tau_i(x_K)$, $i \in \mathcal{I}$ by requiring

$$e_\tau(\tau_i(x_K)) = e_K^{\text{ref}}, \qquad i \in \mathcal{I}. \tag{6.24}$$

This gives a number of alternative local configurations $\{\tau_i(x_K)\}_{i \in \mathcal{I}}$ (and the corresponding anisotropy of elements) with the same continuous error estimate. Naturally, we choose the configuration associated with the smallest local hp-density

$$i_K^{\text{opt}} := \arg\min_{i \in \mathcal{I} \cup \{0\}} \{\tau_i(x_K)(p_K + i + 1) \dots (p_K + i + d)\} \tag{6.25}$$

and set $\mathcal{P}^\star(x_K) := \mathcal{P}(x_K) + i_K^{\text{opt}}$, $K \in \mathcal{T}_h$. The practical implementation of this idea is described in the next section.

Remark 6.7 In the heuristic approach explained above, we determine several candidates with polynomial approximation degree $p_K + i$, $i \in \mathcal{I} \cup \{0\}$ having the same interpolation error bound and we chose the candidate with the smallest local hp-density. Alternatively, we can determine the candidates such that they have the same local hp-density and we select the candidates with the smallest interpolation error bound. Due to the monotonicity of the error bounds w.r.t. τ both techniques are equivalent. The latter one is used in Chap. 8.

Finally, we present the hp-analogue of the equidistribution of the error formulated in Lemma 5.28.

Lemma 6.8 *Let τ be the mesh density distribution given by (6.21) and $\mathcal{P}$ be the polynomial distribution function. Then*

$$\frac{\|u - \pi_h u\|_{L^q(K)}^q}{(p_K + 2) \dots (p_K + d)} \approx const \qquad \forall K \in \mathcal{T}_h, \tag{6.26}$$

where $p_K \approx \mathcal{P}|_K$.

Proof Let τ be the mesh distribution satisfying (6.21) and the constraint (6.15). From Definition 6.4, (6.13), (6.20), and (6.16), we have

$$\|u - \pi_h u\|_{L^q(K)}^q \approx E_K^\star(\tau^\star, \mathcal{P}) := \int_K e_\tau^q \, dx = \int_K b B^q \tau^{-s} \, dx \tag{6.27}$$

$$= \int_K \underbrace{s\, b\, B^q\, \gamma^{-1}\, \tau^{-s-1}}_{=\,s \text{ due to } (6.20)} \frac{\gamma \tau}{s} \, dx = \frac{S\, d}{q} \int_K \frac{\gamma \tau}{\mathcal{P} + 1} \, dx.$$

Further, using $\mathcal{P}(x)|_K \approx p_K$, relation (5.91) implying $\int_K \tau \, dx \approx c_d$ and (6.3), we obtain from (6.27) the relation

$$\|u - \pi_h u\|_{L^q(K)}^q \approx \frac{S\, d}{q} \int_K \frac{\gamma \tau}{(\mathcal{P} + 1)} \, dx = \frac{S\, d}{q\, d!} \int_K \frac{\tau (\mathcal{P} + 1) \dots (\mathcal{P} + d)}{(\mathcal{P} + 1)} \, dx \tag{6.28}$$

$$\approx \frac{S\, c_d}{q\, (d - 1)!} (p_K + 2) \dots (p_K + d),$$

which implies (6.26). ☐

6.3 Anisotropic *hp*-Mesh Adaptation Algorithm

Similarly as in Sect. 5.6, we develop the *hp*-variant of the anisotropic mesh adaptation algorithm for the numerical solution of partial differential equations. The main enrichment is the possibility to locally modify the polynomial approximation degree p_K, $K \in \mathcal{T}_h$ by selecting the degree which minimizes the local *hp*-density, cf. (6.25). As mentioned above, we admit in each mesh adaptation level to modify the local polynomial degree by one, i.e., $\mathcal{I} = \{-1, 1\}$. Therefore, we select among the candidates $p_K - 1$, p_K, $p_K + 1$ which means that we need approximations of the derivatives of the exact solution u of order p_K, $p_K + 1$, $p_K + 2$. Naturally, we approximate these derivatives by higher-order reconstructions of the corresponding approximate solution $u_h \in S_{hp}$ obtained by a suitable numerical method.

Hence, we introduce the spaces (cf. (6.1))

$$S_{hp+1} := \{v_h \in L^2(\Omega); \ v_h|_K \in P^{p_K+1}(K) \ \forall K \in \mathcal{T}_h\}, \qquad (6.29)$$

$$S_{h,p+2} := \{v_h \in L^2(\Omega); \ v_h|_K \in P^{p_K+2}(K) \ \forall K \in \mathcal{T}_h\} \qquad$$

and define functions

$$u_h^{(0)} := u_h \in S_{hp}, \quad u_h^{(1)} := \mathcal{R}^1(u_h) \in S_{hp+1}, \quad u_h^{(2)} := \mathcal{R}^2(u_h) \in S_{h,p+2}, \qquad (6.30)$$

where $\mathcal{R}^1$ and $\mathcal{R}^2$ are suitable (local) higher-order reconstruction techniques, cf. Sect. 9.1. The partial derivatives of order $p_K + i$ of $u_h^{(i)}$, $i = 0, 1, 2$ are constant on K and they are employed for the approximation of the derivatives of u. Similar to (5.5), we define the *interpolation error functions* of degree $p_K + i$, $i = 0, 1, 2$ on element $K \in \mathcal{T}_h$ by

$$w_{x_K,i}^{int} = u_h^{(i)} - \Pi_{x_K,p} u_h^{(i)}, \qquad i = 0, 1, 2. \qquad (6.31)$$

Furthermore, we present explicit relations for the choice of the optimal polynomial degree. We consider here only the case of the L^q-norm for $d = 2$ and $q < \infty$. The other cases can be treated analogously. Having the higher-order reconstructions (6.30), we evaluate the partial derivatives of order $p_K + i$ on K and approximate the anisotropy of the interpolation error functions $w_{x_K,i}^{int}$, $i = p_K - 1, p_K, p_K + 1$:

$$\{A_{w,p_K-1}, \ \rho_{w,p_K-1}, \ \varphi_{w,p_K-1}\} \quad \ldots \text{ anisotropy of } w_{x_K,p_K-1}^{int}, \qquad (6.32)$$

$$\{A_{w,p_K}, \ \rho_{w,p_K}, \ \varphi_{w,p_K}\} \quad \ldots \text{ anisotropy of } w_{x_K,p_K}^{int},$$

$$\{A_{w,p_K+1}, \ \rho_{w,p_K+1}, \ \varphi_{w,p_K+1}\} \quad \ldots \text{ anisotropy of } w_{x_K,p_K+1}^{int}.$$

Let $\tau^\star(x_K)$ be the optimized mesh density given by (6.21) and (6.23), then the reference values e_K^{ref} (cf. (6.24)) follow from (5.93) and (5.95) as

$$e_K^{ref} := \left(\frac{2\pi}{q(p_K+1)+2}\right)^{1/q} A_{w,p_K} \ (\rho_{w,p_K})^{1/2} \ \tau^\star(x_K)^{-(p_K+1)/2}. \qquad (6.33)$$

Then we define the configurations $\tau_i(x_K)$, $i = -1, 0, 1$ such that

$$e_K^{ref} = \left(\frac{2\pi}{qp_K+2}\right)^{1/q} A_{w,p_K-1} \ (\rho_{w,p_K-1})^{1/2} \ \tau_{-1}(x_K)^{-p_K/2}, \qquad (6.34)$$

$$e_K^{ref} = \left(\frac{2\pi}{q(p_K+2)+2}\right)^{1/q} A_{w,p_K+1} \ (\rho_{w,p_K+1})^{1/2} \ \tau_1(x_K)^{-(p_K+2)/2}$$

and $\tau_0(x_K) := \tau^\star(x_K)$. Finally, we select i_K^{opt} using (6.25) and set the new polynomial approximation degree

$$p_K^\star = \mathcal{P}^\star(x_K) := p_K + i_K^{\text{opt}}, \qquad K \in \mathcal{T}_h. \tag{6.35}$$

At the beginning of Sect. 6.2, we mentioned that the optimal hp-mesh can be obtained iteratively by the combination of two steps. However, in practical computations we repeat this procedure only once for each level of mesh refinement $k = 0, 1, \ldots$ and the polynomial degree distribution from the previous level is taken as the initial guess for the next adaptation level.

Similarly as in the h-variant, cf. Sect. 5.6.2.2, we have to prescribe either the sequence of decreasing error tolerances $\bar{\omega}_k$, $k = 0, 1, \ldots$ for the error bound $\mathscr{E}^\star(\tau, \mathcal{P})$ (cf. (6.13)) or the sequence of increasing number of mesh elements N_k, $k = 0, 1, \ldots$, which scales the constant S in (6.23). For the hp-variant, the latter case is easier to implement. The final anisotropic hp-mesh adaptation algorithm based on interpolation error control is shown in Algorithm 6.1 for $d = 2$, the 3D version is analogous.

Algorithm 6.1: Anisotropic hp-mesh adaptation (2D)

1: set tolerance $\omega > 0$ and initial hp-mesh $\{\mathcal{T}_h^0, \boldsymbol{p}^0\}$
2: **for** $k = 0, 1, \ldots$ **do**
3:     compute approximate solution $u_h^k \in S_{hp}^k$
4:     compute higher-order reconstruction $u_h^+ = \mathcal{R}(u_h^k) \in S_{hp+1}^k$
5:     **if** $\|u_h^+ - \pi_h u_h^+\| \le \omega$ **then**
6:         STOP the computation
7:     **else**
8:         **for** each element K **do**
9:             compute the higher order reconstructions $u_h^{(i)}$, $i = 0, 1, 2$ by (6.30)
10:            determine local anisotropies (6.32)
11:            compute optimal mesh density $\tau^\star$ using (6.21)–(6.22)
12:            compute corresponding τ_i, $i = -1, 0, 1$ using (6.33)–(6.34)
13:            choose polynomial order $p_K^\star$ according to (6.25) and (6.35)
14:            construct new local metric $\mathbb{M}_K^\star = \mathcal{M}(x_K)$ using (2.54)
15:        **end for**
16:        construct continuous hp-mesh $\{\mathcal{M}, \mathcal{P}\}$ from $(\mathbb{M}_K^\star, p_K)_{K \in \mathcal{T}_h}$, cf. Section 2.5.2
17:        construct new hp-mesh $\{\mathcal{T}_h^{k+1}, \boldsymbol{p}^{k+1}\}$ from $\{\mathcal{M}, \mathcal{P}\}$, cf. Section 2.5.3
18:    **end if**
19: **end for**

6.4   Numerical Experiments

We demonstrate the performance of the anisotropic hp-mesh adaptation Algorithm 6.1 for several partial differential equations. Some problems were introduced
in Sect. 5.7 for the h-variant of the methods and the others are new. Again, we
employ the *discontinuous Galerkin method*, the corresponding numerical schemes
are described in Sects. 7.3 and 7.4, for the analysis of the method, we refer to [45].

6.4.1 L^2-Projection of Piecewise Polynomial Function

The first example represents just the L^2-projection of piecewise polynomial functions where we can verify the ability of the adaptive algorithm to select the
corresponding optimal polynomial approximation degrees. Let $\Omega := (0, 1)^2$ be a
unit square, we consider two functions

$$
u_1 = \begin{cases} \left(Z - \frac{1}{2}\right) & \text{for } Z \le \frac{1}{2} \\ \left(Z - \frac{1}{2}\right)^2 & \text{for } Z > \frac{1}{2} \end{cases}, \qquad u_2 = \begin{cases} \left(Z - \frac{1}{2}\right) & \text{for } Z \le \frac{1}{2} \\ \left(Z - \frac{1}{2}\right)^3 & \text{for } Z > \frac{1}{2} \end{cases}, \qquad (6.36)
$$

where $Z = x_1^2 + x_2^2$. Obviously, u_1 is a polynomial of degree $p = 2$ in the interior of
the quarter circle with radius $1\sqrt{2}$ and a polynomial of degree $p = 4$ in the exterior
of this quarter circle. Similarly, u_2 is a polynomial of degrees $p = 2$ and $p = 6$ in
the interior and exterior of the quarter circle, respectively. Moreover, both functions
are continuous but not continuously differentiable.

We apply the anisotropic hp-mesh adaptive algorithm (the numerical scheme is
replaced by the L^2-projection of functions u_1 and u_2 to the space S_{hp}) with an
initial isotropic coarse mesh and initial polynomial approximation degree $p = 3$.
Figures 6.1 and 6.2 show the final hp-meshes. We observe a strong anisotropic mesh
capturing of the circle arc with radius $1\sqrt{2}$. Further, there are large elements outside
of this arc with the polynomial approximation degrees corresponding to the degrees
of piecewise polynomial functions u_1 and u_2. In the detail of Fig. 6.2, we observe a
transition region between polynomial degrees 2 and 6.

6.4.2   Boundary Layers Problem

We consider the convection–diffusion problem (5.151) and (5.152) from Chap. 5,
where the solution possess two boundary layers along $x_1 = 0$ and $x_2 = 0$ with
the width proportional to the diffusion coefficient ε, we consider diffusive constants
$\varepsilon = 10^{-2}$ and $\varepsilon = 10^{-3}$, see Fig. 5.4.

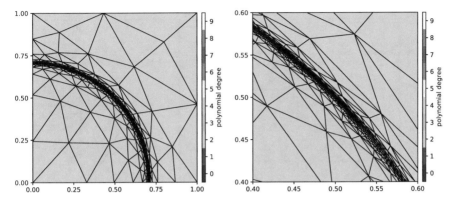

Fig. 6.1 The L^2-projection of u_1 from (6.36), the final hp-mesh, mesh optimization with respect to the $L^2(\Omega)$-norm, total view (left) and the detail near the circle arc (right)

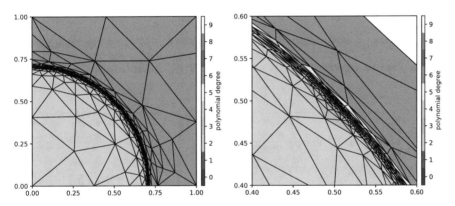

Fig. 6.2 The L^2-projection of u_2 from (6.36), the final hp-mesh, mesh optimization with respect to the $L^2(\Omega)$-norm, total view (left) and the detail near the circle arc (right)

We compare the efficiency of the hp-methods with the h-variants using $p = 2$, $p = 3$, and $p = 4$ polynomial approximations. Figures 6.3 and 6.4 show the convergence of the interpolation error estimator with respect to the square root of the number of degrees of freedom for $\varepsilon = 10^{-2}$ and $\varepsilon = 10^{-3}$, respectively. There are shown the convergences for the mesh optimizations with respect to the $L^2(\Omega)$-norm and the $H^1(\Omega)$-seminorm. We observe that the hp-variant is superior to the h-variants even for the largest tested polynomial degree.

Furthermore, Figs. 6.5 and 6.6 show the histogram of the distribution of the element error estimator η_K, $K \in \mathscr{T}_h^k$ on the middle and the last adaptation levels obtained by Algorithm 6.1. Omitting the left parts of the graph, where the elements have almost vanishing error estimators, we observe the distributions on the last adaptation levels are close to the Gaussian one for $\varepsilon = 10^{-2}$ and a little more far distribution for $\varepsilon = 10^{-3}$. We suppose that additional levels of adaptations (with the same local tolerance ω_k) will equidistribute the error estimator in a better manner.

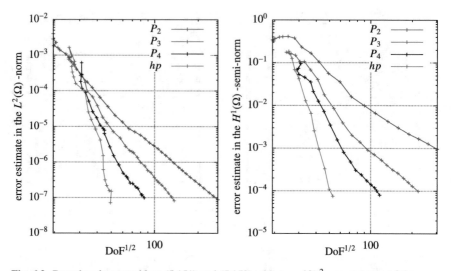

Fig. 6.3 Boundary layer problem (5.151) and (5.152) with $\varepsilon = 10^{-2}$, convergence of the error with respect to $\mathrm{DoF}^{1/2}$ for P_2, P_3, and P_4 polynomial approximation and the *hp*-variant, mesh optimization with respect to the $L^2(\Omega)$-norm (left) and $H^1(\Omega)$-seminorm (right)

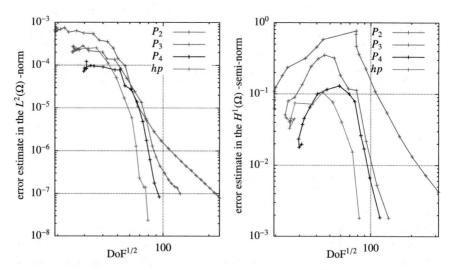

Fig. 6.4 Boundary layer problem (5.151) and (5.152) with $\varepsilon = 10^{-3}$, convergence of the error with respect to $\mathrm{DoF}^{1/2}$ for P_2, P_3, and P_4 polynomial approximation and the *hp*-variant, mesh optimization with respect to the $L^2(\Omega)$-norm (left) and $H^1(\Omega)$-seminorm (right)

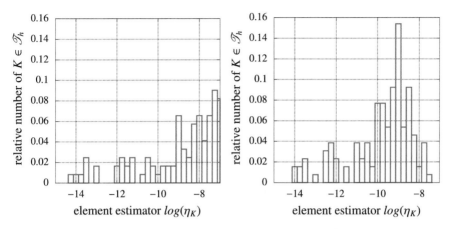

Fig. 6.5 Boundary layer problem (5.151) and (5.152) with $\varepsilon = 10^{-2}$, the histogram of the distribution of η_K, $K \in \mathscr{T}_h^k$ after 9 (left) and 17 (right) adaptation levels obtained by Algorithm 6.1

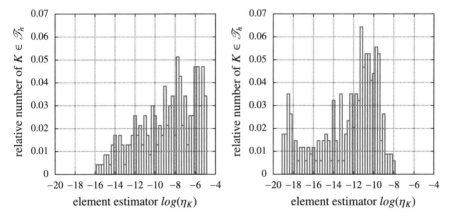

Fig. 6.6 Boundary layer problem (5.151) and (5.152) with $\varepsilon = 10^{-3}$, the histogram of the distribution of η_K, $K \in \mathscr{T}_h^k$ after 8 (left) and 15 (right) adaptation levels obtained by Algorithm 6.1

However, we note that the goal is not to equidistribute the error but achieve the given tolerance as soon as possible.

Finally, Fig. 6.7 shows the comparison of the distributions of η_K, $K \in \mathscr{T}_h^k$ and $\eta_K/(p_K + 2)$, $K \in \mathscr{T}_h^k$. The former corresponds to Lemma 5.28 whereas the latter one to its hp-variant Lemma 6.8. We observe only small differences for both cases. Based on this experience, we employ this principle also in Chap. 8, Sect. 8.2.1.

Finally, Figs. 6.8 and 6.9 show the final hp-mesh obtained by the mesh optimization with respect to the $L^2(\Omega)$-norm. The expected capturing of the boundary layers is easy to observe. Moreover, outside of the boundary layers, the mesh has large

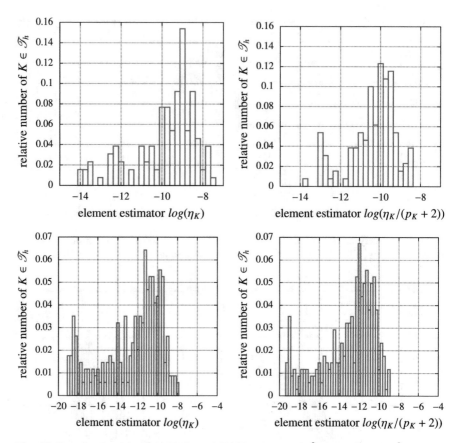

Fig. 6.7 Boundary layer problem (5.151) and (5.152) with $\varepsilon = 10^{-2}$ (top) and $\varepsilon = 10^{-3}$ (bottom), the histogram of the distribution of η_K, $K \in \mathscr{T}_h^k$ (left) and the distribution of $\eta_K/(p_K + 2)$, $K \in \mathscr{T}_h^k$ (right) on the last adaptation level obtained by Algorithm 6.1

elements with degree $p = 2$ since the solution $u \approx (c_1 + c_2(1-x_1))(c_1 + c_2(1-x_2))$ for $x_1 \gg 0$ and $x_2 \gg 0$ is close to a quadratic function in this region, cf. (5.152).

6.4.3 Multiple Difficulties Problem

We consider the Poisson problem (5.153) and (5.154) with multiple difficulties (a point singularity due to a re-entrant corner, a circular wave front, a sharp peak, and a boundary layer).

We compare the efficiency of the hp-methods with the h-variants using $p = 2$, $p = 3$, and $p = 4$ polynomial approximations. Figure 6.10 presents the convergence of the interpolation error estimator with respect to the square root of DoF. There

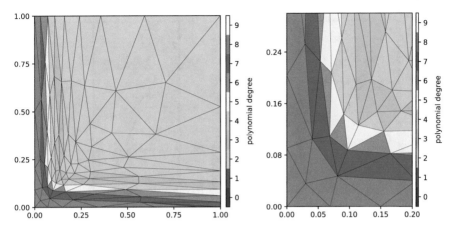

Fig. 6.8 Boundary layer problem (5.151) and (5.152) with $\varepsilon = 10^{-2}$, the final hp-mesh, mesh optimization with respect to the $L^2(\Omega)$-norm, total view (left) and the detail close to the left-bottom corner (right)

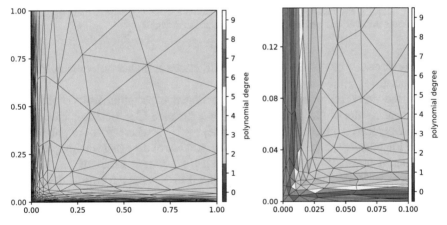

Fig. 6.9 Boundary layer problem (5.151) and (5.152) with $\varepsilon = 10^{-3}$, the final hp-mesh, mesh optimization with respect to the $L^2(\Omega)$-norm, total view (left) and the detail close to the left-bottom corner (right)

are shown the convergences for the mesh optimizations with respect to the $L^2(\Omega)$-norm and the $H^1(\Omega)$-seminorm. We observe that the hp-variant is superior to the h-variants even for the largest tested polynomial degree.

Figure 6.11 shows the final hp-mesh obtained by the mesh optimization with respect to the $L^2(\Omega)$-norm. The expected capturing of the interior and boundary layers is easy to observe. We see also the refinement close to the interior corner but with rather high polynomial approximation degree. It can be surprising since low polynomial approximation degree should be expected due to the corner sin-

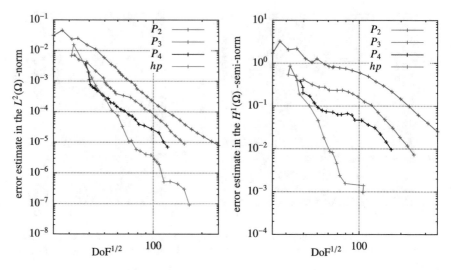

Fig. 6.10 Multiple difficulties problem (5.153) and (5.154), convergence of the error with respect to $\text{DoF}^{1/2}$ for P_2, P_3, and P_4 polynomial approximation and the *hp*-method, mesh optimization with respect to the $L^2(\Omega)$-norm (left) and $H^1(\Omega)$-seminorm (right)

gularity. It can be caused by the insufficiently accurate higher-order reconstruction techniques.

On the other hand, the previously mentioned expectation follows from the theoretical results of the finite element approximation [31, 71, 106]. The rate of the convergence is given by the value $\min(p + 1, s)$, where p is the polynomial approximation degree and s is the Sobolev regularity of the exact solution. Numerical experiments in [45, Chapter 2] show that if $p + 1 > s$, then the experimental rate of convergence is equal to (in agreement with theory) s but the error is decreasing for increasing p. Since the goal of the mesh adaptation is to achieve the given error tolerance and not the optimal asymptotic rate of convergence then the use of the higher polynomial degrees in the vicinity of the singularity may be advantageous in some situations.

Finally, Fig. 6.12 shows the isolines of the approximate solution on the final *hp*-mesh. We observe a nice capturing of all phenomena.

6.4.4 Mixed Hyperbolic-Elliptic Problem

According to [76, Example 2], we consider the problem

$$-\varepsilon \Delta u + \nabla \cdot (\boldsymbol{b}\, u) + cu = f \qquad \text{in } \Omega := (0, 1)^2, \tag{6.37}$$

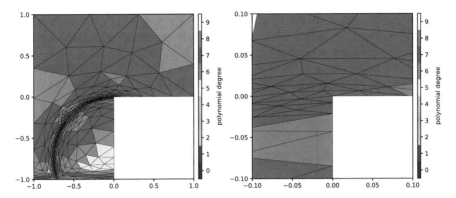

Fig. 6.11 Multiple difficulties problem (5.153) and (5.154), the final hp-mesh, mesh optimization with respect to the $L^2(\Omega)$-norm, total view (left) and the detail close to the interior corner (right)

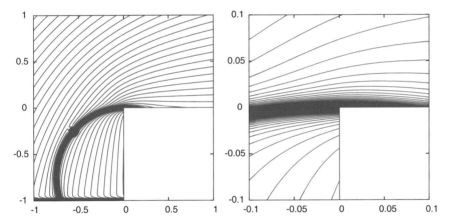

Fig. 6.12 Multiple difficulties problem (5.153) and (5.154), the isolines on the final hp-mesh, mesh optimization with respect to the $L^2(\Omega)$-norm, total view (left) and the detail close to the interior corner (right)

with varying diffusion

$$\varepsilon = \tfrac{\delta}{2}\left(1 - \tanh\left(\tfrac{1}{\gamma}\left(r^2 - \tfrac{1}{16}\right)\right)\right), \quad r = \sqrt{(x_1 - 0.5)^2 + (x_2 - 0.5)^2},$$

where $\delta > 0$, $\gamma > 0$ are constants. The diffusion ε is close to δ in the interior of the circle

$$Z := \{(x_1, x_2);\ (x_1 - 0.5)^2 + (x_2 - 0.5)^2 = 1/16\}$$

having the center at $[1/2, 1/2]$ and diameter $1/4$. Outside of this circle, diffusion ε is quickly vanishing, more precisely $\varepsilon \approx 10^{-16}$ at the boundary $\Gamma = \partial\Omega$. We use the values $\delta = 0.01$ and $\gamma = 0.05$.

Further, the convective field is $b = (2x_2^2 - 4x_1 + 1, x_2 + 1)^\mathsf{T}$, the reaction $c = -\nabla \cdot b = 3$ and the source term $f = 0$. The characteristics associated with the convective field enter the domain Ω through the horizontal edge along $x_2 = 0$ and the vertical edges along $x_1 = 0$ and $x_1 = 1$. Therefore, the Dirichlet boundary conditions are given on this "inflow" part of the boundary $\Gamma_\mathrm{D} = \{(x_1, x_2) \in \Gamma : x_1 = 0 \text{ or } x_1 = 1 \text{ or } x_2 = 0\}$ by

$$
u = \begin{cases} 1 & \text{if } x_1 = 0 \text{ and } 0 < x_2 \le 1, \\ \sin^2(\pi x) & \text{if } 0 \le x_1 \le 1 \text{ and } x_2 = 0, \\ e^{-50x_2^4} & \text{if } x_1 = 1 \text{ and } 0 < x_2 \le 1. \end{cases} \tag{6.38}
$$

The homogeneous Neumann boundary condition is prescribed on $\Gamma_\mathrm{N} := \Gamma \backslash \Gamma_\mathrm{D}$. This problem is elliptic on Ω (since $\varepsilon > 0$) but from the computational point of view, the diffusion is vanishing (the ratio between of the diffusion and convection is under the machine accuracy) in the region with $r > 1/4$ and hence the problem behaves like hyperbolic one.

Figure 6.13 shows the final hp-generated by the adaptive algorithm, the corresponding isolines of the solution are viewed in Fig. 6.14. We observe the capturing of the solution features and the corresponding anisotropic hp-refinement: strong anisotropic adaptation with lower polynomial degrees along the interior layers, large elements with high degree in areas where the solution is smooth and the lowest possible degree where the solution is constant.

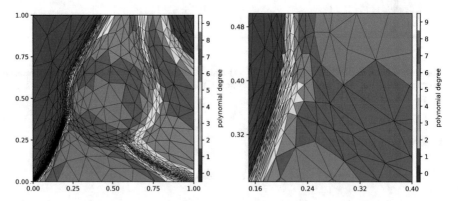

Fig. 6.13 Mixed hyperbolic-elliptic problem (6.37), the final hp-mesh mesh optimization with respect to the $L^2(\Omega)$-norm, total view (left) and the detail close to the left-bottom corner (right)

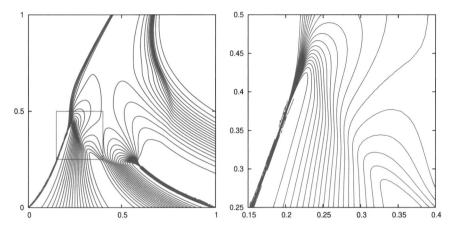

Fig. 6.14 Mixed hyperbolic-elliptic problem (6.37), the solution isolines on the final hp-mesh mesh optimization with respect to the $L^2(\Omega)$-norm, total view (left) and the detail close to the left-bottom corner (right)

6.4.5 Convection-Dominated Problem

This example comes from [61] and [27]. The computational domain is the right-bottom L-shape defined by $\Omega := [0, 4] \times [0, 4] \setminus [0, 2] \times [0, 2]$. We consider the linear convection–diffusion equation

$$-\varepsilon \Delta u + \nabla \cdot (\boldsymbol{b}\, u) = 0 \qquad \text{in } \Omega, \tag{6.39}$$

where the diffusion $\varepsilon = 10^{-3}$ and $\varepsilon = 10^{-6}$ and the convective field $\boldsymbol{b} = (x_2, -x_1)$. We prescribe the combined Dirichlet and Neumann boundary conditions

$$u = 1 \quad \text{on } \{x_1 = 0, \ x_2 \in (2, 4)\}, \tag{6.40}$$

$$\nabla u \cdot \boldsymbol{n} = 0 \quad \text{on } \Gamma_1 := \{x_1 = 4, \ x_2 \in (0, 4)\} \cup \Gamma_2 := \{x_1 \in (2, 4), \ x_2 = 0\},$$

$$u = 0 \quad \text{elsewhere.}$$

The solution of (6.40) exhibits two boundary layers along $\{x_1 \in (0, 4), \ x_2 = 4\}$ and $\{x_1 \in (0, 2), \ x_2 = 2\}$. Moreover, the solution contains two circular-shaped interior layers.

Figure 6.15 shows the final hp-generated by the adaptive algorithm for the case $\varepsilon = 10^{-3}$, the corresponding isolines of the solution are viewed in Fig. 6.16. We observe the capturing of circular-shaped interior layers. The results of the case $\varepsilon = 10^{-6}$ are given in Figs. 6.15 and 6.16. Due to the smaller diffusion the layers are thinner and the mesh refinement is stronger. In the majority of the domain, where the solution is close to a constant, the meshes with the lowest possible polynomial degree ($p = 1$) are generated (Figs. 6.17 and 6.18).

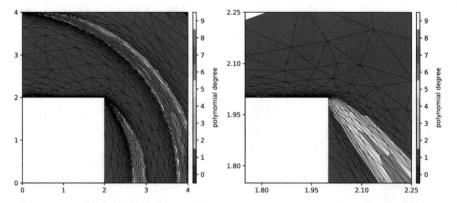

Fig. 6.15 Convection-dominated problem (6.39) and (6.40) with $\varepsilon = 10^{-3}$, the final *hp*-mesh mesh optimization with respect to the $L^2(\Omega)$-norm, total view (left) and the detail close to the left-bottom corner (right)

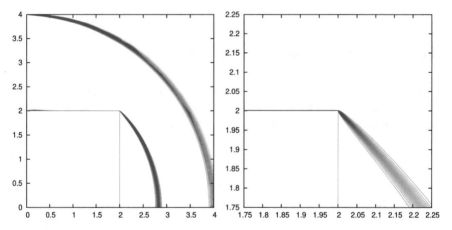

Fig. 6.16 Convection-dominated problem (6.39) and (6.40) with $\varepsilon = 10^{-3}$, the solution isolines on the final *hp*-mesh mesh optimization with respect to the $L^2(\Omega)$-norm, total view (left) and the detail close to the left-bottom corner (right)

Furthermore, Fig. 6.19 shows the histogram of the distribution of the element error estimator η_K, $K \in \mathscr{T}_h^k$ on the last adaptation level obtained by Algorithm 6.1. Omitting the left parts of the graph, where the elements have almost vanishing error estimators, we observe an distribution close to the Gaussian one for $\varepsilon = 10^{-3}$. On the other hand, for smaller diffusion $\varepsilon = 10^{-6}$, the distribution is more far from the Gaussian one, similarly as in the boundary layer problem (5.151) and (5.152) with $\varepsilon = 10^{-3}$, cf. Fig. 6.6.

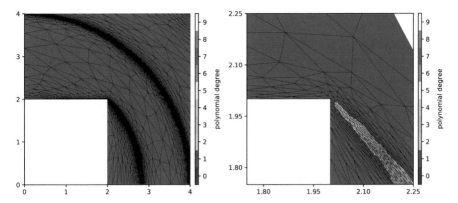

Fig. 6.17 Convection-dominated problem (6.39) and (6.40) with $\varepsilon = 10^{-6}$, the final hp-mesh mesh optimization with respect to the $L^2(\Omega)$-norm, total view (left) and the detail close to the left-bottom corner (right)

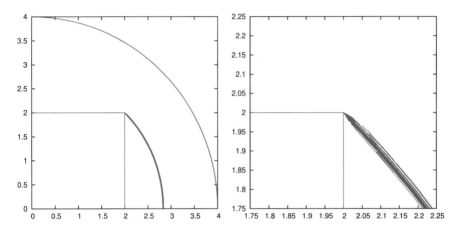

Fig. 6.18 Convection-dominated problem (6.39) and (6.40) with $\varepsilon = 10^{-6}$, the solution isolines on the final hp-mesh mesh optimization with respect to the $L^2(\Omega)$-norm, total view (left) and the detail close to the left-bottom corner (right)

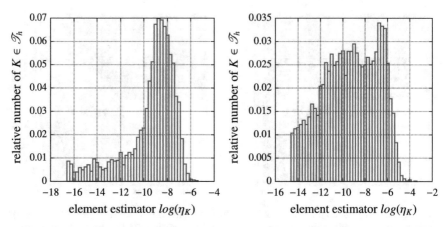

Fig. 6.19 Convection-dominated problem (6.39) and (6.40) with $\varepsilon = 10^{-3}$ (left) and $\varepsilon = 10^{-6}$ (right), the histogram of the distribution of η_K, $K \in \mathscr{T}_h^k$ on the last adaptation level obtained by Algorithm 6.1

Chapter 7
Framework of the Goal-Oriented Error Estimates

In many practical applications, we are not interested in the solution u of the given partial differential equations as such, but in the value of a certain *quantity of interest*, which depends on the solution. This quantity is given by a solution-dependent (target) functional denoted hereafter as $J(u)$. As examples may serve the temperature at a given point of the domain or the heat flux through a part of the boundary for heat conduction problems, the mean stress in elastoplastic deformations, or the aerodynamic coefficients for flow simulation problems (for more examples, see, e.g., [14]). Therefore, the output of the numerical solution is the value $J(u_h)$ and the goal is the estimation of the error $J(u) - J(u_h)$.

In this and next chapters, we develop estimates of the error of the quantity of interest, taking into account the geometry of mesh elements. Moreover, we propose an anisotropic mesh adaptation algorithm, which reduces the error of the target functional. In Sects. 7.1 and 7.2, we briefly recall the general framework of the goal-oriented error estimates for linear and nonlinear problems, respectively. These techniques are based on the solution of the adjoint problem corresponding to the primal one. In Sects. 7.3–7.5, we apply this framework to particular problems. We introduce the estimates of type I and type II, the latter are employed in Chap. 8 where the final error estimates taking into account the geometry of mesh elements are derived.

7.1 Goal-Oriented Error Estimates for Linear PDEs

In this section, we briefly recall the framework of the goal-oriented error estimates for an abstract linear problem. For a more detailed explanation and analysis, we refer to, e.g., [14, 17, 69, 77].

7.1.1 Primal Problem

Let Ω be a computational domain with boundary Γ. We consider the abstract boundary value problem

$$\mathscr{L}u = f \quad \text{in } \Omega, \qquad \mathscr{B}u = g \quad \text{on } \Gamma, \tag{7.1}$$

where $\mathscr{L}$ denotes a linear differential operator on Ω, $\mathscr{B}$ denotes a linear boundary differential operator on Γ, f and g are given functions, and $u : \Omega \to \mathbb{R}$ is the sought solution. Let the *variational formulation* of (7.1) read

$$a(u, v) = \ell(v) \qquad \forall v \in V, \tag{7.2}$$

where $u \in W$ is a weak solution, $a(\cdot, \cdot) : W \times V \to \mathbb{R}$ is a bilinear form, $\ell(\cdot) : V \to \mathbb{R}$ is a linear form, and W and V are Banach spaces. We assume that (7.2) is well-posed, i.e., it admits a unique weak solution, cf. Definition 1.1.

In order to solve (7.2) numerically, we consider the finite element spaces W_h and V_h, see Sect. 1.2.2. According to (1.11), we define the space $W(h) = W + W_h$, and in addition, we consider the space $V(h) = V + V_h$. In order to define an approximate solution of (7.2), we introduce the (bi)linear forms

$$a_h : W(h) \times V(h) \to \mathbb{R} \qquad \text{and} \qquad \ell_h : V(h) \to \mathbb{R}. \tag{7.3}$$

Remark 7.1 We note that in Sect. 1.2.2, we required only $a_h : W(h) \times V_h \to \mathbb{R}$. However, the larger data range in (7.3) is necessary for the definition of the adjoint problem in the following.

The *approximate solution* of the *primal problem* (7.2) $u_h \in W_h$ satisfies

$$a_h(u_h, v_h) = \ell_h(v_h) \qquad \forall v_h \in V_h. \tag{7.4}$$

We assume that the numerical scheme (7.4) is *consistent*, i.e.,

$$a_h(u, v) = \ell_h(v) \qquad \forall v \in V(h), \tag{7.5}$$

where $u \in W$ is the weak solution of (7.2). This implies the *Galerkin orthogonality of the error* of the primal problem

$$a_h(u_h - u, v_h) = 0 \qquad \forall v_h \in V_h. \tag{7.6}$$

Finally, we define the *residual of the primal problem* by

$$r_h(u_h)(v) := \ell_h(v) - a_h(u_h, v) = a_h(u - u_h, v), \quad v \in V(h), \tag{7.7}$$

where the last equality follows from the consistency (7.5) and the linearity of a_h.

Remark 7.2 We note that the consistency of the numerical method given by (7.5) is stronger than property (1.13) from Chap. 1. However, this condition does not represent any restriction for concrete problems, e.g., Examples 1.3 and 1.4 fulfill consistency condition (7.5).

7.1.2 Quantity of Interest and the Adjoint Problem

As mentioned above, we are interested in the *quantity of interest* $J(u) \in \mathbb{R}$ given by the linear functional in the form

$$J(u) = (j_\Omega, u)_\Omega + (j_\Gamma, \mathscr{C}u)_\Gamma, \tag{7.8}$$

where j_Ω and j_Γ are given integrable functions on Ω and Γ, respectively, $\mathscr{C}$ is a boundary differential operator on Γ, and the symbols $(\cdot, \cdot)_\Omega$ and $(\cdot, \cdot)_\Gamma$ denote the $L^2(\Omega)$ and $L^2(\Gamma)$ scalar products, respectively.

In order to estimate the error $J(u) - J(u_h)$, we consider the *adjoint* (or *dual*) *problem* to (7.1) in the form: find $z : \Omega \to \mathbb{R}$ such that

$$\mathscr{L}^*z = j_\Omega \quad \text{in } \Omega, \qquad \mathscr{B}^*z = j_\Gamma \quad \text{on } \Gamma, \tag{7.9}$$

where $\mathscr{L}^*$ and $\mathscr{B}^*$ are adjoint operators to $\mathscr{L}$ and $\mathscr{B}$ from (7.1), respectively, and j_Ω and j_Γ are the functions from (7.8). We note that operators $\mathscr{L}$, $\mathscr{B}$, and $\mathscr{C}$ have to be *compatible*

$$(\mathscr{L}u, z)_\Omega + (\mathscr{B}u, \mathscr{C}^*z)_\Gamma = (u, \mathscr{L}^*z)_\Omega + (\mathscr{C}u, \mathscr{B}^*z)_\Gamma \tag{7.10}$$

for any functions u and z, see [77] for more details.

Furthermore, we introduce the *approximate solution* of the *adjoint problem* (7.9) by $z_h \in V_h$ such that

$$a_h(w_h, z_h) = J(w_h) \qquad \forall w_h \in W_h, \tag{7.11}$$

where a_h and J are given by (7.3) and (7.8), respectively. Moreover, we assume that the numerical scheme (7.11) is *adjoint consistent*, i.e.,

$$a_h(w, z) = J(w) \qquad \forall w \in W(h), \tag{7.12}$$

where $z \in V$ is the weak solution of the adjoint problem (7.9). Relations (7.11)–(7.12) imply the *Galerkin orthogonality of the error* of the adjoint problem

$$a_h(w_h, z_h - z) = 0 \qquad \forall w_h \in W_h. \tag{7.13}$$

Finally, we define the *residual of the adjoint problem* by

$$r_h^*(z_h)(w) := J(w) - a_h(w, z_h) = a_h(w, z - z_h), \quad w \in W(h), \tag{7.14}$$

where the last equality follows from the adjoint consistency (7.12).

Remark 7.3 According to [77], the adjoint consistency requires a specific modification of the target functional in some situations, see Sect. 7.3.5 for an example.

7.1.3   Abstract Goal-Oriented Error Estimates

We are ready to derive abstract error estimates of the quantity of interest, i.e., the difference between the (unknown) exact value $J(u)$ and its approximation $J(u_h)$, which is obtained by computing u_h first and then by the evaluating J at u_h.

Let u and z be the exact solutions of the primal and adjoint problems (7.2) and (7.9), respectively, and similarly, let u_h and z_h be the approximate solutions of the primal and adjoint problems (7.4) and (7.11), respectively. Using the linearity of J and the adjoint consistency (7.12), we have the identity

$$J(u) - J(u_h) = J(u - u_h) = a_h(u - u_h, z). \tag{7.15}$$

The Galerkin orthogonality of the error (7.6) implies

$$a_h(u - u_h, z) = a_h(u - u_h, z - v_h) = r_h(u_h)(z - v_h) \quad \forall v_h \in V_h, \tag{7.16}$$

where the second equality follows from the definition of the primal residual $r_h(u_h)(\cdot)$ (7.7). Hence, we obtain from (7.15)–(7.16) relation

$$J(u) - J(u_h) = r_h(u_h)(z - v_h) \quad \forall v_h \in V_h, \tag{7.17}$$

which is called the *primal error identity*.

Moreover, setting $v_h := z_h$ in (7.17), and exploiting the Galerkin orthogonality of the error of the adjoint problem (7.13) and the definition of the residual of the adjoint problem (7.14), we have

$$J(u) - J(u_h) = a_h(u - u_h, z - z_h) = a_h(u - w_h, z - z_h) \tag{7.18}$$
$$= r_h^*(z_h)(u - w_h) \quad \forall w_h \in W_h,$$

which is called the *adjoint error identity*. Obviously, there is the *residual equivalence* between the primal and adjoint residuals

$$r_h(u_h)(z - v_h) = r_h^*(z_h)(u - w_h) \quad \forall v_h \in V_h \, \forall w_h \in W_h. \tag{7.19}$$

Remark 7.4 Both error identities (7.17) and (7.18) are valid only for exactly computed approximate solutions u_h and z_h. Nevertheless, they are not available in practical computations since the outputs of a practical realization of any numerical method suffer from algebraic (iterative and rounding) errors. Hence these outputs do not satisfy (7.17) and (7.18). Therefore, for efficient and reliable computations, the algebraic errors have to be taken into account, see the review in [6]. The goal-oriented error estimates including algebraic errors were studied and developed, e.g., in [50, 54, 84, 92, 93]. For simplicity, we do not consider the algebraic errors in the forthcoming error estimates since their treatment has only a minor impact on the anisotropy of error estimates and the corresponding mesh adaptation method. We refer to [43, 54] for more detail.

7.1.4 Computable Goal-Oriented Error Estimates

The right-hand sides of (7.17) and (7.18) contain the exact adjoint solution z and the exact primal solution u, respectively, which are unknown. Therefore, in order to have a computable error estimate, we have to use approximations

$$u \approx u_h^+ \in W_h^+ \quad \text{and} \quad z \approx z_h^+ \in V_h^+, \tag{7.20}$$

where $W_h^+ \subset W(h)$ and $V_h^+ \subset V(h)$ are "richer" finite-dimensional spaces than W_h and V_h, respectively. We note that the choice $u_h^+ \in W_h$ and $z_h^+ \in V_h$ leads to nullification of the right-hand sides of (7.17) and (7.18) due to the Galerkin orthogonalities (7.6) and (7.13), respectively.

The spaces W_h^+ and V_h^+ are usually defined by h-, p-, or hp-refinement of spaces W_h and V_h, respectively. The approximations u_h^+ and z_h^+ can be obtained by high-order (global or local) reconstructions of the approximate solutions u_h and z_h of the primal and adjoint problems, respectively. Some reconstruction techniques are described in Sect. 9.1.

Alternatively, u_h^+ and z_h^+ can be evaluated by the solution of the discrete problems (7.4) and (7.11) directly discretized using spaces W_h^+ and V_h^+. This technique is computationally more expensive, but it was used, e.g., in [108]. However, the numerical experiments presented in [50] show that taking u_h^+ and z_h^+ as the solutions of the global problem $u_h^+ \in W_h^+$ and $z_h^+ \in V_h^+$, respectively, does not bring any essential gain in the accuracy.

Here, we use a local reconstruction technique, and we formally set

$$u_h^+ = \mathscr{R}(u_h), \qquad z_h^+ = \mathscr{R}^*(z_h), \tag{7.21}$$

where $\mathscr{R} : W_h \to W_h^+$ and $\mathscr{R}^* : V_h \to V_h^+$, cf. Sect. 9.1.

Then using (7.20)–(7.21), we obtain from (7.17) and (7.18) the computable goal-oriented error estimates

$$J(u - u_h) \approx r_h(u_h)(z_h^+ - v_h), \quad v_h \in V_h \tag{7.22}$$

$$J(u - u_h) \approx r_h^*(z_h)(u_h^+ - w_h), \quad w_h \in W_h.$$

We note that the error equivalence (7.19) does not hold in general when we replace u by u_h^+ and z by z_h^+ (for an exception, see Remark 9.3). Therefore, we employ both relations in (7.22) as the mean value. We suppose that in this way we employ the available information of the primal and adjoint solutions.

Both right-hand sides in (7.22) are independent of the choice of v_h and w_h due to the Galerkin orthogonalities of error. However, in virtue of Remark 7.4, this is not true in practical computations when u_h and z_h suffer from algebraic errors. Then the choice of v_h and w_h has impact on the resulting estimates. In order to obtain accurate error estimates, it is advantageous to minimize the arguments of $r_h(u_h)(\cdot)$ and $r_h^*(z_h)(\cdot)$. There are two natural possibilities:

$$\text{(i)} \quad v_h := z_h, \qquad w_h := u_h, \tag{7.23a}$$

$$\text{(ii)} \quad v_h := \Pi^* z_h^+, \quad w_h := \Pi u_h^+, \tag{7.23b}$$

where $\Pi : W_h^+ \to W_h$ and $\Pi^* : W_h^+ \to W_h$ are suitable projections. Hence, for both choices from (7.23), we obtain from (7.22) the estimates

$$J(u - u_h) \approx \tfrac{1}{2}\left(r_h(u_h)(z_h^+ - z_h) + r_h^*(z_h)(u_h^+ - u_h)\right), \tag{7.24a}$$

$$J(u - u_h) \approx \tfrac{1}{2}\left(r_h(u_h)(z_h^+ - \Pi^* z_h^+) + r_h^*(z_h)(u_h^+ - \Pi u_h^+)\right). \tag{7.24b}$$

The choice (7.23b) (and the corresponding estimate (7.24b)) is more suitable for the forthcoming anisotropic error estimates, which serve as the base of the anisotropic mesh adaptation process.

7.2 Goal-Oriented Error Estimates for Nonlinear PDEs

We extend the framework of the goal-oriented error estimates from Sect. 7.1 to nonlinear problems. The adjoint problem is defined by a linearization of the primal problem, where the linearization is usually carried out by the differentiation of the weak formulation and the target functional at the approximate solution, see, e.g., [14, 17]. In [43], we recently presented an approach where the concept of a general linearization can be employed. Similar to Sect. 7.1, we present the main relations that are important for anisotropic error estimates and mesh adaptation. For a more detailed explanation and analysis, we refer again to, e.g., [14, 17, 69, 77].

7.2.1 Primal Problem

We consider a nonlinear partial differential equation with *variational formulation*

$$a(u, v) = 0 \qquad \forall v \in V, \tag{7.25}$$

where $u \in W$ is a weak solution, $a(\cdot, \cdot) : W \times V \to \mathbb{R}$ is a form linear in its second argument, and W and V are Banach spaces. We assume that (7.25) admits a unique weak solution.

The approximate solution of (7.25) is sought in the finite-dimensional space W_h and the test functions are chosen in the space V_h. Similar to Sect. 7.1.1, we set $W(h) = W + W_h$ and $V(h) = V + V_h$.

Let $a_h : W(h) \times V(h) \to \mathbb{R}$ represent a discretization of (7.25) by a suitable numerical method. Then the *approximate solution* of the *primal problem* (7.25), $u_h \in W_h$, satisfies

$$a_h(u_h, v_h) = 0 \qquad \forall v_h \in V_h. \tag{7.26}$$

We assume that the numerical scheme (7.26) is *consistent*, i.e.,

$$a_h(u, v) = 0 \qquad \forall v \in V(h), \tag{7.27}$$

where $u \in V$ is the solution of (7.25). Finally, we define the *residual of the primal problem* by

$$r_h(u_h)(v) := -a_h(u_h, v), \quad v \in V(h). \tag{7.28}$$

7.2.2 Quantity of Interest and Adjoint Problem Based on Differentiation

We consider again the *quantity of interest* given by a functional $J : W(h) \to \mathbb{R}$, which is nonlinear in general. As mentioned above, the adjoint problem is defined usually by the differentiation of forms a_h and J.

By $J'[v_h](v)$ we denote the Fréchet derivative of J at v_h along the direction v, and similarly, $a'[v](v, \cdot)$ and $a'_h[v_h](v, \cdot)$ denote the derivative of a and a_h with respect to its first argument at v, respectively, v_h, along the direction v.

We define the *adjoint problem* (linearized at $v \in W(h)$): find $z \in V$ such that

$$a'[v](w, z) = J'[v](w) \qquad \forall w \in W. \tag{7.29}$$

The discretization of (7.29) using the differentiation of form a_h from (7.26) gives the *approximate adjoint problem*: find $z_h \in V_h$ such that

$$a_h'[u_h](w_h, z_h) = J'[u_h](w_h) \qquad \forall w_h \in W_h, \tag{7.30}$$

where u_h is the approximate solution of (7.26). We assume that (7.30) is *adjoint consistent*, which means that

$$a_h'[u](w, z) = J'[u](w) \qquad \forall w \in W(h), \tag{7.31}$$

where $u \in W$ is the exact solution of primal problem (7.25) and $z \in V$ is the weak solution of the corresponding adjoint problem (7.29). We note that the well-posedness of the adjoint problem is open, in general.

Furthermore, the *adjoint residual* of (7.30) is given by

$$r_h'(u_h, z_h)(w) := J'[u_h](w) - a_h'[u_h](w, z_h), \quad w \in W(h). \tag{7.32}$$

The fundamental identity, which can be taken from [17, Proposition 2.2] or [14, Proposition 6.2] (see also [101, Proposition 3.1] for the case including algebraic errors), is the following:

Theorem 7.5 *Let $u_h \in W_h$ and $z_h \in V_h$ be the solutions of the problems (7.26) and (7.30), respectively. Then the error of the quantity of interest satisfies*

$$J(u) - J(u_h) = \tfrac{1}{2} r_h(u_h)(z - v_h) + \tfrac{1}{2} r_h'(u_h, z_h)(u - w_h) + \mathcal{R}_h^{(3)} \tag{7.33}$$

for any $v_h \in V_h$, $w_h \in W_h$, where $r_h(\cdot)(\cdot)$ and $r_h'(\cdot, \cdot)(\cdot)$ are the primal and adjoint residuals given by (7.28) and (7.32), respectively, and $\mathcal{R}_h^{(3)}$ is a higher (third)-order term.

Remark 7.6 The remainder term $\mathcal{R}_h^{(3)}$ in (7.33) depends on the second and third derivatives of forms a_h and J whose existence is not guaranteed in practice. In [69] and [78], the authors presented an alternative formulation of the adjoint problem which leads to an estimate similar to (7.33) without the remainder term. Nevertheless, this adjoint problem depends also on the solution of primal problem (7.25) which is not available in practice and it has to be approximated by the approximate solution. Therefore, an additional source of error arises.

Remark 7.7 The relation (7.33) contains the average of the primal and adjoint residuals. According to [17, Proposition 2.3] or [14, Proposition 6.2], if only the primal residual is considered, then the remaining term is the second order term, and hence the estimate is less accurate.

7.2.3   Quantity of Interest and Adjoint Problem Based on Linearization

The discrete primal problem (7.26) represents a system of nonlinear algebraic equations that are solved iteratively. One possibility is the well-known Newton method, but in many situations (e.g., when the form a_h is not differentiable or its derivatives are difficult to compute), it is possible to consider a different iterative scheme.

We assume that there exist forms $a_h^L : W(h) \times W(h) \times V(h) \to \mathbb{R}$ and $\tilde{a}_h : W(h) \times V(h) \to \mathbb{R}$ which are consistent with a_h by

$$a_h(u_h, v_h) = a_h^L(u_h, u_h, v_h) - \tilde{a}_h(u_h, v_h) \qquad \forall u_h \in W(h) \; \forall v_h \in V(h), \tag{7.34}$$

where a_h^L is linear in its second and third arguments and $\tilde{a}_h$ is linear in its second argument. Then problem (7.26) is solved iteratively by computing the sequence $u_h^n \in W_h, n = 0, 1, \ldots$, satisfying

$$a_h^L(u_h^n, u_h^{n+1}, v_h) = \tilde{a}_h(u_h^n, v_h) \quad \forall v_h \in V_h, \; n = 0, 1, \ldots. \tag{7.35}$$

The initial solution u_h^0 is given from the available information. We assume that the iterative scheme (7.35) is convergent. In Sects. 7.3–7.4, we present the concrete definitions of forms a_h^L and $\tilde{a}_h$ of particular problems.

It is shown in [77] that the adjoint consistency requires a modification of the functional J even for linear problems, cf. Sect. 7.3.5 for an example. Therefore, we introduce formally a new functional $J_h : W(h) \to \mathbb{R}$, which is consistent with J by the identity

$$J_h(u) = J(u), \tag{7.36}$$

where u is the solution of (7.25).

Moreover, we assume the existence of a suitable linearization of J_h, particularly forms $J_h^L : W(h) \times W(h) \to \mathbb{R}$ and $\tilde{J}_h : W(h) \to \mathbb{R}$ such that (compare with (7.34))

$$J_h(w_h) = J_h^L(w_h, w_h) + \tilde{J}_h(w_h) \qquad \forall w_h \in W(h), \tag{7.37}$$

where form J_h^L is linear in its second argument. In many situations, form $\tilde{J}_h$ is independent of w_h. Then it does not influence the error since $J_h(u) - J_h(u_h) = J_h^L(u, u) - J_h^L(u_h, u_h)$.

We introduce the adjoint problem using the linearized forms a_h^L and J_h^L. In (7.30), we replace $a_h'[u_h](\cdot, \cdot)$ by $a_h^L(u_h, \cdot, \cdot)$ and $J'[u_h](\cdot)$ by $J_h^L(u_h, \cdot)$. Then we have an *approximate adjoint problem* linearized at $u_h \in W_h$: Find $z_h \in V_h$ such that

$$a_h^L(u_h, w_h, z_h) = J_h^L(u_h, w_h) \qquad \forall w_h \in W_h, \qquad (7.38)$$

where a_h^L and J_h^L are the forms from (7.34) and (7.37), respectively. The corresponding adjoint residual is given by

$$r_h^*(u_h, z_h)(w) := J_h^L(u_h, w) - a_h^L(u_h, w, z_h), \quad u_h, w \in W(h), \ z_h \in V(h). \qquad (7.39)$$

Finally, we say that discretization (7.38) is *adjoint consistent* if

$$a_h^L(u, w_h, z) = J_h^L(u, w_h) \qquad \forall w_h \in W(h), \qquad (7.40)$$

where $u \in W$ is the weak solution of (7.25) and $z \in V$ is the weak solution of the corresponding adjoint problem. Several examples are given in Sect. 7.4.

7.2.4 Goal-Oriented Error Estimates

We present the *goal-oriented error estimates* using the adjoint problem (7.38). In virtue of Theorem 7.5, we set

$$J(u) - J(u_h) \approx \tfrac{1}{2} r_h(u_h)(z - v_h) + \tfrac{1}{2} r_h^*(u_h, z_h)(u - w_h), \quad v_h \in V_h, \ w_h \in W_h, \qquad (7.41)$$

where $r_h(u_h)(\cdot)$ and $r_h^*(z_h)(\cdot)$ are the primal and adjoint residuals given by (7.28) and (7.39), respectively.

The validity of estimate (7.41) is questionable, and additional assumptions for the linearized forms a_h^L and J_h^L would be required. We refer to [43] where this aspect is discussed and numerically justified.

The estimate (7.41) depends on the exact solutions u and z of primal and adjoint problems, respectively. Moreover, we have to specify the choice of v_h and w_h in (7.41) since the right-hand side of (7.41) is independent of v_h and w_h only for exactly computed primal and adjoint approximate solutions u_h and z_h, cf. Remark 7.4.

We proceed similarly as in Sect. 7.1.4. Using the local reconstructions (7.20)–(7.21) and the choice (7.23b), we obtain the following computable error estimate:

$$J(u) - J(u_h) \approx \tfrac{1}{2} r_h(u_h)(z_h^+ - \Pi^* z_h^+) + \tfrac{1}{2} r_h^*(u_h, z_h)(u_h^+ - \Pi u_h^+). \qquad (7.42)$$

Finally, we note that if the form a from (7.25), the form a_h from (7.26), and the functional J are affine, then all results from Sect. 7.2 reduce to those ones from Sect. 7.1.

7.3   Error Estimates for Linear Convection–Diffusion Equation

In this section, we apply the abstract goal-oriented error estimate framework from Sect. 7.1 to the numerical solution of a linear convection–diffusion–reaction equation which is discretized by the discontinuous Galerkin method (DGM). We recall that we denote by $(u, v)_M := \int_M u\, v\, dx$, $u, v \in L^2(M)$ the L^2-scalar product over domain $M \subset \mathbb{R}^m$, $m = 1, 2, 3$. For simplicity, we write $(\cdot, \cdot)$ instead of $(\cdot, \cdot)_\Omega$. Moreover, $\|\cdot\|_M$ denotes the L^2-norm over $M \subset \mathbb{R}^m$, $d = 1, 2, 3$.

7.3.1   Problem Formulation

We deal with the solution of a linear convection–diffusion–reaction equation in a bounded polygonal domain $\Omega \subset \mathbb{R}^d$, $d = 2, 3$, with Lipschitz boundary $\Gamma := \partial\Omega$. At each $x \in \Gamma$, there exists the unit outward normal vector to Γ denoted by $n(x)$.

We consider

(i) A linear convective field $b : \Omega \to \mathbb{R}^d$ such that b has Lipschitz continuous components
(ii) A function $c \in L^\infty(\Omega)$ representing the reaction
(iii) A symmetric matrix $\mathbb{A} = \{a_{i,j}\}_{i,j=1}^d$ representing diffusion such that $\mathbb{A}$ has bounded piecewise continuous entries satisfying

$$\zeta^T \mathbb{A}(x)\zeta \geq 0 \ \forall \zeta \in \mathbb{R}^d, \ \text{a.e. } x \in \Omega.$$

In order to define well-posed problem, it is possible to prescribe the Dirichlet boundary condition only on the parts of Γ where either $n(x)^T \mathbb{A}(x)n(x) > 0$ (non-vanishing diffusion) or $n(x) \cdot b(x) < 0$ (inflow convective field). Therefore, in the same way as in [81], we decompose Γ onto three disjoint parts

$$\Gamma^0 := \{x \in \Gamma : n(x)^T \mathbb{A}(x)n(x) > 0\},$$

$$\Gamma^- := \{x \in \Gamma \backslash \Gamma^0 : n(x) \cdot b(x) < 0\},$$

$$\Gamma^+ := \{x \in \Gamma \backslash \Gamma^0 : n(x) \cdot b(x) \geq 0\}.$$

Moreover, we split Γ^0 into two disjoint parts Γ_D and Γ_N, where the Dirichlet and Neumann boundary conditions are considered, respectively. We assume that $\Gamma^- \cup \Gamma_D \neq \emptyset$ and that $b \cdot n \geq 0$ on Γ_N whenever $\Gamma_N \neq \emptyset$.

We consider the following linear convection–diffusion–reaction problem: find $u : \Omega \rightarrow \mathbb{R}$ such that

$$-\nabla \cdot \mathbb{A}\nabla u + \nabla \cdot (\boldsymbol{b}\, u) + cu = f \qquad \text{in } \Omega, \tag{7.43a}$$

$$u = u_D \qquad \text{on } \Gamma_D \cup \Gamma^-, \tag{7.43b}$$

$$\mathbb{A}\nabla u \cdot \boldsymbol{n} = g_N \qquad \text{on } \Gamma_N, \tag{7.43c}$$

where the data satisfy $f \in L^2(\Omega)$, u_D is trace of some $u^* \in H^1(\Omega)$ on $\Gamma_D \cup \Gamma^-$, and $g_N \in L^2(\Gamma_N)$. In the standard way, we introduce the variational formulation of problem (7.43).

Definition 7.8 The function $u \in H^1(\Omega)$ is called the *weak solution* of (7.43) if $u - u^* \in H_D^1(\Omega) := \{v \in H^1(\Omega); \ v|_{\Gamma_D \cup \Gamma^-} = 0\}$ and

$$a(u, v) = \ell(v) \quad \forall v \in H_D^1(\Omega), \tag{7.44}$$

where

$$a(u, v) := (\mathbb{A}\nabla u, \nabla v) - (\boldsymbol{b}\, u, \nabla v) + (c\, u, v) + (\boldsymbol{b} \cdot \boldsymbol{n}\, u, v)_{\Gamma^+ \cup \Gamma_N}, \ u, v \in H^1(\Omega),$$

$$\ell(v) := (f, v) + (g_N, v)_{\Gamma_N}, \ v \in H^1(\Omega).$$

The well-posedness of (7.44) is shown in [80, Appendix] for the homogeneous boundary conditions. The case with the non-homogeneous conditions can be proved using a similar technique by the decomposition $u = u_0 + u^*$, where $u_0 \in H_D^1(\Omega)$ and solving problem $a(u_0, v) = \ell(v) - a(u^*, v) \ \forall v \in H_D^1(\Omega)$.

7.3.2 Triangulation and Finite Element Spaces

Let $\mathcal{T}_h$ be a conforming triangular grid, cf. Definition 1.12 consisting of triangular elements K. By $\boldsymbol{n}_K$ we denote the unit outer normal to $K \in \mathcal{T}_h$. Over the triangulation $\mathcal{T}_h$, we define the *broken Sobolev space* by

$$H^s(\mathcal{T}_h) = \{v \in L^2(\Omega), \ v|_K \in H^s(K) \ \forall K \in \mathcal{T}_h\}, s = 1, 2, \ldots, \tag{7.45}$$

where $H^s(K)$ is the Sobolev space of functions having square integrable weak derivatives up to order s over $K \in \mathcal{T}_h$. Hence, the space $H^s(\mathcal{T}_h)$ consists of piecewise Sobolev functions with possible discontinuities across the boundaries of elements $K \in \mathcal{T}_h$.

Let $v \in H^1(\mathcal{T}_h)$, $K \in \mathcal{T}_h$. Moreover, by $v|_{\partial K}^{(\text{int})}$ and $v|_{\partial K}^{(\text{ext})}$, we denote the trace of v on ∂K from the interior and exterior of K, respectively, i.e.,

$$v|_{\partial K}^{(\text{int})}(x) = \lim_{t \to 0^+} v(x - t\boldsymbol{n}_K), \quad v|_{\partial K}^{(\text{ext})}(x) = \lim_{t \to 0^-} v(x - t\boldsymbol{n}_K), \quad x \in \partial K.$$

$$(7.46)$$

The same notation is used for a vector-valued function $\boldsymbol{g} \in [H^1(\mathcal{T}_h)]^d$. Furthermore, symbols $\langle \cdot \rangle$ and $[\![\cdot]\!]$ denote the *mean value* and the *jump* on $\partial K \setminus \Gamma$ given by

$$\langle v \rangle_{\partial K} = \tfrac{1}{2}\big(v|_{\partial K}^{(\text{int})} + v|_{\partial K}^{(\text{ext})}\big), \quad [\![v]\!]_{\partial K} = \big(v|_{\partial K}^{(\text{int})} - v|_{\partial K}^{(\text{ext})}\big)\boldsymbol{n}_k, \quad v \in H^1(\mathcal{T}_h)$$

$$(7.47)$$

$$\langle \boldsymbol{g} \rangle_{\partial K} = \tfrac{1}{2}\big(\boldsymbol{g}|_{\partial K}^{(\text{int})} + \boldsymbol{g}|_{\partial K}^{(\text{ext})}\big), \quad [\![\boldsymbol{g}]\!]_{\partial K} = \big(\boldsymbol{g}|_{\partial K}^{(\text{int})} - \boldsymbol{g}|_{\partial K}^{(\text{ext})}\big) \cdot \boldsymbol{n}_k, \quad \boldsymbol{g} \in [H^1(\mathcal{T}_h)]^d.$$

Therefore, $[\![v]\!]_{\partial K}$ is a vector for a scalar function v and $[\![\boldsymbol{g}]\!]_{\partial K}$ is a scalar function for a vector-valued function $\boldsymbol{g}$. Obviously, if $K, K' \in \mathcal{T}_h$ share an edge for $d = 2$ or a face for $d = 3$, then

$$[\![v]\!]_{\partial K} = [\![v]\!]_{\partial K'} \quad v \in H^1(\mathcal{T}_h), \qquad [\![\boldsymbol{g}]\!]_{\partial K} = [\![\boldsymbol{g}]\!]_{\partial K'} \quad \boldsymbol{g} \in [H^1(\mathcal{T}_h)]^d.$$

Finally, for $\partial K \cap \Gamma \neq 0$, we set

$$\langle v \rangle_{\partial K \cap \Gamma} = v|_{\partial K}^{(\text{int})}, \qquad [\![v]\!]_{\partial K \cap \Gamma} = v|_{\partial K}^{(\text{int})}\boldsymbol{n}_K, \quad v \in H^1(\mathcal{T}_h) \qquad (7.48)$$

$$\langle \boldsymbol{g} \rangle_{\partial K \cap \Gamma} = \boldsymbol{g}|_{\partial K}^{(\text{int})}, \qquad [\![\boldsymbol{g}]\!]_{\partial K \cap \Gamma} = \boldsymbol{g}|_{\partial K}^{(\text{int})} \cdot \boldsymbol{n}_K, \quad \boldsymbol{g} \in [H^1(\mathcal{T}_h)]^d, \qquad (7.49)$$

where the traces are given by (7.46).

In the following, we suppress the subscripts K and ∂K in $\boldsymbol{n}_K$, $[\![\cdot]\!]_{\partial K}$ and $\langle \cdot \rangle_{\partial K}$, since it will always be clear from the context which $K \in \mathcal{T}_h$ is considered. Similarly, we write v instead of $v|_{\partial K}^{(\text{int})}$ (interior trace) and $v^{(\text{ext})}$ instead of $v|_{\partial K}^{(\text{ext})}$ (exterior trace), for simplicity.

Furthermore, for $K \in \mathcal{T}_h$, we set

$$\partial K^- = \{x \in \partial K; \ \boldsymbol{b} \cdot \boldsymbol{n}_{\partial K}(x) < 0\},$$

$$\partial K^+ = \{x \in \partial K; \ \boldsymbol{b} \cdot \boldsymbol{n}_{\partial K}(x) \geq 0\}.$$

Finally, the approximate solution is sought in the space of piecewise polynomial function (cf. (1.23)). We define spaces

$$S_{hp} := \{v \in L^2(\Omega); \ v|_K \in P^{p_K}(K) \, \forall K \in \mathcal{T}_h\}, \qquad (7.50)$$

$$S_{hp+1} := \{v \in L^2(\Omega); \ v|_K \in P^{p_K+1}(K) \, \forall K \in \mathcal{T}_h\},$$

where p_K local polynomial approximation degree on $K \in \mathscr{T}_h$. Obviously, $S_{hp} \subset S_{hp+1} \subset H^s(\Omega)$ for all $s = 1, 2, \ldots$. The approximate solution is sought in the space S_{hp}, whereas the space S_{hp+1} is used for the higher-order reconstructions (7.21). The dimension of S_{hp}, which is called the number of *degrees of freedom* DoF, fulfills

$$\text{DoF} := \dim S_{hp} = \begin{cases} \frac{1}{2} \sum_{K \in \mathscr{T}_h} (p_K + 1)(p_K + 2) & \text{for } d = 2, \\ \frac{1}{6} \sum_{K \in \mathscr{T}_h} (p_K + 1)(p_K + 2)(p_K + 3) & \text{for } d = 3. \end{cases} \tag{7.51}$$

Obviously, if p is an integer, setting $p_K = p$ for all $K \in \mathscr{T}_h$, the finite element space S_{hp} given by (7.50) reduces to space (1.22) and we write S_{hp} instead.

7.3.3 Discretization of the Primal Problem

We discretize (7.44) by the discontinuous Galerkin method (DGM) seeking the approximate solution in space S_{hp}, cf. (7.50). The convective term is discretized by the *upwinding* technique, whereas the diffusive terms are treated by the *symmetric interior penalty Galerkin* (SIPG) variant of DGM. The SIPG technique ensures (under some assumptions) the *adjoint consistency* of the methods, cf. [7, 45, 77]. It has been demonstrated in [75] that the lack of adjoint consistency leads to non-smooth adjoint solutions and consequently to a suboptimal convergence rate of the quantity of interest. We present only the final discretization of (7.44), and the detailed derivation can be found, e.g., in [45, Section 4.6] or [81].

For $u, v \in H^2(\mathscr{T}_h)$, we define forms

$$a_h(u, v) := \sum_{K \in \mathscr{T}_h} \{ (\mathbb{A}\nabla u, \nabla v)_K - (u\, \boldsymbol{b}, \nabla v)_K + (c\, u, v)_K \} \tag{7.52}$$

$$- \sum_{K \in \mathscr{T}_h} \left\{ \left((\langle \mathbb{A}\nabla u \rangle - \sigma [\![u]\!]) \, \boldsymbol{n}_K, v \right)_{\partial K \setminus \Gamma_{\text{N}}} + e \left(\mathbb{A}[\![u]\!], \nabla v \right)_{\partial K \setminus \Gamma_{\text{N}}} \right\}$$

$$+ \sum_{K \in \mathscr{T}_h} \left\{ (\boldsymbol{b} \cdot \boldsymbol{n}_K\, u, v)_{\partial K^+} + \left(\boldsymbol{b} \cdot \boldsymbol{n}_K\, u^{(\text{ext})}, v \right)_{\partial K^- \setminus (\Gamma_{\text{D}} \cup \Gamma^-)} \right\},$$

$$\ell_h(v) := (f, v) + (g_N, v)_{\Gamma_{\text{N}}} - \sum_{K \in \mathscr{T}_h} (\boldsymbol{b} \cdot \boldsymbol{n}_K\, u_D, v)_{\partial K^- \cap (\Gamma_{\text{D}} \cup \Gamma^-)}$$

$$+ \sum_{K \in \mathscr{T}_h} (\sigma v - \mathbb{A}\nabla v \cdot \boldsymbol{n}, u_D)_{\partial K \cap \Gamma_{\text{D}}},$$

where $e = 1/2$ on $\partial K \setminus \Gamma_D$ and $e = 1$ on $\partial K \cap \Gamma_D$. The penalty parameter $\sigma > 0$ is chosen by

$$\sigma|_{\partial K} = \begin{cases} \varepsilon C_W \max(\frac{p_K^2}{h_K}, \frac{p_{K'}^2}{h_{K'}}) & \text{if } \partial K \text{ is shared with } \partial K', \ K \neq K' \in \mathscr{T}_h, \\ \varepsilon C_W \frac{p_K^2}{h_K} & \text{if } \partial K \subset \Gamma, \end{cases}$$

(7.53)

where ε denotes the amount of diffusivity ($\approx |\mathbb{A}|$), and $C_W > 0$ has to be chosen sufficiently large to guarantee the existence of an approximate solution and the convergence of the method, see [45, Chapter 2].

Definition 7.9 We say that $u_h \in S_{hp}$ is the *approximate solution* of (7.44) if

$$a_h(u_h, v_h) = \ell_h(v_h) \qquad \forall v_h \in S_{hp},$$

(7.54)

where a_h and ℓ_h are given by (7.52).

7.3.4   Consistency

We prove that the discretization of (7.44) by DGM is *consistent*. Moreover, we introduce some residual forms which are employed later for the anisotropic goal-oriented error estimates.

Lemma 7.10 *The numerical scheme (7.54) is consistent, i.e.,*

$$a_h(u, v_h) = \ell_h(v_h) \qquad \forall v_h \in H^2(\mathscr{T}_h),$$

(7.55)

where u is the solution of (7.44).

Proof In virtue of (7.7), we define the residual of the problem (7.54) by

$$r_h(u_h)(v_h) := \ell_h(v_h) - a_h(u_h, v_h), \qquad u_h, v_h \in H^2(\mathscr{T}_h).$$

(7.56)

We have to prove that $r_h(u)(v_h) = 0 \ \forall v_h \in H^2(\mathscr{T}_h)$, where u is the weak solution of (7.44). Taking into account (7.52), after some manipulation, we obtain

$$r_h(u_h)(v_h) = \sum_{K \in \mathscr{T}_h} r_K(u_h)(v_h),$$

(7.57)

where $r_K(u_h)(\cdot)$ denotes the local residuals for $K \in \mathscr{T}_h$ having the form

$$r_K(u_h)(v_h) = \big(r_{K,V}(u_h), v_h\big)_K + \big(r_{K,B}(u_h), v_h\big)_{\partial K} + \big(r_{K,D}(u_h), \nabla v_h\big)_{\partial K}$$

(7.58)

and

$$r_{K,\mathrm{V}}(u_h) := f + \nabla \cdot \mathbb{A}\nabla u_h - \nabla \cdot (\boldsymbol{b}u_h) - cu_h, \tag{7.59}$$

$$r_{K,\mathrm{B}}(u_h) := \begin{cases} -\sigma \llbracket u_h \rrbracket \cdot \boldsymbol{n}_K - \frac{1}{2}\llbracket \mathbb{A}\nabla u_h \rrbracket & \text{on } \partial K^+ \setminus \Gamma, \\ -\sigma \llbracket u_h \rrbracket \cdot \boldsymbol{n}_K - \frac{1}{2}\llbracket \mathbb{A}\nabla u_h \rrbracket + \boldsymbol{b} \cdot \llbracket u_h \rrbracket & \text{on } \partial K^- \setminus \Gamma, \\ g_N - \mathbb{A}\nabla u_h \cdot \boldsymbol{n} & \text{on } \partial K \cap \Gamma_{\mathrm{N}}, \\ \sigma(u_D - u_h) & \text{on } \partial K^+ \cap \Gamma_{\mathrm{D}}, \\ (\sigma - \boldsymbol{b} \cdot \boldsymbol{n})(u_D - u_h) & \text{on } \partial K^- \cap \Gamma_{\mathrm{D}}, \\ -\boldsymbol{b} \cdot \boldsymbol{n}(u_D - u_h) & \text{on } \partial K \cap \Gamma^-, \\ 0 & \text{on } \partial K \cap \Gamma^+, \end{cases}$$

$$r_{K,\mathrm{D}}(u_h) := \begin{cases} \frac{1}{2}\mathbb{A}\llbracket u_h \rrbracket & \text{on } \partial K \setminus \Gamma, \\ (u_h - u_D)\mathbb{A}\boldsymbol{n} & \text{on } \partial K \cap \Gamma_{\mathrm{D}}, \\ 0 & \text{on } \partial K \cap \Gamma_{\mathrm{N}}. \end{cases}$$

Obviously, if $u \in H^2(\Omega)$ is the exact solution of (7.44), then $\bigl(r_{K,\mathrm{V}}(u_h), v_h\bigr)_K = 0$ for all $v_h \in H^2(K)$, $K \in \mathscr{T}_h$. Moreover, $\llbracket u \rrbracket = \llbracket \mathbb{A}\nabla u \rrbracket = 0$ for all interior edges ($d = 2$) or faces ($d = 3$), $u = u_D$ on Γ_{D}, $g_N = \mathbb{A}\nabla u_h \cdot \boldsymbol{n}$ on Γ_{N} in the sense of traces, and consequently, $\bigl(r_{K,\mathrm{B}}(u_h), v_h\bigr)_{\partial K} + \bigl(r_{K,\mathrm{D}}(u_h), \nabla v_h\bigr)_{\partial K} = 0$ for all $v_h \in H^2(K)$, $K \in \mathscr{T}_h$. Hence, (7.58) implies that

$$r_h(u)(v_h) = 0 \quad \forall v_h \in H^2(\mathscr{T}_h),$$

which proves the lemma.   □

The consequence of (7.54)–(7.56) is the *Galerkin orthogonality* of the error

$$r_h(u_h)(v_h) = a_h(u_h - u, v_h) = 0 \qquad \forall v_h \in S_{hp}. \tag{7.60}$$

Finally, using the notation

$$R_{K,\mathrm{V}} := \bigl\| r_{K,\mathrm{V}}(u_h) \bigr\|_K, \quad R_{K,\mathrm{B}} := \bigl\| r_{K,\mathrm{B}}(u_h) \bigr\|_{\partial K}, \quad R_{K,\mathrm{D}} := \bigl\| r_{K,\mathrm{D}}(u_h) \bigr\|_{\partial K} \tag{7.61}$$

and the Cauchy inequality, we obtain from (7.58) the estimate

$$|r_K(u_h)(v_h)| \le R_{K,\mathrm{V}}\|v_h\|_K + R_{K,\mathrm{B}}\|v_h\|_{\partial K} + R_{K,\mathrm{D}}\|\nabla v_h\|_{\partial K}, \quad v_h \in H^2(\mathscr{T}_h), \tag{7.62}$$

which will be used later.

7.3.5 Quantity of Interest and Adjoint Problem

In this section, we introduce the adjoint problem to (7.43) and show that for some type of target functional, the adjoint consistency of the method requires a modification of the target functional. This modification, which does not affect the original quantity of interest, already ensures the adjoint consistency of the method.

We consider the *quantity of interest* of type (7.8), which represents a combination of the weighted mean values of the solution over the computational domain, the diffusive flux through the Dirichlet part of the boundary, and the weighted mean value of the solution on the Neumann part of the boundary. Particularly, we define $J(u)$ by

$$J(v) = (j_\Omega, v) + \left(j_{\Gamma_D}, \mathbb{A}\nabla v \cdot \boldsymbol{n}\right)_{\Gamma_D} + \left(j_{\Gamma_N}, v\right)_{\Gamma^+ \cup \Gamma_N}, \quad v \in H^1(\mathcal{T}_h), \qquad (7.63)$$

where $j_{\Gamma_D}, j_{\Gamma_N} \in L^2(\Gamma)$ and $j_\Omega \in L^2(\Omega)$ are given weight functions. Very often, these are characteristic functions of some subdomain of Γ or Ω, respectively. We assume that j_{Γ_D} is trace of some $z^* \in H^1(\Omega)$ on Γ_D.

We introduce the *adjoint problem* to the original problem (7.43): find a function $z : \Omega \to \mathbb{R}$ such that

$$-\nabla \cdot \mathbb{A}\nabla z - \boldsymbol{b} \cdot \nabla z + cz = j_\Omega \qquad \text{in } \Omega, \qquad (7.64a)$$

$$z = -j_{\Gamma_D} \qquad \text{on } \Gamma_D, \qquad (7.64b)$$

$$\mathbb{A}\nabla z \cdot \boldsymbol{n} + \boldsymbol{b} \cdot \boldsymbol{n} z = j_{\Gamma_N} \qquad \text{on } \Gamma_N, \qquad (7.64c)$$

$$\boldsymbol{b} \cdot \boldsymbol{n} z = j_{\Gamma_N} \qquad \text{on } \Gamma^+. \qquad (7.64d)$$

It is proved in [105] that primal problem (7.43), the adjoint problem (7.64), and the target functional (7.63) are compatible in the sense of (7.10), see also [77].

The discretization of (7.64) by the discontinuous Galerkin method leads to the following problem.

Definition 7.11 Function $z_h \in S_{hp}$ is the solution of the *discrete adjoint problem* if

$$a_h(w_h, z_h) = J(w_h) \qquad \forall w_h \in S_{hp}, \qquad (7.65)$$

where a_h is given by (7.52) and J by (7.63).

It is shown in [77] that (7.65) is not adjoint consistent (i.e., weak solution of (7.64) does not fulfill (7.65)) unless $j_{\Gamma_D} = 0$. This problem can be overcome by a modification of the target functional according to the method from [77] (see also [50] for details); namely, we set

$$J_h(v) := J(v) - \mathcal{N}(v), \qquad (7.66)$$

where $\mathcal{N}(v) := \sum_{K \in \mathcal{T}_h} \mathcal{N}_K(v)$ and

$$\mathcal{N}_K(v) := \big(\sigma(v - u_D), j_{\Gamma_D}\big)_{\partial K - \cap \Gamma_D} + \big((\sigma + \boldsymbol{b} \cdot \boldsymbol{n})(v - u_D), j_{\Gamma_D}\big)_{\partial K + \cap \Gamma_D}. \tag{7.67}$$

This modification is designed such that $J_h(u) = J(u)$ for u being the exact solution of (7.43). We define the directional derivative of J_h at u in the direction v by

$$J_h'[u](v) := \lim_{\tau \to 0} \frac{1}{\tau} \big(J_h(u + \tau v) - J_h(u)\big) \tag{7.68}$$

$$= J(v) - \sum_{K \in \mathcal{T}_h} \left\{ \big(\sigma v, j_{\Gamma_D}\big)_{\partial K - \cap \Gamma_D} + \big((\sigma + \boldsymbol{b} \cdot \boldsymbol{n})v, j_{\Gamma_D}\big)_{\partial K + \cap \Gamma_D} \right\}.$$

Since J_h is affine, $J_h'[u](v)$ does not depend on u, and hence we will use the notation $J_h'(v) = J_h'[u](v)$.

Finally, we replace the adjoint problem (7.65) by the following one.

Definition 7.12 Function $z_h \in S_{hp}$ is the solution of the *discrete adjoint problem* if

$$a_h(w_h, z_h) = J_h'(w_h) \qquad \forall w_h \in S_{hp}, \tag{7.69}$$

where a_h is given by (7.52) and J_h' by (7.68).

From the definition of J_h and J_h' (cf. (7.66) and (7.68)), we have

$$J(u) - J(u_h) = J_h'(u - u_h) - \mathcal{N}(u_h), \tag{7.70}$$

and $\mathcal{N}(u_h)$ (cf. (7.67)) can be understood as the "amount of violation" of the Dirichlet boundary condition (7.43b).

7.3.6 Adjoint Consistency

We show that the discretization of (7.44) by DGM with the modification of the target functional (7.66) is *adjoint consistent*. Moreover, we introduce the adjoint residual forms which will be used later.

Lemma 7.13 *The numerical scheme (7.54) is adjoint consistent, i.e.,*

$$a_h(w_h, z) = J_h'(w_h) \qquad \forall w_h \in H^2(\mathcal{T}_h), \tag{7.71}$$

where z is the weak solution of (7.64).

Proof We define the *adjoint residual* of problem (7.69) by

$$r_h^*(z_h)(w_h) := J_h'(w_h) - a_h(w_h, z_h), \qquad z_h, w_h \in H^2(\mathcal{T}_h) \tag{7.72}$$

and show that $r_h^*(z)(w_h) = 0$ for all $w_h \in H^2(\mathcal{T}_h)$, where z is the weak solution of (7.64). Taking into account (7.52) and (7.63) and the integration by parts, we get using a similar technique as in Sect. 7.3.4,

$$r_h^*(z_h)(w_h) = \sum_{K \in \mathcal{T}_h} r_K^*(z_h)(w_h), \tag{7.73}$$

where

$$r_K^*(z_h)(w_h) := \left(r_{K,V}^*(z_h), w_h\right)_K + \left(r_{K,B}^*(z_h), w_h\right)_{\partial K} + \left(r_{K,D}^*(z_h), \nabla w_h\right)_{\partial K}, \tag{7.74}$$

and

$$r_{K,V}^*(z_h) := j_\Omega + \nabla \cdot (\mathbb{A}\nabla z_h) + \boldsymbol{b} \cdot \nabla z_h - c z_h, \tag{7.75}$$

$$r_{K,B}^*(z_h) := \begin{cases} -\frac{1}{2}[\![\mathbb{A}\nabla z_h]\!] - (\sigma \boldsymbol{n}_K + \boldsymbol{b}) \cdot [\![z_h]\!] & \text{on } \partial K^+ \setminus \Gamma, \\ -\frac{1}{2}[\![\mathbb{A}\nabla z_h]\!] - \sigma \boldsymbol{n}_K \cdot [\![z_h]\!] & \text{on } \partial K^- \setminus \Gamma, \\ -\sigma(z_h + j_{\Gamma_D}) & \text{on } \partial K^- \cap \Gamma_D, \\ -(\sigma + \boldsymbol{b} \cdot \boldsymbol{n})(z_h + j_{\Gamma_D}) & \text{on } \partial K^+ \cap \Gamma_D, \\ j_{\Gamma_N} - \mathbb{A}\nabla z_h \cdot \boldsymbol{n} - \boldsymbol{b} \cdot \boldsymbol{n} z_h & \text{on } \partial K \cap \Gamma_N, \\ j_{\Gamma_N} - \boldsymbol{b} \cdot \boldsymbol{n} z_h & \text{on } \partial K \cap \Gamma^+, \\ 0 & \text{on } \partial K \cap \Gamma^-, \end{cases}$$

$$r_{K,D}^*(z_h) := \begin{cases} \frac{1}{2}\mathbb{A}[\![z_h]\!] & \text{on } \partial K \setminus \Gamma, \\ (j_{\Gamma_D} + z_h)\mathbb{A}\boldsymbol{n} & \text{on } \partial K \cap \Gamma_D, \\ 0 & \text{on } \partial K \cap (\Gamma \setminus \Gamma_D). \end{cases}$$

If $z \in H^2(\Omega)$ is the weak solution of (7.64), then $\left(r_{K,V}^*(z_h), w_h\right)_K = 0$ for all $w_h \in H^2(\mathcal{T}_h)$ and $K \in \mathcal{T}_h$. Moreover, $[\![z]\!] = 0$ and $[\![\mathbb{A}\nabla z_h]\!] = 0$ on all interior edges $(d = 2)$ or faces $(d = 3)$ of $\mathcal{T}_h$ and (7.64b)–(7.64d) are valid in the sense of traces. Therefore, $\left(r_{K,B}^*(z_h), w_h\right)_{\partial K} = \left(r_{K,D}^*(z_h), \nabla w_h\right)_{\partial K} = 0$ for all $w_h \in H^2(\mathcal{T}_h)$ and $K \in \mathcal{T}_h$. Hence (7.74) implies that

$$r_h^*(z)(w_h) = 0 \quad \forall w_h \in H^2(\mathcal{T}_h),$$

which proves the lemma. $\square$

Remark 7.14 The two underlined terms in the definition of $r^*_{K,B}$ have arisen due to the modification of the target functional by (7.66). Therefore, they are missing when considering (7.65) instead of (7.69) and the adjoint consistency is not valid in this case.

A simple consequence of (7.69), (7.71), and (7.72) is the *adjoint Galerkin orthogonality* of the error

$$r^*_h(z_h)(w_h) = a_h(w_h, z_h - z) = 0 \qquad \forall w_h \in S_{hp}. \tag{7.76}$$

Finally, applying the Cauchy inequality, we obtain from (7.74) the estimate

$$|r^*_K(z_h)(w_h)| \le R^*_{K,V}\|w_h\|_K + R^*_{K,B}\|w_h\|_{\partial K} + R^*_{K,D}\|\nabla w_h\|_{\partial K}, \quad w_h \in H^2(\mathscr{T}_h), \tag{7.77}$$

where

$$R^*_{K,V} := \left\|r^*_{K,V}(z_h)\right\|_K, \quad R^*_{K,B} := \left\|r^*_{K,B}(z_h)\right\|_{\partial K}, \quad R^*_{K,D} := \left\|r^*_{K,D}(z_h)\right\|_{\partial K}, \tag{7.78}$$

and $r^*_{K,V}$, $r^*_{K,B}$, and $r^*_{K,D}$ are defined by (7.75).

7.3.7 Goal-Oriented Error Estimates

We apply the abstract framework of the goal-oriented error estimates from Sect. 7.1.3 to problem (7.43) and its adjoint counterpart (7.64). Using (7.69), (7.70), and the Galerkin orthogonality (7.60), we get the *primal error identity*

$$J(u - u_h) = \mathscr{N}(u_h) + a_h(u - u_h, z) = \mathscr{N}(u_h) + \ell_h(z) - a_h(u_h, z) \tag{7.79}$$
$$= \mathscr{N}(u_h) + r_h(u_h)(z) = \mathscr{N}(u_h) + r_h(u_h)(z - v_h) \qquad \forall v_h \in S_{hp},$$

where $r_h(u_h)(\cdot)$ denotes the residual of the problem (7.54) given by (7.56).

Similarly, exploiting in addition the adjoint Galerkin orthogonality (7.76), we get the *adjoint error identity*

$$J(u - u_h) = \mathscr{N}(u_h) + a_h(u - u_h, z - z_h) = \mathscr{N}(u_h) + a_h(u - w_h, z - z_h)$$
$$= \mathscr{N}(u_h) + J'_h(u - w_h) - a_h(u - w_h, z_h) \tag{7.80}$$
$$= \mathscr{N}(u_h) + r^*_h(z_h)(u - w_h) \qquad \forall w_h \in S_{hp},$$

where $r^*_h(z_h)(\cdot)$ denotes the residual of the adjoint problem (7.69).

7.3.7.1   Error Estimate of Type I

The error estimates (7.79) and (7.80) depend on the unknown exact solutions z and u of the adjoint and primal problems, respectively. As mentioned in Sect. 7.1.4, the usual way is to approximate them by higher-order reconstructions, cf. (7.20)–(7.21). Therefore, we set

$$u \approx u_h^+ := \mathscr{R}(u_h), \quad z \approx z_h^+ := \mathscr{R}(z_h), \tag{7.81}$$

where $\mathscr{R} : S_{hp} \to S_{hp+1}$ denotes a reconstruction operator to a higher-polynomial degree space (cf. (7.50)). The possible variants of the higher-order reconstructions are discussed in Sect. 9.1.

Moreover, we have to specify the choice of v_h and w_h in (7.79) and (7.80), respectively. We use the choice (7.23b) discussed in Sect. 7.1.4. Moreover, although identities (7.79) and (7.80) are equivalent for the exact primal and adjoint solutions u and z, respectively, this equivalence is violated when u and z are replaced by reconstructed functions u_h^+ and z_h^+. Therefore, in order to employ all available information, we employ both of them using the arithmetic average. We note that the use of both primal and adjoint identities is important for nonlinear problems, see Remark 7.7.

Finally, using (7.79)–(7.80), (7.81), and (7.23b), we define the *goal-oriented error estimate of type I*

$$J(u - u_h) \tag{7.82}$$

$$\approx \eta^{\mathrm{I}}(u_h, z_h) := \mathscr{N}(u_h) + \tfrac{1}{2}\left[r_h(u_h)(z_h^+ - \Pi z_h^+) + r_h^*(z_h)(u_h^+ - \Pi u_h^+)\right],$$

where $\Pi : S_{hp+1} \to S_{hp}$ is a projection which will be specified in Sect. 8.1.

Remark 7.15 In virtue of (7.79)–(7.82), the exact representation of the error is

$$J(u - u_h) = \eta^{\mathrm{I}}(u_h, z_h) + \eta^{(2)}, \tag{7.83}$$

where $\eta^{(2)} := \tfrac{1}{2}\left[r_h(u_h)(z - z_h^+) + r_h^*(z_h)(u - u_h^+)\right]$. Therefore the error estimator $\eta^{\mathrm{I}}(u_h, z_h)$ is not the upper bound of the error $J(u - u_h)$. Techniques that guarantee the upper bound of the error were developed, e.g., in [92, 98].

7.3.7.2   Localization of the Error Estimate

For the purpose of the mesh adaptation, it is necessary to localize the error estimator, i.e., express it by a sum of local estimators over all mesh elements. A typical localization is based on the classical form of the error estimators proposed by [17], see Sect. 7.3.7.3. In [19], the localization based on filtering the variational formulation was proposed. Here, we follow the approach proposed in [103], which is based on the use of a partition of unity. This approach is very easy to use in the framework of discontinuous Galerkin method since the basis functions are discontinuous and localized for one mesh element.

We *localize* the error estimate $\eta^{\mathrm{I}}(u_h, z_h)$ from (7.82) by

$$\eta^{\mathrm{I}}(u_h, z_h) = \sum_{K \in \mathscr{T}_h} \eta_K^{\mathrm{I}}, \tag{7.84}$$

$$\eta_K^{\mathrm{I}} := \mathscr{N}_K(u_h) + \tfrac{1}{2}\left[r_h(u_h)(\chi_K(z_h^+ - \Pi z_h^+)) + r_h^*(z_h)(\chi_K(u_h^+ - \Pi u_h^+)) \right],$$

where $\mathscr{N}_K$ is given by (7.67) and χ_K is the characteristic function of $K \in \mathscr{T}_h$, i.e., $\chi_K = 1$ on K and $\chi_K = 0$ on $\Omega \setminus K$. The localization of η^{I}, given by (7.82), is employed for the setting of the "optimal" size of mesh elements. Roughly speaking, the size of elements with a large $|\eta_K^{\mathrm{I}}|$ is reduced and the size of $K \in \mathscr{T}_h$ with a small η_K^{I} is enlarged. However, estimate (7.84) does not provide information concerning the optimal shape of mesh elements. Therefore, in the next section, we introduce another form of goal-oriented error estimates, which separate the terms containing the primal and adjoint solutions (the classical form of the error estimators from [17]).

7.3.7.3 Error Estimate of Type II

For the purpose of the anisotropic mesh adaptation, we present another variant of the goal-oriented error estimates where we directly employed the results of Lemmas 7.10 and 7.13. Inserting (7.57), (7.62), (7.73), and (7.77) into (7.82), we obtain

$$|\eta^{\mathrm{I}}(u_h, z_h)| \le \eta^{\mathrm{II}}(u_h, z_h), \qquad \eta^{\mathrm{II}}(u_h, z_h) := \sum_{K \in \mathscr{T}_h} \eta_K^{\mathrm{II}}(u_h, z_h), \tag{7.85}$$

where

$$\eta_K^{\mathrm{II}} := \mathscr{N}_K(u_h) + \tfrac{1}{2}\big[R_{K,\mathrm{V}} \| z_h^+ - \Pi z_h^+ \|_K + R_{K,\mathrm{V}}^* \| u_h^+ - \Pi u_h^+ \|_K \tag{7.86}$$
$$+ R_{K,\mathrm{B}} \| z_h^+ - \Pi z_h^+ \|_{\partial K} + R_{K,\mathrm{B}}^* \| u_h^+ - \Pi u_h^+ \|_{\partial K}$$
$$+ R_{K,\mathrm{D}} \| \nabla(z_h^+ - \Pi z_h^+) \|_{\partial K} + R_{K,\mathrm{D}}^* \| \nabla(u_h^+ - \Pi u_h^+) \|_{\partial K} \big],$$

and residuals $R_{K,\mathrm{V}}$, $R_{K,\mathrm{B}}$, $R_{K,\mathrm{D}}$ and $R_{K,\mathrm{V}}^*$, $R_{K,\mathrm{B}}^*$, $R_{K,\mathrm{D}}^*$ are given by (7.61) and (7.78), respectively. With respect to notation in [17], estimate (7.85)–(7.86) is called *dual weighted residual* estimates since it contains the sum of residuals multiplied by the weights (= $\| u_h^+ - \Pi u_h^+ \|$ and $\| z_h^+ - \Pi z_h^+ \|$). We call (7.85)–(7.86) the *goal-*

oriented error estimate of type II. Obviously, we have also the local variant of this estimate

$$|\eta_K^{\mathrm{I}}(u_h, z_h)| \leq \eta_K^{\mathrm{II}}(u_h, z_h) \qquad \forall K \in \mathscr{T}_h. \tag{7.87}$$

The relation (7.86) is the base of the anisotropic mesh adaptation algorithm which is developed in Sect. 8.1. Unfortunately, we are not able to obtain information of the anisotropy nature of the errors from the residual terms $R_{K,\mathrm{V}}$, $R_{K,\mathrm{B}}$, $R_{K,\mathrm{D}}$, $R_{K,\mathrm{V}}^*$, $R_{K,\mathrm{B}}^*$, $R_{K,\mathrm{D}}^*$, but we can extract such information from the weight terms $\left\| u_h^+ - \Pi u_h^+ \right\|$ and $\left\| z_h^+ - \Pi z_h^+ \right\|$, see Remark 8.2.

7.4   Error Estimates for Nonlinear Convection–Diffusion Equations

In this section, we apply the abstract goal-oriented error estimate framework from Sect. 7.2 to the numerical solution of a nonlinear convection–diffusion equation which is discretized by the discontinuous Galerkin method (DGM).

7.4.1   Problem Formulation

We consider the nonlinear convection–diffusion problem: find $u : \Omega \to \mathbb{R}$ such that

$$\nabla \cdot \mathbf{f}(u) - \nabla \cdot (\mu(|\nabla u|)\nabla u) = g \qquad \text{in } \Omega, \tag{7.88}$$

$$u = u_D \qquad \text{on } \Gamma_{\mathrm{D}},$$

$$\nabla u \cdot \boldsymbol{n} = 0 \qquad \text{on } \Gamma_{\mathrm{N}} = \Gamma \setminus \Gamma_{\mathrm{D}},$$

where $g \in L^2(\Omega)$, u_D is a trace of a function from $H^1(\Omega)$ prescribed on the Dirichlet part of boundary Γ_{D}, and μ is a nonlinear function such that $0 < \mu_0 \leq \mu(s) \leq \mu_1$ for all $s \geq 0$, where μ_0 and μ_1 are positive constants. The vector-valued function $\mathbf{f}(u) \in [C^1(\mathbb{R})]^d$ is the given convective field and we assume that it can be written in the form

$$\mathbf{f}(u) = (f_1^{\mathrm{L}}(u)u, \dots, f_d^{\mathrm{L}}(u)u)^{\mathsf{T}}, \tag{7.89}$$

where $f_s^{\mathrm{L}} \in C(\mathbb{R})$, $s = 1, \dots, d$. For example, if $f_s(u) = u^2/2$ (Burger's equation), we put $f_s^{\mathrm{L}}(u) = u/2$, $s = 1, \dots, d$.

We employ the notation from Sect. 7.3.2 and discretize (7.88) again by the discontinuous Galerkin method. We present only the main relation for completeness. For more detail, we refer to [45].

The diffusive terms are treated by the SIPG variant of DGM. The convective flux through the boundary element ∂K, $K \in \mathcal{T}_h$ is approximated by

$$\mathbf{f}(u) \cdot \boldsymbol{n}_K|_{\partial K} \approx \mathrm{H}(u|_{\partial K}^{(\mathrm{int})}, u|_{\partial K}^{(\mathrm{ext})}, \boldsymbol{n}_K), \qquad K \in \mathcal{T}_h, \tag{7.90}$$

where function $\mathrm{H} : \mathbb{R} \times \mathbb{R} \times B_1 \to \mathbb{R}$ is called the *numerical flux*, see (7.46) for the notation. The numerical flux has to be *conservative*

$$\mathrm{H}(u_1, u_2, \boldsymbol{n}) = -\mathrm{H}(u_2, u_1, -\boldsymbol{n}) \qquad \forall u_1, u_2 \in \mathbb{R}, \; \boldsymbol{n} \in B_1 \tag{7.91}$$

and *consistent*, i.e., if u is continuous across ∂K (i.e., $u|_{\partial K}^{(\mathrm{int})} = u|_{\partial K}^{(\mathrm{ext})}$) then $\mathbf{f}(u) \cdot \boldsymbol{n}|_{\partial K} = \mathrm{H}(u, u, \boldsymbol{n}_\gamma)$. Furthermore, for the boundary elements, we set

$$u|_{\partial K}^{(\mathrm{ext})} := u_D \text{ on } \partial K \cap \Gamma_{\mathrm{D}}, \qquad u|_{\partial K}^{(\mathrm{ext})} := u|_{\partial K}^{(\mathrm{int})} \text{ on } \partial K \cap \Gamma_{\mathrm{N}}. \tag{7.92}$$

We employ the numerical flux which is based on the upwinding technique, namely

$$\mathrm{H}(u|_{\partial K}^{(\mathrm{int})}, u|_{\partial K}^{(\mathrm{ext})}, \boldsymbol{n}_K) := \mathrm{P}^+(\bar{u}, \boldsymbol{n}_K)u|_{\partial K}^{(\mathrm{int})} + \mathrm{P}^-(\bar{u}, \boldsymbol{n}_K)u|_{\partial K}^{(\mathrm{ext})}, \quad K \in \mathcal{T}_h, \tag{7.93}$$

where

$$\mathrm{P}^\pm(\bar{u}, \boldsymbol{n}_K) = \left(\mathbf{f}^{\mathrm{L}}(\langle \bar{u} \rangle) \cdot \boldsymbol{n}_K\right)^\pm, \qquad \bar{u} \in H^1(\mathcal{T}_h), \; K \in \mathcal{T}_h, \tag{7.94}$$

the symbol $\langle \cdot \rangle$ is the mean value given by (7.47), and $\xi^+ = \max(\xi, 0)$ and $\xi^- = \min(\xi, 0)$ are the positive and negative parts of $\xi \in \mathbb{R}$, respectively. A possible modification to, e.g., the Lax–Friedrichs numerical flux is straightforward. Again, we write u_h instead of $u_h|_{\partial K}^{(\mathrm{int})}$ (interior trace) and $u_h^{(\mathrm{ext})}$ instead of $u_h|_{\partial K}^{(\mathrm{ext})}$ (exterior trace), for simplicity.

Then the form a_h from (7.26) corresponding to the discretization of (7.88) by DGM reads

$$a_h(u_h, v_h) := a_h^{\mathrm{L}}(u_h, u_h, v_h) - \tilde{a}_h(u_h, v_h), \qquad u_h, v_h \in H^2(\mathcal{T}_h), \tag{7.95}$$

where

$$a_h^{\mathrm{L}}(\bar{u}_h, u_h, v_h) := \sum_{K \in \mathcal{T}_h} \left(\mu(|\nabla \bar{u}_h|)\nabla u_h - \mathbf{f}^{\mathrm{L}}(\bar{u}_h)u_h, \nabla v_h\right)_K \tag{7.96}$$

$$+ \sum_{K \in \mathcal{T}_h} \left((\mathrm{P}^+(\bar{u}_h, \boldsymbol{n}_K)u_h, v_h)_{\partial K} + (\mathrm{P}^-(\bar{u}_h, \boldsymbol{n}_K)u_h^{(\mathrm{ext})}, v_h)_{\partial K \setminus \Gamma_{\mathrm{D}}}\right)$$

$$- \sum_{K \in \mathcal{T}_h} \left((\langle \mu(|\nabla \bar{u}_h|)\nabla u_h \rangle - \sigma[\![u_h]\!])\boldsymbol{n}_K, v_h\right)_{\partial K \setminus \Gamma_{\mathrm{N}}}$$

$$- e \sum_{K \in \mathscr{T}_h} \left([\![\mu(|\nabla \bar{u}_h|) u_h]\!] \boldsymbol{n}_K, \nabla v_h \right)_{\partial K \setminus \Gamma_N}$$

$$\tilde{a}_h(\bar{u}_h, v_h) := - \sum_{K \in \mathscr{T}_h} \left(\mathrm{P}^-(\bar{u}_h, \boldsymbol{n}_K) u_D, v_h \right)_{\partial K \cap \Gamma_D} + (g, v_h)$$

$$- \sum_{K \in \mathscr{T}_h} \left(\mu(|\nabla \bar{u}_h|) \nabla v_h, u_D \boldsymbol{n}_K \right)_{K \cap \Gamma_D} + \sum_{K \in \mathscr{T}_h} (\sigma u_D, v_h)_{K \cap \Gamma_D},$$

and $\sigma > 0$ is the penalty given by (7.53), $e = 1/2$ on $\partial K \setminus \Gamma_D$ and $e = 1$ on $\partial K \cap \Gamma_D$.

Function $u_h \in S_{hp}$ (cf. (7.50)) is the *approximate solution* of (7.88) if it satisfies

$$a_h(u_h, v_h) = 0 \qquad \forall v_h \in S_{hp} \tag{7.97}$$

with a_h given by (7.95)–(7.96). Finally, we set the *primal residual* as

$$r_h(u_h)(v_h) := -a_h(u_h, v_h), \qquad u_h, v_h \in H^2(\mathscr{T}_h). \tag{7.98}$$

Lemma 7.16 *The discretization* (7.97) *is consistent, i.e.,* (7.97) *is valid for* $u \in H^2(\Omega)$ *being the exact primal solution of* (7.88) *and* $v_h \in H^2(\mathscr{T}_h)$.

Proof Using the relations (7.95)–(7.96) and some manipulation, we express residual (7.98) as the sum of element residuals

$$r_h(u_h)(v_h) = \tilde{a}_h(u_h, v_h) - a_h^{\mathrm{L}}(u_h, u_h, v_h) = \sum_{K \in \mathscr{T}_h} r_K(u_h)(v_h), \tag{7.99}$$

where

$$r_K(u_h)(v_h) = \left(r_{K,\mathrm{V}}(u_h), v_h \right)_K + \left(r_{K,\mathrm{B}}(u_h), v_h \right)_{\partial K} + \left(r_{K,\mathrm{D}}(u_h), \nabla v_h \right)_{\partial K}, \tag{7.100}$$

and

$$r_{K,\mathrm{V}}(u_h) = g + \nabla \cdot (\mu(|\nabla u_h|) \nabla u_h) - \nabla \cdot \mathbf{f}(u_h), \tag{7.101}$$

$$r_{K,\mathrm{B}}(u_h) = \begin{cases} \mathbf{f}(u_h) \cdot \boldsymbol{n}_K - \mathrm{P}^+(u_h, \boldsymbol{n}_K) u_h - \mathrm{P}^-(u_h, \boldsymbol{n}_K) u_h^{(\mathrm{ext})} \\ \quad + \left(\frac{1}{2} [\![\mu(|\nabla u_h|) \nabla u_h]\!] - \sigma [\![u_h]\!] \right) \boldsymbol{n}_K & \text{on } \partial K \setminus \Gamma, \\[2mm] \mathbf{f}(u_h) \cdot \boldsymbol{n}_K - \mathrm{P}^+(u_h, \boldsymbol{n}_K) u_h - \mathrm{P}^-(u_h, \boldsymbol{n}_K) u_D \\ \quad + \frac{1}{2} [\![\mu(|\nabla u_h|) \nabla u_h]\!] \boldsymbol{n}_K - \sigma(u_h - u_D) & \text{on } \partial K \cap \Gamma_D, \\[2mm] \mathbf{f}(u_h) \cdot \boldsymbol{n}_K - \mathrm{P}^+(u_h, \boldsymbol{n}_K) u_h - \mathrm{P}^-(u_h, \boldsymbol{n}_K) u_h \\ \quad + \mu(|\nabla u_h|) \nabla u_h \cdot \boldsymbol{n}_K & \text{on } \partial K \cap \Gamma_N, \end{cases}$$

$$r_{K,D}(u_h) = \begin{cases} \frac{1}{2}[\![\mu(|\nabla u_h|)u_h]\!]\boldsymbol{n}_K & \text{on } \partial K \setminus \Gamma, \\ \mu(|\nabla u_h|)(u_h - u_D) & \text{on } \partial K \cap \Gamma_D, \\ 0 & \text{on } \partial K \cap \Gamma_N. \end{cases}$$

Let $u \in H^2(\Omega)$ be the solution of (7.88), and we insert $u_h := u$ in (7.101) and use the fact that $u = u^{(\text{ext})}$ and $[\![u]\!] = [\![\mu(|\nabla u_h|)u_h]\!] = [\![\mu(|\nabla u_h|)\nabla u_h]\!] = 0$ in the sense of traces. Moreover the consistency of numerical flux implies $\mathbf{f}(u) \cdot \boldsymbol{n}_K = \mathrm{P}^+(u, \boldsymbol{n}_K)u + \mathrm{P}^-(u, \boldsymbol{n}_K)u^{(\text{ext})}$, which together with the fact that u satisfies the boundary conditions in (7.88) implies $r_h(u_h)(v_h) = 0$ for all $v_h \in H^2(\mathscr{T}_h)$. $\square$

Corollary 7.17 *Let $u_h \in S_{hp}$ and*

$$R_{K,\mathrm{V}} := \left\| r_{K,\mathrm{V}}(u_h) \right\|_K, \quad R_{K,\mathrm{B}} := \left\| r_{K,\mathrm{B}}(u_h) \right\|_{\partial K}, \quad R_{K,\mathrm{D}} := \left\| r_{K,\mathrm{D}}(u_h) \right\|_{\partial K}, \tag{7.102}$$

where $r_{K,\mathrm{V}}$, $r_{K,\mathrm{B}}$, and $r_{K,\mathrm{D}}$ are given by (7.101). Then the Cauchy inequality and (7.100) imply

$$|r_K(u_h)(v_h)| \le R_{K,\mathrm{V}}\|v_h\|_K + R_{K,\mathrm{B}}\|v_h\|_{\partial K} + R_{K,\mathrm{D}}\|\nabla v_h\|_{\partial K}, \quad v_h \in H^2(\mathscr{T}_h). \tag{7.103}$$

7.4.2 Target Functional and Adjoint Problem

We consider the *quantity of interest* as the possible combination of the diffusive flux through the boundary Γ_D and the convective flux through the Neumann part of the boundary Γ_N. Then the target functional has the form

$$J(u) = \int_{\Gamma_D} \chi_1 \mu(|\nabla u|)\nabla u \cdot \boldsymbol{n}\, dS + \int_{\Gamma_N} \chi_2\, \mathbf{f}(u) \cdot \boldsymbol{n}\, dS, \tag{7.104}$$

where $\chi_1, \chi_2 : \Gamma \to \mathbb{R}$ are given weight functions.

Employing the technique from [77], we define the strong form of the adjoint problem, linearized at $u \in H^2(\mathscr{T}_h)$, by

$$-\mathbf{f}^{\mathrm{L}}(u) \cdot \nabla z - \nabla \cdot (\mu(|\nabla u|)\nabla z) = 0 \qquad\qquad \text{in } \Omega, \tag{7.105}$$

$$z = -\chi_1 \qquad\qquad \text{on } \Gamma_D,$$

$$\mu(|\nabla u|)\nabla z \cdot \boldsymbol{n} + \mathbf{f}^{\mathrm{L}}(u) \cdot \boldsymbol{n}\, z = \chi_2\, \mathbf{f}^{\mathrm{L}}(u) \cdot \boldsymbol{n} \qquad \text{on } \Gamma_N.$$

In order to guarantee the adjoint consistency of (7.97) with the form given by (7.95), we modify the target functional (7.104) as

$$J_h(u) = \int_{\Gamma_D} \chi_1 \mu(|\nabla u|) \nabla u \cdot \boldsymbol{n} \, dS - \int_{\Gamma_D} \chi_1 \left(\sigma + \left(\mathbf{f}^L(u_D) \cdot \boldsymbol{n}\right)^+\right) (u - u_D) \, dS$$

$$+ \int_{\Gamma_N} \chi_2 \mathbf{f}(u) \cdot \boldsymbol{n} \, dS, \tag{7.106}$$

where $\sigma > 0$ is the penalty parameter appearing in (7.96) and symbol $a^+ = \max(a, 0)$ for $a \in \mathbb{R}$. Obviously, $J(u) = J_h(u)$ for u being the exact solution of (7.88), which verifies (7.36). Moreover, in order to define the variational form of the adjoint problem, we linearize J_h at $u_h \in H^2(\mathcal{T}_h)$ as

$$J_h^L(u_h, v_h) := \int_{\Gamma_D} \chi_1 \mu(|\nabla u_h|) \nabla v_h \cdot \boldsymbol{n} \, dS - \int_{\Gamma_D} \chi_1 \left(\sigma + \left(\mathbf{f}^L(u_D) \cdot \boldsymbol{n}\right)^+\right) v_h \, dS$$

$$+ \int_{\Gamma_N} \chi_2 \mathbf{f}^L(u_h) v_h \cdot \boldsymbol{n} \, dS, \qquad u_h, v_h \in H^2(\mathcal{T}_h). \tag{7.107}$$

Then expression (7.37) is valid with $\tilde{J}_h(v_h) = \int_{\Gamma} \chi_1 \sigma u_D \, dS = $ const, and then it is independent of v_h and $J_h(u) - J_h(u_h) = J_h^L(u, u) - J_h^L(u_h, u_h)$.

Definition 7.18 Let $u_h \in S_{hp}$ be given, and the function $z_h \in S_{hp}$ is the solution of the *discrete adjoint problem* if

$$a_h^L(u_h, w_h, z_h) = J_h^L(u_h, w_h) \qquad \forall w_h \in S_{hp}, \tag{7.108}$$

where a_h^L and J_h^L are given by (7.96) and (7.107), respectively.

Finally, we define the *adjoint residual* (cf. (7.39)) as

$$r_h^*(u_h, z_h)(w_h) := J_h^L(u_h, w_h) - a_h^L(u_h, w_h, z_h), \qquad w_h \in H^2(\mathcal{T}_h). \tag{7.109}$$

Lemma 7.19 *The discretization (7.97) is adjoint consistent, i.e., (7.108) is valid for $u \in H^2(\Omega)$ and $z \in H^2(\Omega)$ being the exact primal and adjoint solutions of (7.88) and (7.105), respectively, and any $w_h \in H^2(\mathcal{T}_h)$.*

Proof Using the relations (7.96) and (7.107) defining forms a_h^L and J_h^L, respectively, and some manipulation, we express the adjoint residual (7.109) as the sum of element residuals

$$r_h^*(u_h, z_h)(w_h) = \sum_{K \in \mathcal{T}_h} r_K^*(u_h, z_h)(w_h), \tag{7.110}$$

where

$$r_K^*(u_h, z_h)(w_h) = \left(r_{K,V}^*(u_h, z_h), w_h\right)_K + \left(r_{K,B}^*(u_h, z_h), w_h\right)_{\partial K} \tag{7.111}$$
$$+ \left(r_{K,D}^*(u_h, z_h), \nabla w_h\right)_{\partial K},$$

and

$$r_{K,V}^*(u_h, z_h) = \nabla \cdot (\mu(|\nabla u_h|)\nabla z_h) + \mathbf{f}^L(u_h) \cdot \nabla z_h, \tag{7.112}$$

$$r_{K,B}^*(u_h, z_h) = \begin{cases} -\frac{1}{2}\llbracket \mu(|\nabla u_h|)\nabla z_h \rrbracket - (\sigma + \mathrm{P}^+(u_h, \mathbf{n}_K))\llbracket z_h \rrbracket \cdot \mathbf{n}_K & \text{on } \partial K \setminus \Gamma, \\ -(\sigma + \mathrm{P}^+(u_h, \mathbf{n}_K))z_h - \left(\sigma + (\mathbf{f}^L(u_D) \cdot \mathbf{n}_K)^+\right)\chi_1 & \text{on } \partial K \cap \Gamma_D, \\ (\chi_2 - z_h)\,\mathbf{f}^L(u_h) \cdot \mathbf{n}_K - \mu(|\nabla u_h|)\nabla z_h \cdot \mathbf{n}_K & \text{on } \partial K \cap \Gamma_N, \end{cases}$$

$$r_{K,D}^*(u_h, z_h) = \begin{cases} \frac{1}{2}\mu(|\nabla u_h|)\llbracket z_h \rrbracket & \text{on } \partial K \setminus \Gamma, \\ \mu(|\nabla u_h|)\mathbf{n}_K(z_h + \chi_1) & \text{on } \partial K \cap \Gamma_D, \\ 0 & \text{on } \partial K \cap \Gamma_N. \end{cases}$$

A more detailed derivation can be found in [43].

Let $u, z \in H^2(\Omega)$ be the solutions of primal and adjoint problems (7.88) and (7.105), respectively. Putting $u_h := u$ and $z_h := z$ in (7.112), using the identities $\llbracket z \rrbracket = 0$ and $\llbracket \mu(|\nabla u|)\nabla z \rrbracket = 0$ valid on all inter-element boundaries, the consistency of numerical flux (7.94), and the fact that z fulfills (7.105), we obtain $r_h^*(u, z)(w_h) = 0$ for any $w_h \in H^2(\mathcal{T}_h)$. $\square$

Corollary 7.20 *Let* $u_h, z_h \in S_{hp}$ *and*

$$R_{K,V}^* := \left\| r_{K,V}^*(u_h, z_h) \right\|_K, \quad R_{K,B}^* := \left\| r_{K,B}^*(u_h, z_h) \right\|_{\partial K}, \quad R_{K,D}^* := \left\| r_{K,D}^*(u_h, z_h) \right\|_{\partial K}, \tag{7.113}$$

where $r_{K,V}^*$, $r_{K,B}^*$, *and* $r_{K,D}^*$ *are given by* (7.112). *Then the Cauchy inequality and* (7.111) *imply*

$$|r_K^*(u_h, z_h)(w_h)| \le R_{K,V}^*\|w_h\|_K + R_{K,B}^*\|w_h\|_{\partial K} + R_{K,D}^*\|\nabla w_h\|_{\partial K}, \quad w_h \in H^2(\mathcal{T}_h). \tag{7.114}$$

7.4.3 Goal-Oriented Error Estimates

We apply the abstract framework of the goal-oriented error estimates from Sect. 7.2 to problem (7.88) and its adjoint counterpart (7.105). Following the strategy for the linear problems given in Sect. 7.3.7, we employ (7.42) and introduce the *goal-oriented error estimate of type I* in the form

$$J(u) - J(u_h) \tag{7.115}$$

$$\approx \eta^{\mathrm{I}}(u_h, z_h) := \tfrac{1}{2} \left(r_h(u_h)(z_h^+ - \Pi z_h^+) + r_h^*(u_h, z_h)(u_h^+ - \Pi u_h^+) \right),$$

where primal and adjoint residuals $r_h(\cdot)(\cdot)$ and $r_h^*(\cdot)(\cdot)$ are given by (7.98) and (7.109), respectively, functions $u_h^+, z_h^+ \in S_{hp+1}$ are the higher-order reconstructions of the approximate solutions of primal and adjoint problems, and $\Pi u_h^+, \Pi z_h^+ \in S_{hp}$ are the corresponding projections.

Similar to Sect. 7.3.7.2, we *localize* the error estimates by the partition of unity and have

$$\eta^{\mathrm{I}}(u_h, z_h) = \sum_{K \in \mathscr{T}_h} \eta_K^{\mathrm{I}}, \tag{7.116}$$

$$\eta_K^{\mathrm{I}} := \tfrac{1}{2} \left[r_h(u_h)(\chi_K(z_h^+ - \Pi z_h^+)) + r_h^*(u_h, z_h)(\chi_K(u_h^+ - \Pi u_h^+)) \right],$$

where χ_K is the characteristic function of $K \in \mathscr{T}_h$.

Finally, as in Sect. 7.3.7.3, we employ (7.99), (7.103), (7.110), and (7.114) and obtain the *goal-oriented error estimate of type II* in the form

$$|\eta^{\mathrm{I}}(u_h, z_h)| \leq \eta^{\mathrm{II}}(u_h, z_h), \qquad \eta^{\mathrm{II}}(u_h, z_h) := \sum_{K \in \mathscr{T}_h} \eta_K^{\mathrm{II}}(u_h, z_h), \tag{7.117}$$

where

$$\eta_K^{\mathrm{II}} := \tfrac{1}{2} \big[R_{K,\mathrm{V}} \| z_h^+ - \Pi z_h^+ \|_K + R_{K,\mathrm{V}}^* \| u_h^+ - \Pi u_h^+ \|_K \tag{7.118}$$

$$+ R_{K,\mathrm{B}} \| z_h^+ - \Pi z_h^+ \|_{\partial K} + R_{K,\mathrm{B}}^* \| u_h^+ - \Pi u_h^+ \|_{\partial K}$$

$$+ R_{K,\mathrm{D}} \| \nabla(z_h^+ - \Pi z_h^+) \|_{\partial K} + R_{K,\mathrm{D}}^* \| \nabla(u_h^+ - \Pi u_h^+) \|_{\partial K} \big],$$

and residuals $R_{K,\mathrm{V}}$, $R_{K,\mathrm{B}}$, $R_{K,\mathrm{D}}$ and $R_{K,\mathrm{V}}^*$, $R_{K,\mathrm{B}}^*$, $R_{K,\mathrm{D}}^*$ are given by (7.102) and (7.113), respectively. Obviously, we also have the local variant of this estimate

$$|\eta_K^{\mathrm{I}}(u_h, z_h)| \leq \eta_K^{\mathrm{II}}(u_h, z_h) \qquad \forall K \in \mathscr{T}_h. \tag{7.119}$$

7.5 Compressible Euler Equations

In this book we present also numerical experiments dealing with the simulation of steady inviscid compressible flow around an isolated profile. This problem is described by the system of the *Euler equations* with appropriate boundary conditions which can be written as

$$\sum_{s=1}^{d} \frac{\partial f_s(\boldsymbol{u})}{\partial x_s} = 0, \tag{7.120}$$

where $\boldsymbol{u} : \Omega \to \mathbb{R}^{d+2}$ is the unknown vector-valued function whose components are the density, components of momentum, and energy. Furthermore, $\boldsymbol{f}_s : \mathbb{R}^{d+2} \to \mathbb{R}^{d+2}$, $s = 1, \ldots, d$, denote the physical fluxes.

The quantity of interest is frequently the drag c_D or the lift c_L coefficient which represents the normalized components of the force acting on the profile. In order to have an adjoint consistent discretization, the corresponding target functional and adjoint problem have to be carefully formulated. We analyzed this problem in detail in [51] for $d = 2$ and applied the framework from Sect. 7.2.

For the anisotropic mesh adaptation technique, it is sufficient to introduce that primal and adjoint residuals can be written in the form appearing in Corollaries 7.17 and 7.20, namely

$$r_h(\boldsymbol{u}_h)(\boldsymbol{\varphi}) = \left(\boldsymbol{r}_{K,\text{V}}(\boldsymbol{u}_h), \boldsymbol{\varphi}\right)_K + \left(\boldsymbol{r}_{K,\text{B}}(\boldsymbol{u}_h), \boldsymbol{\varphi}\right)_{\partial K}, \tag{7.121}$$

and

$$r_h^*(\boldsymbol{u}_h, \boldsymbol{z}_h)(\boldsymbol{\varphi}) = \left(\boldsymbol{r}_{K,\text{V}}^*(\boldsymbol{u}_h), \boldsymbol{\varphi}\right)_K + \left(\boldsymbol{r}_{K,\text{B}}^*(\boldsymbol{u}_h), \boldsymbol{\varphi}\right)_{\partial K}, \tag{7.122}$$

respectively, where $\boldsymbol{u}_h \in [S_{hp}]^{d+2}$ and $\boldsymbol{z}_h \in [S_{hp}]^{d+2}$ are the approximate solutions of the discrete primal and adjoint problems, and $\boldsymbol{\varphi} \in [H^1(\mathscr{T}_h)]^{d+2}$ is the test function. The residuals $\boldsymbol{r}_{K,\text{V}}$, $\boldsymbol{r}_{K,\text{B}}$, $\boldsymbol{r}_{K,\text{V}}^*$, and $\boldsymbol{r}_{K,\text{B}}^*$ are vector-valued functions analogous to those appearing in (7.101) and (7.112).

It is possible to formulate the error estimate of types I and II as well as localize the error estimates as in Sects. 7.3 and 7.4. Since the derivation is straightforward, we omit it. We only mention that the error estimate of type II has $4(d + 2)$ terms, since it contains residuals of type $R_{K,\text{V}}$, $R_{K,\text{B}}$, $R_{K,\text{V}}^*$, and $R_{K,\text{B}}^*$ for each of $d + 2$ equations.

Chapter 8
Goal-Oriented Anisotropic Mesh Adaptation

Employing the results from the previous chapter, we derive goal-oriented error estimates including the geometry of mesh elements. These estimates are based on interpolation error bounds. Furthermore, we present a goal-oriented anisotropic hp-mesh adaptation algorithm as well as several numerical examples demonstrating its accuracy, efficiency, and robustness.

Goal-oriented mesh adaptation, including anisotropic (metric-based) techniques, was developed over many years and there are numerous publications, e.g., [28, 29, 48, 100, 111, 112, 116], dealing with theoretical as well as practical aspects. Some ideas presented in these publications are employed in this chapter.

8.1 Goal-Oriented Estimates Including the Geometry of Elements

We summarize the goal-oriented error estimates from Chap. 7 for linear as well as nonlinear problems. Let u and z be the exact solutions of the primal and adjoint problems, respectively, and $u_h, z_h \in S_{hp}$ be the corresponding approximate solutions. Furthermore, $u_h^+, z_h^+ \in S_{hp+1}$ denotes the higher-order reconstructions given by (7.81) and $\Pi : S_{hp+1} \rightarrow S_{hp}$ is a projection.

(i) The error of the quantity of interest is estimated by

$$J(u) - J(u_h) \tag{8.1}$$

$$\approx \eta^I(u_h, z_h) := \mathcal{N}(u_h) + \tfrac{1}{2}r_h(u_h)(z_h^+ - \Pi z_h^+) + \tfrac{1}{2}r_h^*(u_h, z_h)(u_h^+ - \Pi u_h^+),$$

cf. (7.82), (7.115), which is the *goal-oriented error estimate of type I*.

V. Dolejší, G. May, *Anisotropic* hp-*Mesh Adaptation Methods*, Nečas Center Series, https://doi.org/10.1007/978-3-031-04279-9_8

(ii) The localization of estimate (8.1) gives

$$\eta^{\mathrm{I}}(u_h, z_h) = \sum_{K \in \mathscr{T}_h} \eta_K^{\mathrm{I}} \qquad \text{with} \tag{8.2}$$

$$\eta_K^{\mathrm{I}} := \mathscr{N}_K(u_h) + \frac{1}{2} \left[r_h(u_h)(\chi_K(z_h^+ - \Pi z_h^+)) + r_h^*(u_h, z_h)(\chi_K(u_h^+ - \Pi u_h^+)) \right],$$

cf. (7.84), (7.116).

(iii) Separating the residuals and the weights, we obtain the *goal-oriented error estimate of type II*

$$|J(u - u_h)| \approx |\eta^{\mathrm{I}}(u_h, z_h)| \le \eta^{\mathrm{II}}(u_h, z_h) = \sum_{K \in \mathscr{T}_h} \eta_K^{\mathrm{II}}(u_h, z_h), \tag{8.3}$$

$$\eta_K^{\mathrm{II}} := \mathscr{N}_K + \frac{1}{2} \sum_{\star} \left(R_{K,\star} \|z_h^+ - \Pi z_h^+\|_{K,\star} + R_{K,\star}^* \|u_h^+ - \Pi u_h^+\|_{K,\star} \right),$$

where $R_{K,\star} \in \{R_{K,\mathrm{V}}, R_{K,\mathrm{B}}, R_{K,\mathrm{D}}\}$, $R_{K,\star}^* \in \{R_{K,\mathrm{V}}^*, R_{K,\mathrm{B}}^*, R_{K,\mathrm{D}}^*\}$ and $\|w\|_{K,\star}$ denotes the corresponding norms $\{\|w\|_K, \|w\|_{\partial K}, \|\nabla w\|_{\partial K}\}$, cf. (7.85) and (7.86) and (7.117) and (7.118).

(iv) The *local variant* of the estimate of type II is valid as well, i.e.,

$$|\eta_K^{\mathrm{I}}(u_h, z_h)| \le \eta_K^{\mathrm{II}}(u_h, z_h) \qquad \forall K \in \mathscr{T}_h. \tag{8.4}$$

cf. (7.87) and (7.119).

The reconstructed solutions $u_h^+, z_h^+ \in S_{hp+1}$ are obtained by a technique from Sect. 9.1 (preferably by that of Sect. 9.1.2). Concerning the projection Π appearing in estimates η^{I} and η^{II} above, it is advantageous to use the interpolation operator from Sect. 3.2 given by Definition 3.2. In this case the differences $u_h^+ - \Pi u_h^+$ and $z_h^+ - \Pi z_h^+$ are s-homogeneous functions in the sense of Definition 3.6 and, therefore, the estimates from Sect. 3.3 can be directly employed. We consider separately the two- and three-dimensional cases.

8.1.1 Estimates Including the Geometry of Elements for 2D

Let u_h^+ and z_h^+ be the reconstructed functions from S_{hp+1}, cf. (7.50). Let $K \in \mathscr{T}_h$ be a triangle with the geometry $\{\mu_K, \sigma_K, \phi_K\}$, cf. Definition 3.1, and p_K be the corresponding polynomial approximation degree. Applying Theorem 3.17 and Definition 3.18, there exist values defining the anisotropic bounds of $p_K + 1$-homogeneous functions $u_h^+ - \Pi u_h^+$, $z_h^+ - \Pi z_h^+$ and $2p_K$-homogeneous functions

$|\nabla(u_h^+ - \Pi u_h^+)|^2$, $|\nabla(z_h^+ - \Pi z_h^+)|^2$ restricted to K, denoted as

$$\{A_u, \rho_u, \varphi_u\} \ldots \text{the anisotropic bound of } (u_h^+ - \Pi u_h^+)|_K, \tag{8.5}$$

$$\{A_z, \rho_z, \varphi_z\} \ldots \text{the anisotropic bound of } (z_h^+ - \Pi z_h^+)|_K,$$

$$\{A_{u'}, \rho_{u'}, \varphi_{u'}\} \ldots \text{the anisotropic bound of } |\nabla(u_h^+ - \Pi u_h^+)|_K|^2,$$

$$\{A_{z'}, \rho_{z'}, \varphi_{z'}\} \ldots \text{the anisotropic bound of } |\nabla(z_h^+ - \Pi z_h^+)|_K|^2,$$

which depend on $(p+1)$th-derivatives of u_h^+ and z_h^+.

We estimate terms $\|z_h^+ - \Pi z_h^+\|_{K,\star}$ and $\|u_h^+ - \Pi u_h^+\|_{K,\star}$ by the results from Sect. 3.3 (particularly estimates (3.69), (3.84), (3.91)). Therefore, from (8.5), we obtain

$$\|u_h^+ - \Pi u_h^+\|_K \leq \left(A_u^2 \frac{\mu_K^{2(p_K+2)}}{2p_K+4} G(p_K+1, p_K+1, \rho_u, \varphi_u; \sigma_K, \phi_K)\right)^{1/2} =: \theta_{K,\mathrm{V}},$$

$$\|u_h^+ - \Pi u_h^+\|_{\partial K} \leq \left(A_u^2 \mu_K^{2p_K+3} \sigma_K\, G(p_K+1, p_K+1, \rho_u, \varphi_u; \sigma_K, \phi_K)\right)^{1/2} =: \theta_{K,\mathrm{B}},$$

$$\|\nabla(u_h^+ - \Pi u_h^+)\|_{\partial K} \leq \left(A_{u'} \mu_K^{2p_K+1} \sigma_K\, G(p_K, 2p_K, \rho_{u'}, \varphi_{u'}; \sigma_K, \phi_K)\right)^{1/2} =: \theta_{K,\mathrm{D}},$$

$$\|z_h^+ - \Pi z_h^+\|_K \leq \left(A_z^2 \frac{\mu_K^{2(p_K+2)}}{2p_K+4} G(p_K+1, p_K+1, \rho_z, \varphi_z; \sigma_K, \phi_K)\right)^{1/2} =: \theta_{K,\mathrm{V}}^*,$$

$$\|z_h^+ - \Pi z_h^+\|_{\partial K} \leq \left(A_z^2 \mu_K^{2p_K+3} \sigma_K\, G(p_K+1, p_K+1, \rho_z, \varphi_z; \sigma_K, \phi_K)\right)^{1/2} =: \theta_{K,\mathrm{B}}^*,$$

$$\|\nabla(z_h^+ - \Pi z_h^+)\|_{\partial K} \leq \left(A_{z'} \mu_K^{2p_K+1} \sigma_K\, G(p_K, 2p_K, \rho_{z'}, \varphi_{z'}; \sigma_K, \phi_K)\right)^{1/2} =: \theta_{K,\mathrm{D}}^*,$$

$$\tag{8.6}$$

where function G is given by (3.57).

Applying (8.6) on the terms appearing in (8.3), we get

$$\eta_K^{\mathrm{I\!I\!I}} \leq \eta_K^{\mathrm{I\!I\!I}}(\mu_K, \sigma_K, \phi_K, p_K), \quad K \in \mathscr{T}_h, \tag{8.7}$$

where

$$\eta_K^{\mathrm{I\!I\!I}}(\mu_K, \sigma_K, \phi_K, p_K) := \mathscr{N}_K + \frac{1}{2}\sum_\star \left(R_{K,\star}\theta_{K,\star} + R_{K,\star}^*\theta_{K,\star}^*\right), \tag{8.8}$$

where $\theta_{K,\star} \in \{\theta_{K,\mathrm{V}}, \theta_{K,\mathrm{B}}, \theta_{K,\mathrm{D}}\}$, $\theta_{K,\star}^* \in \{\theta_{K,\mathrm{V}}^*, \theta_{K,\mathrm{B}}^*, \theta_{K,\mathrm{D}}^*\}$, cf. (8.6). Obviously, $\eta_K^{\mathrm{I\!I\!I}}$ depends on all of the anisotropy parameters $A_u, \rho_u, \ldots$ from (8.5) but for shorter notation, we write only the dependence on the element geometry parameters μ_K, σ_K, ϕ_K (cf. Definition 3.1) and the polynomial degree p_K. This determination leads to the final *goal-oriented error estimate of type III*.

Theorem 8.1 *Let u_h and z_h be the approximate solutions of the primal and adjoint problems given by (7.4) and (7.11), respectively. Furthermore, let u_h^+ and z_h^+ be the higher-order approximations of the primal and adjoint solutions reconstructed from u_h and z_h, respectively. Finally, let $\eta^{\mathrm{I}}(u_h, z_h) \approx J(u - u_h)$ be given by (8.1). Then*

$$|\eta^{\mathrm{I}}(u_h, z_h)| \leq \eta^{\mathrm{II}}(u_h, z_h) \leq \sum_{K \in \mathscr{T}_h} \eta_K^{\mathrm{III}}(\mu_K, \sigma_K, \phi_K, p_K), \qquad (8.9)$$

where η^{II} is given by (8.3) and η_K^{III}, $K \in \mathscr{T}_h$ are computable by (8.5)–(8.8).

Remark 8.2 The estimator η_K^{III}, $K \in \mathscr{T}_h$ is the sum of residuals $R_{K,\star}$, $R_{K,\star}^*$ multiplied by the estimators of the weights $\theta_{K,\star}, \theta_{K,\star}^*$. Obviously, each of these terms depends on the geometry of element K. Due to the interpolation error estimates from Sect. 3.3, we are able to extract the information of the element geometry for the terms $\theta_{K,\star}, \theta_{K,\star}^*$, cf. (8.6). On the other hand, we do not know how to extract such information from the residuals $R_{K,\star}$, $R_{K,\star}^*$. Therefore, in the subsequent optimization process, we fix the residuals $R_{K,\star}$, $R_{K,\star}^*$ and optimize only the parameters coming from the terms $\theta_{K,\star}, \theta_{K,\star}^*$.

8.1.2 Goal-Oriented Optimization of the Anisotropy of Triangles

The aim is to employ estimate (8.9) to find the optimal anisotropy (shape and orientation) of element K. In particular, for given $\mu_K > 0$ and $p_K \in \mathbb{N}$, we seek $\sigma_K \geq 1$ and $\phi_K \in [0, \pi)$ such that η_K^{III} is minimal, i.e.,

$$\{\sigma_K^{\mathrm{new}}, \phi_K^{\mathrm{new}}\} = \underset{\sigma \geq 1, \phi \in [0,\pi)}{\arg\min} \ \eta_K^{\mathrm{III}}(\mu_K, \sigma, \phi, p_K), \qquad K \in \mathscr{T}_h, \qquad (8.10)$$

where η_K^{III} is given by (8.8).

Each $\theta_{K,\star}$ and $\theta_{K,\star}^*$ in (8.8) is proportional to function G given by (3.57) with varying arguments, cf. (8.6)–(8.8). In Sect. 5.1, we derived the analytical formulas for the minimization of a single function G. Unfortunately, we are not able to derive similar relations for the weighted sum of several functions G so we have to find the minimum numerically by an iterative process.

Lemma 3.20 states that function G is π-periodic and blowing up for $\sigma \to \infty$. Therefore, the same properties are valid also for η_K^{III}. Hence, the continuity of η_K^{III} with respect to σ and ϕ implies the existence of at least one minimum of (8.10).

The uniqueness of the minimum is not guaranteed. However, we performed thousands of numerical experiments for realistic problems and we did not observe existence of more than one local minimum. Obviously, an exception is the case when the minimum is attained for $\sigma = 1$, since $\eta_K^{\mathrm{III}}(\cdot, 1, \phi, \cdot) = \mathrm{const}$ for any $\phi \in [0, \pi]$.

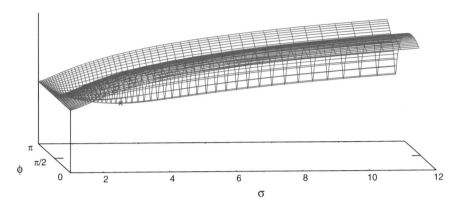

Fig. 8.1 Example of function $\eta_K^{I\!I\!I}(\cdot, \sigma, \phi, \cdot)$ from (8.8) (red graph) and its minimum (blue cross). The vertical (x_3) axis is in the logarithmic scale

However, in this case, the corresponding triangle is isotropic and its orientation ϕ is not relevant. A typical example of $\eta_K^{I\!I\!I}(\cdot, \sigma, \phi, \cdot)$ is shown in Fig. 8.1.

The minimum of (8.10) can be found by standard minimization methods, e.g., the Newton method when the derivatives are approximated by finite differences. However, we used a simple iterative process. We start with the value $\{\sigma_K, \phi_K\}$ corresponding to the anisotropy of $K \in \mathscr{T}_h^k$. With the chosen steps $\delta_\sigma > 1$ and $\delta_\phi \geq 0$, we test successively the values of $\eta_K^{I\!I\!I}(\cdot, \sigma, \phi \pm \delta_\phi, \cdot)$ and $\eta_K^{I\!I\!I}(\cdot, \sigma \delta_\sigma^{\pm 1}, \phi, \cdot)$. From these values, we choose the pair σ, ϕ giving the smallest value of $\eta_K^{I\!I\!I}$ and repeat the search. When we find σ, ϕ such that

$$\eta_K^{I\!I\!I}(\cdot, \sigma, \phi, \cdot) \leq \eta_K^{I\!I\!I}(\cdot, \sigma, \phi \pm \delta_\phi, \cdot) \quad \text{and} \quad \eta_K^{I\!I\!I}(\cdot, \sigma, \phi, \cdot) \leq \eta_K^{I\!I\!I}(\cdot, \sigma \delta_\sigma^{\pm 1}, \phi, \cdot),$$

we decrease the steps $\delta_\sigma := \sqrt{\delta_\sigma}$ and $\delta_\phi := \delta_\phi/2$ and repeat the searching until the required tolerances $\omega_\sigma > 0$ and $\omega_\phi > 0$ for the accuracy of σ and ϕ are achieved, respectively. The decrease of step δ_ϕ is done arithmetically since it represents an absolute tolerance for $\phi > 0$. On the other hand, the decrease of $\delta_\sigma > 1$ is done geometrically since the difference $\sigma - 1$ represents the relative tolerance for σ. In numerical examples presented in this monograph, we use the relative tolerance for σ equal to 1% ($\omega_\sigma = 1.01$) and the absolute tolerance for ϕ equal to $1°$ ($\omega_\phi = \pi/180$). For the initial steps, we use the values $\delta_\sigma = 1.5$ and $\delta_\phi = \pi/10$. This technique is illustrated by Fig. 8.2 and the exact algorithmization is given by Algorithm 8.1.

The presented algorithm is robust, accurate and its total computational costs is significantly smaller in comparison to the solution of linear algebraic systems and/or creating new meshes.

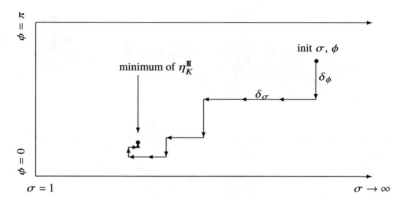

Fig. 8.2 Illustration of the iterative process seeking the minimum of $\eta_K^{I\!I\!I}$

8.1.3   Estimates Including the Geometry of Elements for 3D

For the sake of completeness, we extend the results from Sect. 8.1.1 to the three-dimensional case. Let u_h^+ and z_h^+ be the reconstructed functions from S_{hp+1}, cf. (7.50). Let $K \in \mathcal{T}_h$ be a tetrahedron with the geometry $\{\mu_K, \sigma_K, \varsigma_K, \phi_K\}$, cf. Definition 4.1 and p_K be the corresponding polynomial approximation degree. Similar to (8.5), applying Theorem 4.5 and Definition 4.7, there exist values defining the anisotropic bounds of $p_K + 1$-homogeneous functions $u_h^+ - \Pi u_h^+$, $z_h^+ - \Pi z_h^+$ and $2p_K$-homogeneous functions $|\nabla(u_h^+ - \Pi u_h^+)|^2$, $|\nabla(z_h^+ - \Pi z_h^+)|^2$, restricted to K, denoted as

$$\{A_u, \rho_u, \varrho_u, \varphi_u\} \dots \text{the anisotropic bound of } (u_h^+ - \Pi u_h^+)|_K, \tag{8.11}$$

$$\{A_z, \rho_z, \varrho_z, \varphi_z\} \dots \text{the anisotropic bound of } (z_h^+ - \Pi z_h^+)|_K,$$

$$\{A_{u'}, \rho_{u'}, \varrho_{u'}, \varphi_{u'}\} \dots \text{the anisotropic bound of } |\nabla(u_h^+ - \Pi u_h^+)|_K|^2,$$

$$\{A_{z'}, \rho_{z'}, \varrho_{z'}, \varphi_{z'}\} \dots \text{the anisotropic bound of } |\nabla(z_h^+ - \Pi z_h^+)|_K|^2,$$

which depend on $(p + 1)$th-derivatives of u_h^+ and z_h^+.

We estimate terms $\left\| z_h^+ - \Pi z_h^+ \right\|_{K,\star}$ and $\left\| u_h^+ - \Pi u_h^+ \right\|_{K,\star}$ by the results from Sect. 4.3 (particularly estimates (4.37), (4.38), (4.45)). Therefore, from (8.11), we obtain

$$\left\| u_h^+ - \Pi u_h^+ \right\|_K \leq \left(\frac{A_u^2 \mu_K^{2p_K+5}}{2p_K + 5} G(p_K + 1, p_K + 1, \rho_u, \varrho_u, \varphi_u; \sigma_K, \varsigma_K, \phi_K) \right)^{1/2}$$

$$\left\| u_h^+ - \Pi u_h^+ \right\|_{\partial K} \leq \left(A_u^2 \mu_K^{2p_K+3} \sigma_K \varsigma_K \, G(p_K + 1, p_K + 1, \rho_u, \varrho_u, \varphi_u; \sigma_K, \varsigma_K, \phi_K) \right)^{1/2},$$

$$\left\| \nabla(u_h^+ - \Pi u_h^+) \right\|_{\partial K} \leq \left(A_{u'} \mu_K^{2p_K+1} \sigma_K \varsigma_K \, G(p_K, 2p_K, \rho_{u'}, \varrho_{u'}, \varphi_{u'}; \sigma_K, \varsigma_K, \phi_K) \right)^{1/2},$$

Algorithm 8.1: Minimization of $\eta_K^{I\!I\!I}(\cdot, \sigma, \phi, \cdot)$, cf. (8.10)

1: set the required tolerances $\omega_\sigma > 0$ and $\omega_\phi > 0$
2: set initial steps $\delta_\sigma > 1$ and $\delta_\phi \geq 0$
3: set initial pair $\sigma := \sigma_K, \phi := \phi_K$
4: **repeat**
5:    **repeat**
6:       NSTEP = 0
7:       **loop**
8:          $g^0 \leftarrow \eta_K^{I\!I\!I}(\cdot, \sigma, \phi, \cdot)$,
9:          $g^+ \leftarrow \eta_K^{I\!I\!I}(\cdot, \sigma, \phi + \delta_\phi, \cdot)$,
10:         $g^- \leftarrow \eta_K^{I\!I\!I}(\cdot, \sigma, \phi - \delta_\phi, \cdot)$,
11:         **if** $g^0 \leq g^+$ and $g^0 \leq g^-$ **then**
12:            exit loop
13:         **else if** $g^0 > g^+$ and $g^- > g^+$ **then**
14:            $\phi \leftarrow \phi + \delta_\phi$; NSTEP = NSTEP+1
15:         **else**
16:            $\phi \leftarrow \phi - \delta_\phi$; NSTEP = NSTEP+1
17:         **end if**
18:       **end loop**
19:       **loop**
20:          $g^0 \leftarrow \eta_K^{I\!I\!I}(\cdot, \sigma, \phi, \cdot)$,
21:          $g^+ \leftarrow \eta_K^{I\!I\!I}(\cdot, \sigma \delta_\sigma, \phi, \cdot)$,
22:          $g^- \leftarrow \eta_K^{I\!I\!I}(\cdot, \sigma / \delta_\sigma, \phi, \cdot)$,
23:         **if** $g^0 \leq g^+$ and $g^0 \leq g^-$ **then**
24:            exit loop
25:         **else if** $g^0 > g^+$ and $g^- > g^+$ **then**
26:            $\sigma \leftarrow \sigma \cdot \delta_\sigma$; NSTEP = NSTEP+1
27:         **else**
28:            $\sigma \leftarrow \sigma / \delta_\sigma$; NSTEP = NSTEP+1
29:         **end if**
30:       **end loop**
31:    **until** NSTEP > 0
32:    $\delta_\phi \leftarrow \delta_\phi/2; \delta_\sigma \leftarrow \sqrt{\delta_\sigma}$
33: **until** $\delta_\sigma > \omega_\sigma$ or $\delta\phi > \omega_\phi$
34: set $\sigma_K^{\text{new}} \leftarrow \sigma, \phi_K^{\text{new}} \leftarrow \phi$

$$\left\| z_h^+ - \Pi z_h^+ \right\|_K \leq \left(\frac{A_z^2 \mu_K^{2p_K+5}}{2p_K + 5} \, G(p_K + 1, p_K + 1, \rho_z, \varrho_z, \boldsymbol{\varphi}_z; \sigma_K, \varsigma_K, \boldsymbol{\phi}_K) \right)^{1/2},$$

$$\left\| z_h^+ - \Pi z_h^+ \right\|_{\partial K} \leq \left(A_z^2 \mu_K^{2p_K+3} \sigma_K \varsigma_K \, G(p_K + 1, p_K + 1, \rho_z, \varrho_z, \boldsymbol{\varphi}_z; \sigma_K, \varsigma_K, \boldsymbol{\phi}_K) \right)^{1/2},$$

$$\left\| \nabla(z_h^+ - \Pi z_h^+) \right\|_{\partial K} \leq \left(A_{z'} \mu_K^{2p_K+1} \sigma_K \varsigma_K \, G(p_K, 2p_K, \rho_{z'}, \varrho_{z'}, \boldsymbol{\varphi}_{z'}; \sigma_K, \varsigma_K, \boldsymbol{\phi}_K) \right)^{1/2},$$

$$(8.12)$$

where function G is given by (4.29). From first to last, we denote the terms on the right-hand sides of (8.12) by $\theta_{K,\mathrm{V}}$, $\theta_{K,\mathrm{B}}$, $\theta_{K,\mathrm{D}}$, $\theta_{K,\mathrm{V}}^*$, $\theta_{K,\mathrm{B}}^*$, $\theta_{K,\mathrm{D}}^*$. (Compare with the 2D case.)

Applying (8.12) on the terms appearing in (8.3), we obtain

$$\eta_K^{\mathrm{II}} \leq \eta_K^{\mathrm{III}}(\mu_K, \sigma_K, \varsigma_K, \boldsymbol{\phi}_K, p_K), \quad K \in \mathscr{T}_h, \tag{8.13}$$

where

$$\eta_K^{\mathrm{III}}(\mu_K, \sigma_K, \varsigma_K, \boldsymbol{\phi}_K, p_K) := \mathscr{N}_K + \frac{1}{2}\sum_\star \left(R_{K,\star}\theta_{K,\star} + R_{K,\star}^*\theta_{K,\star}^* \right), \tag{8.14}$$

where $\theta_{K,\star} \in \{\theta_{K,\mathrm{V}}, \theta_{K,\mathrm{B}}, \theta_{K,\mathrm{D}}\}$, $\theta_{K,\star}^* \in \{\theta_{K,\mathrm{V}}^*, \theta_{K,\mathrm{B}}^*, \theta_{K,\mathrm{D}}^*\}$, cf. (8.12). This determination leads to the final *goal-oriented error estimate of type III*.

Theorem 8.3 *Let u_h and z_h be the approximate solutions of the primal and adjoint problems given by (7.4) and (7.11), respectively. Furthermore, let u_h^+ and z_h^+ be the higher-order approximations of the primal and adjoint solutions reconstructed from u_h and z_h, respectively. Finally, let $\eta^{\mathrm{I}}(u_h, z_h) \approx J(u - u_h)$ be given by (8.1). Then*

$$|\eta^{\mathrm{I}}(u_h, z_h)| \leq \eta^{\mathrm{II}}(u_h, z_h) \leq \sum_{K \in \mathscr{T}_h} \eta_K^{\mathrm{III}}(\mu_K, \sigma_K, \varsigma_K, \boldsymbol{\phi}_K, p_K), \tag{8.15}$$

where η^{II} is given by (8.3) and η_K^{III}, $K \in \mathscr{T}_h$ are computable by (8.11)–(8.14).

8.1.4 Goal-Oriented Optimization of the Anisotropy of Tetrahedra

The aim is to employ estimate (8.15) to find the optimal anisotropy (shape and orientation) of element K. Particularly, for given $\mu_K > 0$ and $p_K \in \mathbb{N}$, we seek $\sigma_K \geq 1$, $\varsigma_K \geq 1$ and $\boldsymbol{\phi}_K \in \mathscr{U}$ such that η_K^{III} is minimal, i.e.,

$$\{\sigma_K^{\mathrm{new}}, \varsigma_K^{\mathrm{new}}, \boldsymbol{\phi}_K^{\mathrm{new}}\} = \underset{\sigma \geq 1, \varsigma \geq 1, \boldsymbol{\phi} \in \mathscr{U}}{\arg\min} \; \eta_K^{\mathrm{III}}(\mu_K, \sigma, \varsigma, \boldsymbol{\phi}, p_K), \quad K \in \mathscr{T}_h, \tag{8.16}$$

where η_K^{III} is given by (8.14) and $\mathscr{U}$ is given by (4.3).

The minimization problem (8.16) is more difficult to solve in comparison to the 2D case since the number of unknown parameters is five. We expect that the minimum of (8.16) can be found by known minimization methods. In principle, the 3D generalization of Algorithm 8.1 is possible, but it is technically difficult and potentially much more expensive.

8.2   Goal-Oriented Anisotropic hp-Mesh Adaptive Algorithm

We develop the goal-oriented anisotropic hp-mesh adaptive algorithm which generates a sequence of simplicial meshes and polynomial approximation degrees such that the corresponding error estimate η^{I} of the target functional is under the given tolerance.

Theoretical results of the hp-methods initiated in [10] and further developed for the goal-oriented estimates, e.g., in [68, Corollary 4.8], lead to the expectation that the error estimator converges *exponentially* with respect to DoF in the form (for 2D as well as 3D)

$$|\eta^{\mathrm{I}}| \approx C \exp\left(-b\,\mathrm{DoF}^{1/3}\right), \tag{8.17}$$

where $C > 0$ and $b > 0$ are constants independent of DoF.

Therefore, in order to compare the efficiency of the proposed techniques, we investigate the decay of the error estimator $|\eta^{\mathrm{I}}|$ with respect to $\mathrm{DoF}^{1/3}$ in the logarithmic-linear scale since the exponential rate corresponds to a straight line in figures. Moreover, we plot also the decay of $|\eta^{\mathrm{I}}|$ with respect to the computational time in seconds. We note that the speed of computation depends on many factors, hence the measurements of the convergence with respect to the computational time have only an informative character.

We denote by $\mathscr{T}_{hp} := \{\mathscr{T}_h, \boldsymbol{p}\}$ the hp-*mesh*, which consists of the simplicial mesh $\mathscr{T}_h$ and the set of the corresponding polynomial approximation degrees $\boldsymbol{p} := \{p_K,\ K \in \mathscr{T}_h\}$, cf. Definition 6.1. Obviously, for each $\mathscr{T}_{hp}$, there exists a uniquely defined space of piecewise polynomial functions S_{hp}, given by (7.50).

The mesh adaptive algorithm generates successively a sequence of hp-meshes $\mathscr{T}_{hp}^k$, $k = 0, 1, \dots$ together with the corresponding spaces S_{hp}^k, $k = 0, 1, \dots$, cf. Algorithm 1.1. In particular, we employ the knowledge of the approximate solutions $u_h^k \in S_{hp}^k$ and $z_h^k \in S_{hp}^k$ of the primal and adjoint problems, respectively, for the construction of the next hp-mesh $\mathscr{T}_{hp}^{k+1}$. The initial hp-mesh $\mathscr{T}_{hp}^0$ is usually chosen as a uniform mesh with a constant (and low) polynomial approximation degree.

Let $\omega > 0$ be the prescribed tolerance for the error of the quantity of interest, then the mesh adaptive process is stopped if the condition

$$|J(u) - J(u_h^k)| \approx |\eta^{\mathrm{I}}(u_h^k, z_h^k)| \le \omega \tag{8.18}$$

is achieved for some k. If the stopping criterion (8.18) is not fulfilled, we have to construct the new hp-mesh $\mathscr{T}_{hp}^{k+1}$. In particular, for each $K \in \mathscr{T}_{hp}^k$ we have to define a new ("optimal") size $\mu_K^\star$, a new shape and orientation, $\{\sigma_K^\star, \phi_K^\star\}$ (for 2D) or $\{\sigma_K^\star, \varsigma_K^\star, \boldsymbol{\phi}_K^\star\}$ (for 3D), and the new polynomial degree $p_K^\star$. Then, having these values for each $K \in \mathscr{T}_h$, we define the metric field $\mathcal{M}$ and polynomial distribution function $\mathcal{P}$ and construct the new hp-mesh $\mathscr{T}_{hp}^{k+1}$ (similar to Algorithm 6.1). The whole mesh optimization process for 2D is shown in Algorithm 8.2. The steps 11–13 are described in details in the following sections. The 3D variant can be formulated with minor modifications.

Algorithm 8.2: Goal-oriented anisotropic mesh adaptation (2D)

1: set the desired tolerance $\omega > 0$, $k \leftarrow 0$
2: set the initial coarse mesh $\mathcal{T}_{hp}^0$ and the corresponding space S_{hp}^0, cf. (7.50)
3: **for** $k = 0, 1, \ldots$ **do**
4:     evaluate $u_h^k \in S_{hp}^k$ and $z_h^k \in S_{hp}^k$ by solving (7.26) and (7.38) on $\mathcal{T}_{hp}^k$, respectively
5:     compute η_K^I, $K \in \mathcal{T}_h^k$ by (8.2)
6:     set $\eta_k^I := \eta^I(u_h^k, z_h^k) = \sum_{K \in \mathcal{T}_h} \eta_K^I$, cf. (8.1)
7:     **if** $|\eta_k^I| \leq \omega$ **then**
8:         STOP the computational process
9:     **else**
10:         **for** all $K \in \mathcal{T}_h^k$ **do**
11:             set the new element size $\mu_K^\star$ using η_K^I, cf. Section 8.2.1
12:             set the new shape and orientation $\{\sigma_K^\star, \phi_K^\star\}$, $K \in \mathcal{T}_h^k$, cf. Section 8.2.2
13:             set the new polynomial degree $p_K^\star$, cf. Section 8.2.2
14:             construct new local metric $\mathbb{M}_K^\star = \mathcal{M}(x_K)$ using (2.54)
15:         **end for**
16:         construct continuous hp-mesh $\{\mathcal{M}, \mathcal{P}\}$ from$(\mathbb{M}_K^\star, p_K)_{K \in \mathcal{T}_h}$, cf. Section 2.5.2
17:         construct new hp-mesh $\{\mathcal{T}_h^{k+1}, p^{k+1}\}$ from $\{\mathcal{M}, \mathcal{P}\}$
18:     **end if**
19: **end for**

8.2.1   Setting of Element Size

We employ the ideas developed in Chaps. 5 and 6, namely we equilibrate the error estimator, i.e., $|\eta_K^I| \approx \text{const} \ \forall K \in \mathcal{T}_h^k$. Similar to Sect. 5.6.2.2, we define a strictly decreasing sequence of tolerances for mesh adaptation ω_k, $k = 0, 1, 2, \ldots$ and prescribe the criterion for the local error estimators

$$|\eta_K^I| \approx \frac{\omega_k}{\#\mathcal{T}_h^k}, \qquad k = 0, 1, 2 \ldots, \tag{8.19}$$

where $\#\mathcal{T}_h^k$ denotes the number of elements of $\mathcal{T}_h^k$. We note that the right-hand side of (8.19) does not dependent on K. A simple natural choice ω_k, $k = 0, 1, \ldots$ is

$$\omega_k = \begin{cases} \zeta |\eta^I(u_h^0, z_h^0)| & \text{for } k = 0 \\ \zeta \omega_{k-1} & \text{for } k > 0, \end{cases} \tag{8.20}$$

where $\eta^I(u_h^0, z_h^0)$ is the error estimator on the initial hp-mesh $\mathcal{T}_{hp}^0$ and $\zeta \in (0, 1)$ is a suitably chosen factor. Too small ζ leads to too high number of elements for low

k and on the other hand, choosing ζ close to 1 requires too many mesh adaptation levels to achieve (8.18). Typically, $\zeta = \frac{1}{2}$ or $\zeta = \frac{3}{4}$ are good choices.

Let $K \in \mathscr{T}_h^k$ be a triangle having the size μ_K and set $\bar{\omega}_k := \frac{\omega_k}{\#\mathscr{T}_h^k}$. Then, in virtue of (8.19), if $|\eta_K^l| > \bar{\omega}_k$ we have to decrease the size of K and if $|\eta_K^l| < \bar{\omega}_k$ it is possible to increase the size of K. Hence we set (cf. (5.143))

$$\mu_K^\star = \alpha_K \mu_K, \qquad K \in \mathscr{T}_h^k, \tag{8.21}$$

where α_K is a positive parameter which has to be chosen based on the values η_K^l and $\bar{\omega}_k$.

In order to choose α_K, we adopt and test numerically the techniques introduced in Sect. 5.6.2.2.

- rate—Similar to (5.146), We assume that the error estimator depends on the element size as $|\eta_K^l| \approx O(\mu_K^{\beta_K})$, where $\beta_K > 0$ is the rate of convergence. Then we set

$$\alpha_K = \left(\frac{\bar{\omega}_k}{|\eta_K^l|}\right)^{1/\beta_K}, \qquad K \in \mathscr{T}_h^k. \tag{8.22}$$

However, having knowledge of β_K is questionable. For sufficiently regular solutions, results from [68, Theorem 4.5] lead to $\beta_K = 2p_K$, $K \in \mathscr{T}_h^k$. Although this technique was employed in [48] for the h-variant ($p_K = \mathrm{const}\ \forall K \in \mathscr{T}_h^k$), its application to hp-method is not advisable. This follows from the fact that α_K, as a function of η_K^l, given by (8.22), is changing rapidly for $|\eta_K^l|$ close to $\bar{\omega}_k$, see Fig. 8.3. This leads to frequent repetition of refinement and coarsening which increase the number of mesh adaptation levels. This effect is illustrated in Fig. 8.4 which shows four consecutive mesh adaptation levels of the forthcoming

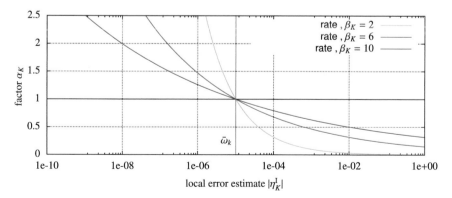

Fig. 8.3 The dependence of the factor α_K on $|\eta_K^l|$ for the rate technique (relation (8.22)) with $\bar{\omega}_k = 10^{-5}$ and the rates of convergence $\beta_K = 2$, $\beta_K = 6$, and $\beta_K = 10$

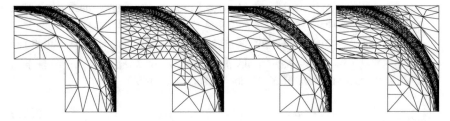

Fig. 8.4 Illustration of the refinement/coarsening of the rate technique close to the interior angle (red-border region) for the mesh adaptation levels $k = 11, 12, 13, 14$ (from the left to the right)

example (8.38)–(8.40) (with target functional J_B). We observe a repetition of the refinement and coarsening in the vicinity of the interior angle.

- **e-elem***—Following (5.147) and (5.148), we employ a relation such that α_K is changing only slowly for $|\eta_K^I| \approx \bar{\omega}_k$. Hence, we define the maximal and minimal value of the estimator η_K^I, $K \in \mathscr{T}_h$ by

$$\eta_{\max}^I = \max_{K \in \mathscr{T}_h^k} |\eta_K^I|, \quad \eta_{\min}^I = \min_{K \in \mathscr{T}_h^k} |\eta_K^I|, \tag{8.23}$$

and two additional parameters, the *maximal refinement factor* $r_{\max} \in (0, 1)$ and the *maximal coarsening factor* $c_{\max} > 1$. Then we set

$$\alpha_K = \begin{cases} 1 + (r_{\max} - 1)\xi_K^2, \ \xi_K := \dfrac{\log(|\eta_K^I|) - \log(\bar{\omega}_k)}{\log(\eta_{\max}^I) - \log(\bar{\omega}_k)} & \text{for } |\eta_K^I| \geq \bar{\omega}_k, \\[3mm] 1 + (c_{\max} - 1)\xi_K^2, \ \xi_K := \dfrac{\log(|\eta_K^I|) - \log(\bar{\omega}_k)}{\log(\eta_{\min}^I) - \log(\bar{\omega}_k)} & \text{for } |\eta_K^I| < \bar{\omega}_k. \end{cases} \tag{8.24}$$

Figure 8.5 shows an example of function $\alpha_K = \alpha_K(|\eta_K^I|)$. Obviously, α_K given by (8.24) is almost equal to 1 in vicinity of $|\eta_K^I| \approx \bar{\omega}_k$ and there this technique does not behave as erroneously as the rate technique.

- **e-elem**—The efficiency of the whole algorithm can be still slightly improved by a modification of (8.24) in such a way that α_K is close to a constant in the vicinity of the limit values $\eta_{\min}^I$ and $\eta_{\max}^I$, cf. (5.149). Namely, we set

$$\alpha_K = \begin{cases} \frac{1}{2}(1 - r_{\max})(\cos(\pi\xi_K) + 1) + r_{\max}, \\ \quad\quad \text{with } \xi_K := \dfrac{\log(|\eta_K^I|) - \log(\bar{\omega}_k)}{\log(\eta_{\max}^I) - \log(\bar{\omega}_k)} \ \text{ for } |\eta_K^I| \geq \bar{\omega}_k, \\[3mm] \frac{1}{2}(c_{\max} - 1)(\cos(\pi(\xi_K + 1)) + 1) + 1, \\ \quad\quad \text{with } \xi_K := \dfrac{\log(|\eta_K^I|) - \log(\bar{\omega}_k)}{\log(\eta_{\min}^I) - \log(\bar{\omega}_k)} \ \text{ for } |\eta_K^I| < \bar{\omega}_k. \end{cases} \tag{8.25}$$

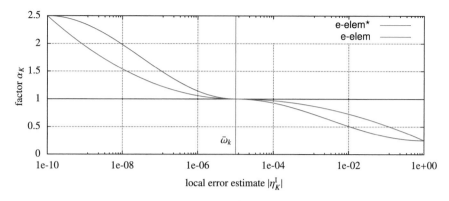

Fig. 8.5 The dependence of the factor α_K on $|\eta_K^l|$ for the e-elem* and e-elem techniques (relations (8.24) and (8.25)) with $\eta_{max}^l = 1$, $\eta_{min}^l = 10^{-10}$, $\bar{\omega}_k = 10^{-5}$, $r_{max} = 0.25$, and $c_{max} = 2.5$

Figure 8.5 compares the e-elem* and e-elem techniques (relations (8.24) and (8.25)).

For illustration, we present a comparison of techniques rate, e-elem*, and e-elem for example (8.38)–(8.40) (with target functional J_B). Figure 8.6 compares the convergences of the error estimator $|\eta_K^l|$ and the error $|J(u) - J(u_h)|$ with respect to DoF$^{1/3}$ and the computational time in seconds. We observe that rate is not very efficient due to the aforementioned reasons. Moreover, the e-elem technique is slightly better than e-elem*.

On the other hand, the disadvantage of e-elem* and e-elem techniques is the necessity of the choice of the parameters $r_{max} \in (0, 1)$, $c_{max} > 1$ in (8.24) and (8.25), respectively. Numerical experiments indicate that the performance of Algorithm 8.2 is not sensitive to the choice of these parameters. We demonstrate this property by Figs. 8.7 and 8.8, where we plot the convergence of the error estimator $|\eta^l|$ with respect to DoF$^{1/3}$ and the computational time for the example (8.33) from Sect. 8.3.1 using the e-elem technique. Figure 8.7 shows the convergence with the fixed coarsening parameter $c_{max} = 2.5$ and varying refinement parameter $r_{max} \in \{0.5, 0.25, 0.15, 0.1, 0.05, 0.03, 0.01\}$. The largest tested values of r_{max} give the slowest rate of convergence (but still exponential). On the other hand there are almost negligible differences for $r_{max} \in \{0.1, 0.05, 0.03, 0.01\}$. Therefore, the sensitivity of the e-elem technique with respect to the value of r_{max} is low for the choice of $r_{max} \lesssim 0.1$.

Figure 8.8 shows the convergence for fixed $r_{max} = 0.1$ and varying coarsening parameter $c_{max} \in \{10, 5, 2.5, 1.5\}$. The differences among all tested values of the parameter are very small which demonstrates the insensitivity of e-elem techniques with respect to the choice of c_{max}. Based on these numerical experiments, we set $r_{max} = 0.1$ and $c_{max} = 2.5$ in the forthcoming computations.

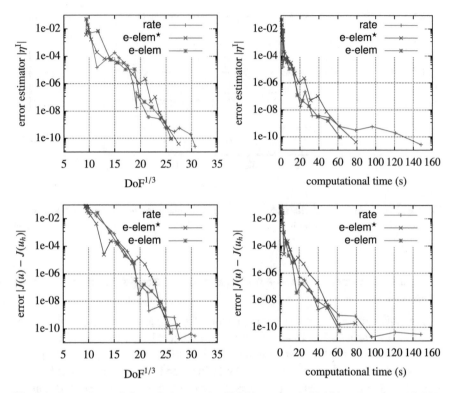

Fig. 8.6 Comparison of the techniques rate ((8.22)), e-elem* ((8.24)) and e-elem ((8.25)), convergences of the error estimator $|\eta^{\mathrm{I}}|$ (top) and the error $|J(u) - J(u_h)|$ (bottom) with respect to $\mathrm{DoF}^{1/3}$ (left) and the computational time in seconds (right)

8.2.2   Setting of Polynomial Approximation Degree and Element Shape

Whereas the new (optimal) size of element $\mu_K^\star$ has been set by the techniques from Sect. 8.2.1 employing the error estimate of type I (η_K^{I}), the setting of the optimal polynomial approximation degree and optimal element shape requires the use of the error estimate of type II (η_K^{II}), cf. (8.2) and (8.3). However, these settings are done locally as in Chaps. 5 and 6.

Let $K \in \mathscr{T}_h^k$ be a triangle with the corresponding polynomial approximation degree p_K. The idea of the optimization of the polynomial approximation degree is testing the possible increase or decrease of p_K by one. It means that for each $K \in \mathscr{T}_h^k$, we choose among the candidates $p_K - 1$, p_K, and $p_K + 1$ the polynomial degree (with the optimal element shapes) which gives the smallest value of the estimator η_K^{II} (with fixed residuals), see (8.3). However, in order to have a fair comparison, we need to keep the *density of the number of degrees of freedom* (= DoF per unit area/volume).

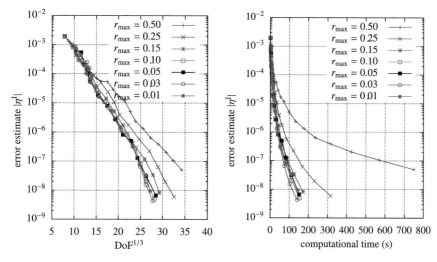

Fig. 8.7 e-elem technique, convergence of the error estimate $|\eta^I|$ with respect to DoF (left) and the computational time (right) with $c_{max} = 2.5$ and $r_{max} \in \{0.5, 0.25, 0.15, 0.1, 0.05, 0.03, 0.01\}$

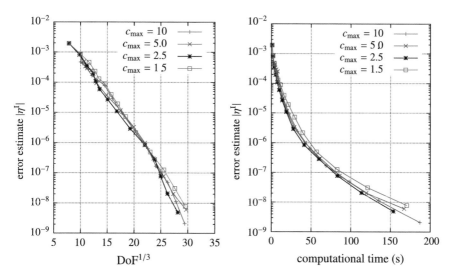

Fig. 8.8 e-elem technique, convergence of the error estimate $|\eta^I|$ with respect to DoF (left) and the computational time (right) with $r_{max} = 0.1$ and $c_{max} \in \{10, 5, 2.5, 1.5\}$

Remark 8.4 In Sect. 6.1, we introduced the density of degrees of freedom as the product $\gamma(x)\tau(x)$, where $\tau(x)$ is the mesh density and $\gamma(x)$ is the local number of degrees of freedom. Here, we use an alternative definition $\gamma(x)\mu_K^{-d}$ which is equivalent up to the multiplicative constant due to (5.87).

Let $K \in \mathcal{T}_h^k$ and $\mu_K^\star$ be the optimal size given by the techniques from Sect. 8.2.1. We recall that the area/volume of K is proportional to $(\mu_K)^d$, $d = 2, 3$. Therefore, in virtue of (7.51), for 2D case, we define the values

$$\mu_K^{(p_K-1)} := \mu_K^\star \left(\frac{p_K(p_K+1)}{(p_K+1)(p_K+2)} \right)^{1/2}, \tag{8.26}$$

$$\mu_K^{(p_K)} := \mu_K^\star,$$

$$\mu_K^{(p_K+1)} := \mu_K^\star \left(\frac{(p_K+2)(p_K+3)}{(p_K+1)(p_K+2)} \right)^{1/2}.$$

Similarly, for the 3D case, these values take the form

$$\mu_K^{(p_K-1)} := \mu_K^\star \left(\frac{p_K(p_K+1)(p_K+2)}{(p_K+1)(p_K+2)(p_K+3)} \right)^{1/3}, \tag{8.27}$$

$$\mu_K^{(p_K)} := \mu_K^\star,$$

$$\mu_K^{(p_K+1)} := \mu_K^\star \left(\frac{(p_K+2)(p_K+3)(p_K+4)}{(p_K+1)(p_K+2)(p_K+3)} \right)^{1/3}.$$

For each of the three candidates $p_K - 1$, p_K, and $p_K + 1$, we seek the corresponding optimal shape using the technique from Sect. 8.1.2 for 2D and the technique from Sect. 8.1.4 for 3D. Therefore, for the 2D case, we set using (8.10)

$$\{\sigma_K^{(p_K+i)}, \phi_K^{(p_K+i)}\} = \operatorname*{arg\,min}_{\sigma \geq 1, \phi \in [0,\pi)} \eta_K^{\text{III}}(\mu_K^{(p_K+i)}, \sigma, \phi, p_K + i), \quad i \in \{-1, 0, 1\} \tag{8.28}$$

and from three candidates $p_K + i$, $i \in \{-1, 0, 1\}$ we choose the one which has the minimal value of η_K^{III}, i.e.,

$$\{\sigma_K^\star, \phi_K^\star, p_K^\star\} = \operatorname*{arg\,min}_{i \in \{-1,0,1\}} \eta_K^{\text{III}}(\mu_K^{(p_K+i)}, \sigma_K^{(p_K+i)}, \phi_K^{(p_K+i)}, p_K + i). \tag{8.29}$$

The 3D version of (8.28) and (8.29) follows from (8.16) and reads

$$\{\sigma_K^{(p_K+i)}, \varsigma_K^{(p_K+i)}, \boldsymbol{\phi}_K^{(p_K+i)}\} = \operatorname*{arg\,min}_{\sigma \geq 1, \varsigma \geq 1, \phi_k \in \phi} \eta_K^{\text{III}}(\mu_K^{(p_K+i)}, \sigma, \varsigma, \boldsymbol{\phi}, p_K + i), \tag{8.30}$$

for $i \in \{-1, 0, 1\}$ and

$$\{\sigma_K^\star, \varsigma_K^\star, \phi_K^\star, p_K^\star\} = \underset{i \in \{-1,0,1\}}{\arg\min} \; \eta_K^{\mathrm{III}}(\mu_K^{(p_K+i)}, \sigma_K^{(p_K+i)}, \varsigma_K^{(p_K+i)}, \phi_K^{(p_K+i)}, p_K + i). \tag{8.31}$$

Finally, we set

$$\mu_K^\star = \mu_K^{(p_K+j)}, \tag{8.32}$$

where $j \in \{-1, 0, 1\}$ for which the minimum in (8.29) or (8.31) is achieved.

8.3   Numerical Experiments

We present several test examples which demonstrate the efficiency, accuracy, and robustness of the anisotropic hp-mesh adaptation Algorithm 8.2 with all the settings presented in Sect. 8.2. We solve the following variants of the linear problem (7.43).

(i) Elliptic problem with a constant diffusion on a "cross" domain with interior corner singularities, cf. Sect. 8.3.1,
(ii) Elliptic problem with vanishing viscosity degenerated to a hyperbolic equation, cf. Sect. 8.3.2,
(iii) Convection-dominated problem with three variants of the target functionals, cf. Sect. 8.3.3.

Additional examples are given in Sect. 10.1. For each case, we compare the following mesh adaptation methods.

(a) hp-AMA: full hp-anisotropic mesh adaptation Algorithm 8.2 with the e-elem technique, cf. (8.25).
(b) h-AMA: h-anisotropic mesh adaptation, we use Algorithm 8.2 and fix the polynomial approximation degrees $p_K = 3$ for all $K \in \mathscr{T}_h^k, k = 0, 1, 2, \dots$.
(c) hp-ISO: hp-isotropic mesh adaptation, which is the generalization of technique (A2) from Sect. 1.4.2, where we mark a fixed ratio of elements having the highest $|\eta_K^1|$ and the marked elements are either split into 4 similar subtriangles or the corresponding polynomial degree is increased by 1. We employ the decision criterion between h- or p-refinement from [44] where we adopted the technique from [71]. Following the numerical study in [16], we select 20% of elements for the refinement at each adaptation level.
(d) hp-ideal: for the purely elliptic problem, we combined the hp-ISO approach with a priori knowledge of the solution regularity. This technique is used only for the purely elliptic problem 8.3 in such a way that if a marked mesh element

touches a corner singularity, h-refinement is applied. Otherwise, p-adaptation is employed.

8.3.1 Elliptic Problem on a "Cross" Domain

This problem was proposed in [3, Example 2] as a benchmark for the goal-oriented mesh adaptation. The solution does not contain anisotropic features but the presence of corner singularities makes this problem as a challenging test example for hp-adaptation.

Let $\Omega = (-2, 2) \times (-1, 1) \cup (-1, 1) \times (-2, 2)$ be a "cross" domain with boundary Γ. We consider the Poisson problem

$$-\Delta u = 1 \quad \text{in } \Omega, \qquad u = 0 \quad \text{on } \Gamma. \tag{8.33}$$

The target functional is the mean value of u over the square subdomain $\Omega_J = [1.2, 1.4] \times [0.2, 0.4]$, i.e.,

$$J(u) = \frac{1}{|\Omega_J|} \int_{\Omega_J} u(x) \, dx. \tag{8.34}$$

Figure 8.9 shows the computational domain with Ω_J and the initial mesh as well as the corresponding primal and adjoint solutions. As the exact value of $J(u)$ we take the reference value 0.407617863684 from [3], which was computed numerically on an adaptively refined mesh with more than 15 million triangles.

Figure 8.10 shows the decrease of the error estimator $|\eta^1|$ with respect to $\text{DoF}^{1/3}$ and the computational time in seconds, respectively, for all tested adaptive methods. We observe that all hp-variants give an exponential rate of convergence and they are superior to the h-variant. Of interest is the comparison between the hp-AMA and the hp-ideal variants. The latter achieves the tolerance with smaller number of DoF (cf.

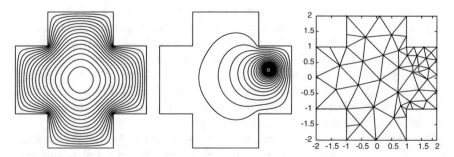

Fig. 8.9 Elliptic problem on a "cross" domain (8.33) and (8.34), the primal (left) and adjoint (center) solutions, domain Ω with the domain of interest Ω_J highlighted by red-bold and the initial triangular grid (right)

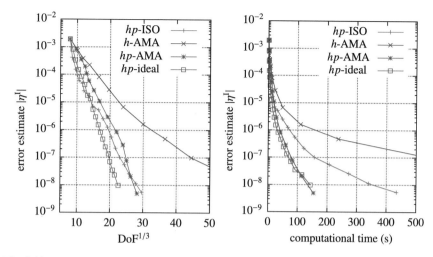

Fig. 8.10 Elliptic problem on a "cross" domain (8.33) and (8.34), convergence of the error estimator $|\eta^{\mathrm{I}}|$ with respect to $\mathrm{DoF}^{1/3}$ (left) and the computational time (right) for all tested adapted strategies

Fig. 8.10, left) but the computational time is practically the same for both techniques (cf. Fig. 8.10, right). The explanation is that the hp-ideal variant requires 18 levels of mesh adaptations to achieve the error tolerance whereas hp-AMA needs only 12 levels. This follows from the observation that the hp-ideal variant admits only a halving of local mesh size in one level of mesh adaptation but the hp-AMA method allows stronger refinement in one adaptation level, as controlled by the parameter $r_{\max}$ in (8.25). The same argument explains why the hp-ISO method requires more than double computational time in comparison to the hp-AMA method, whereas the rate of the convergence with respect to DoF is very similar.

Furthermore, Fig. 8.11, left shows the decay of the error estimators $|\eta^{\mathrm{I}}|$, η^{II} and the error $|J(u - u_h)|$ for the hp-AMA method. We observe that the error is underestimated (cf. Remark 7.15) but both estimates give a reasonable approximation of the error. Figure 8.11, right shows the histogram of the distribution of the element error estimator η_K^{I}, $K \in \mathscr{T}_h^k$ on the last adaptation level. We observe the Gaussian distribution corresponding to the error equidistribution, cf. (8.19).

Finally, Fig. 8.12 shows the final hp-mesh achieved at the last level of mesh adaptation. We observe high polynomial degrees in the major part of Ω. Only in the vicinity of the corner singularities a strong h-refinement with a low polynomial degree is presented, see the detail of the mesh. Therefore, the setting of appropriate polynomial approximation degree from Sect. 8.2.2 is able to identify the singularities of the solution.

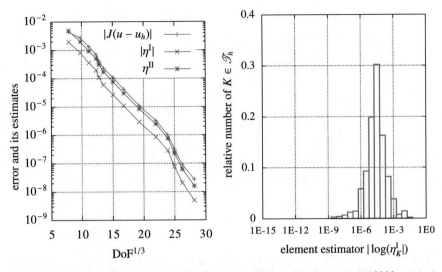

Fig. 8.11 Elliptic problem on a "cross" domain (8.33) and (8.34), the hp-**AMA** method, convergence of estimators $|\eta^{\mathrm{I}}|$, η^{II} and error $|J(u - u_h)|$ (left) and the histogram of the distribution of η_K^{I}, $K \in \mathscr{T}_h^k$ on the last adaptation level

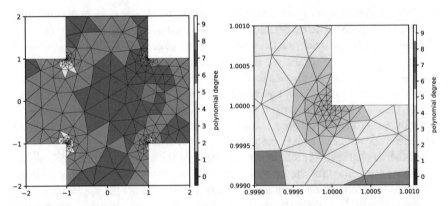

Fig. 8.12 Elliptic problem on a "cross" domain (8.33) and (8.34), the final hp-mesh obtained by the hp-**AMA** method, total view (left) and 1000x zoom near the singularity at $x = (1, 1)$ (right)

8.3.2 Mixed Hyperbolic-Elliptic Problem

This problem comes from [76, Example 2] and it was treated in Sect. 6.4.4 in the context of the interpolation error estimates. We consider the problem

$$-\varepsilon \Delta u + \nabla \cdot (\boldsymbol{b}\, u) + c u = 0 \qquad \text{in } \Omega := (0, 1)^2, \tag{8.35}$$

with varying diffusion $\varepsilon = \frac{\delta}{2}(1 - \tanh((r^2 - \frac{1}{16})/\gamma))$, where $r = ((x_1 - 0.5)^2 + (x_2 - 0.5)^2)^{1/2}$ and $\delta > 0$, $\gamma > 0$ are constants. The diffusion ε is close to δ in the interior of the circle $R = \{(x_1, x_2); (x_1 - 0.5)^2 + (x_2 - 0.5)^2 = 1/16\}$ having the center at $[1/2, 1/2]$ and diameter $1/4$. Outside of this circle, diffusion ε is quickly vanishing, more precisely $\varepsilon \approx 10^{-16}$ at the boundary $\Gamma = \partial\Omega$. We use the values $\delta = 0.01$ and $\gamma = 0.05$.

Furthermore, the convective field is $b = (2x_2^2 - 4x_1 + 1, x_2 + 1)^\mathsf{T}$ and the reaction $c = -\nabla \cdot b = 3$. The characteristics associated with the convective field enter the domain Ω through $\Gamma_\mathrm{D} = \{(x_1, x_2) \in \Gamma : x_1 = 0 \text{ or } x_1 = 1 \text{ or } x_2 = 0\}$, where we prescribe the Dirichlet boundary conditions

$$u = \begin{cases} 1 & \text{if } x_1 = 0 \text{ and } 0 < x_2 \leq 1 \\ \sin^2(\pi x) & \text{if } 0 \leq x_1 \leq 1 \text{ and } x_2 = 0 \\ e^{-50x_2^4} & \text{if } x_1 = 1 \text{ and } 0 < x_2 \leq 1 \end{cases} \quad . \tag{8.36}$$

The homogeneous Neumann boundary condition is prescribed on $\Gamma_\mathrm{N} := \Gamma \setminus \Gamma_\mathrm{D}$.

Finally, we consider the target functional

$$J(u) = \int_{0.25}^{0.625} u(x_1, 1) \, \mathrm{d}x_1. \tag{8.37}$$

The reference value, corresponding to the exact solution, is $J(u) = 0.3240267695611$ and it was obtained by a strong hp-refinement.

This primal solution has a middle anisotropic feature around the circle R. Moreover, the adjoint solution has two line singularities starting at $(x_1, x_2) = (0.25, 1)$ and $(x_1, x_2) = (0.625, 1)$ and going in opposite direction of the convective field b, see Fig. 8.13 showing the reference primal and adjoint solutions, blue lines mark the lines singularities and the red line is the domain of interest of J. These lines are smeared by the diffusion but it is negligible outside of the circle R.

Figure 8.14 shows the error decay of the error estimator $|\eta^\mathrm{I}|$ with respect to $\mathrm{DoF}^{1/3}$ and the computational time for all tested adaptive methods. The dominance of the hp-AMA approach is obvious, namely from the point of view of the computational time. Moreover, we observe that the h-AMA approach has a very fast decay for $\eta^\mathrm{I} \geq 5 \cdot 10^{-9}$. It follows from the presence of discontinuities where the h-adaptation has a large impact. However, for lower values of $|\eta^\mathrm{I}|$, also the influence of the smooth part of the solution is important and then the methods allowing p-adaptation start to dominate.

Figure 8.15, left shows the convergence of the error estimators $|\eta^\mathrm{I}|$, η^II and the error $|J(u - u_h)|$ for the hp-AMA method. We observe the different decay in comparison to the elliptic example from Sect. 8.3.1. The error estimator η^II significantly overestimates the estimator $|\eta^\mathrm{I}|$ and moreover, the overestimation is increasing. It is caused by the estimation in (8.3) where the arguments of the corresponding scalar products on the left-hand side are almost orthogonal and

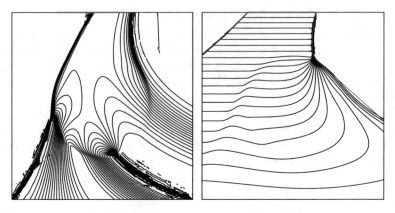

Fig. 8.13 Mixed hyperbolic-elliptic problem (8.35)–(8.37), the isolines of the reference primal (left) and the adjoint (right) solutions, the red line is the "domain of interest," the blue lines mark the singularities of the adjoint solution

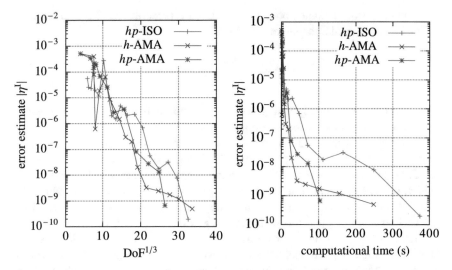

Fig. 8.14 Mixed hyperbolic-elliptic problem (8.35)–(8.37), comparison of the convergence of the estimator $|\eta^{\mathrm{I}}|$ with respect to DoF (left) and the computational time (right)

the Cauchy inequality overestimates $|\eta^{\mathrm{I}}|$. A decrease the error estimate η^{II} can be achieved by a modification of Algorithm 8.2 by using η^{II}_K instead of η^{I}_K, $K \in \mathscr{T}^k_h$ in step 11. Figure 8.15, right shows the corresponding convergence. In this case, η^{II} converges much better.

Furthermore, Fig. 8.16 shows the histogram of the distribution of the element error estimator η^{I}_K, $K \in \mathscr{T}^k_h$ on the last adaptation level obtained by the original Algorithm 8.2 and by the replacing η^{I}_K by η^{II}_K, $K \in \mathscr{T}^k_h$ in step 11 of Algorithm 8.2. The original method gives again the Gaussian distribution corresponding to the error

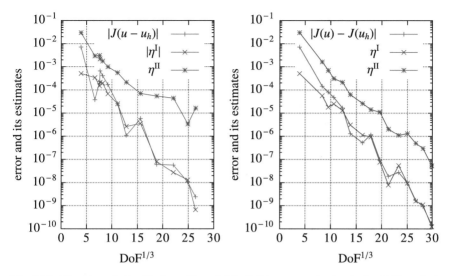

Fig. 8.15 Mixed hyperbolic-elliptic problem (8.35)–(8.37), computation by the hp-AMA method, the convergence of the error estimators $|\eta^{\mathrm{I}}|$, η^{II} and the error $|J(u - u_h)|$ (left). The same convergence obtained by the replacing η^{I}_K by η^{II}_K, $K \in \mathcal{T}_h^k$ in step 11 of Algorithm 8.2 (right)

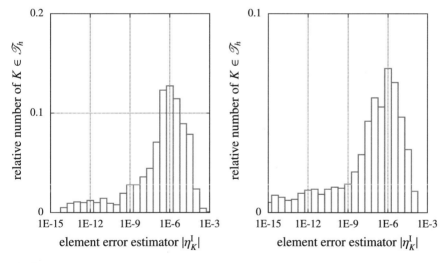

Fig. 8.16 Elliptic problem on a "cross" domain (8.33) and (8.34), the histogram of the distribution of η^{I}_K, $K \in \mathcal{T}_h^k$ on the last adaptation level obtained by the original Algorithm 8.2 (left) and by replacing η^{I}_K with η^{II}_K, $K \in \mathcal{T}_h^k$ in step 11 of Algorithm 8.2 (right)

equidistribution, cf. (8.19). However, this is not true for the modified Algorithm which attempts to equidistribute η^{II}_K, $K \in \mathcal{T}_h$. This explains the discrepancy between both histograms in Fig. 8.16.

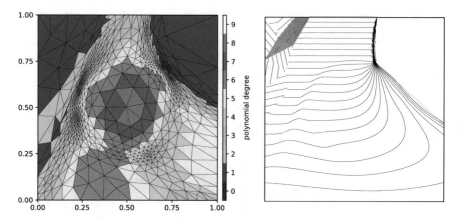

Fig. 8.17 Mixed hyperbolic-elliptic problem (8.35)–(8.37), the final hp-mesh (left) and the isolines of the adjoint (right) solution obtained with η_K^{I}, $K \in \mathscr{T}_h^k$ in step 11 of Algorithm 8.2

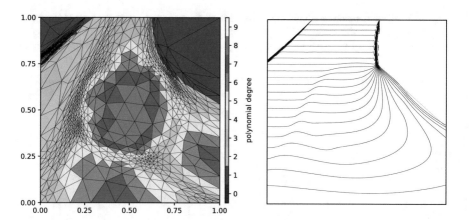

Fig. 8.18 Mixed hyperbolic-elliptic problem (8.35)–(8.37), the final hp-mesh (left) and the isolines of the adjoint (right) solution obtained with η_K^{II}, $K \in \mathscr{T}_h^k$ in step 11 of Algorithm 8.2

Moreover, Figs. 8.17 and 8.18 compare the outputs of Algorithm 8.2 using η_K^{I} and η_K^{II}, respectively, in step 11. Namely, the final hp-mesh and the isolines of the adjoint solution are shown. We see that using η_K^{II}, $K \in \mathscr{T}_h$, the mesh is adapted also along the left singularity line (colored in blue) of the adjoint solution and then this singularity is well resolved (compare the isolines plotted). On the other hand, Fig. 8.15 shows that the use of η_K^{I}, $K \in \mathscr{T}_h^k$ requires a few more DoF for achieving the same error tolerance. Therefore, the use of η^{II} and η_K^{II}, $K \in \mathscr{T}_h^k$ in Algorithm 8.2 is possible but it can be less efficient.

8.3.3 Convection-Dominated Problem

This problem comes from [61] and [27] and it was treated in Sect. 6.4.5 in the context of the interpolation error estimates. The computational domain is the right-bottom L-shape defined by $\Omega := [0, 4] \times [0, 4] \setminus [0, 2] \times [0, 2]$. We consider the linear convection–diffusion equation

$$-\varepsilon \Delta u + \nabla \cdot (\boldsymbol{b}\, u) = 0 \qquad \text{in } \Omega, \tag{8.38}$$

where the diffusion $\varepsilon = 10^{-3}$ and the convective field $\boldsymbol{b} = (x_2, -x_1)$. We prescribe the combined Dirichlet and Neumann boundary conditions

$$u = 1 \quad \text{on } \{x_1 = 0,\ x_2 \in (2, 4)\}, \tag{8.39}$$

$$\nabla u \cdot \boldsymbol{n} = 0 \quad \text{on } \Gamma_1 := \{x_1 = 4,\ x_2 \in (0, 4)\} \cup \Gamma_2 := \{x_1 \in (2, 4),\ x_2 = 0\},$$

$$u = 0 \quad \text{elsewhere.}$$

The solution of (8.39) exhibits two boundary layers along $\{x_1 \in (0, 4),\ x_2 = 4\}$ and $\{x_1 \in (0, 2),\ x_2 = 2\}$. Moreover, the solution contains two circular-shaped interior layers, see Fig. 8.19, top left.

We consider three functionals $J_V(u)$, $J_B(u)$, and $J_D(u)$ given by

$$J_V(u) = \int_E u(x)\, dx, \qquad E := (2.5, 3.5) \times (2.5, 3.5), \tag{8.40}$$

$$J_B(u) = \int_{G_B} \boldsymbol{b} \cdot \boldsymbol{n}\, u\, dS, \qquad G_B := \Gamma_1,$$

$$J_D(u) = \int_{G_D} \boldsymbol{b} \cdot \boldsymbol{n}\, u\, dS, \qquad G_D := \Gamma_1 \cup \Gamma_2.$$

The functional J_V corresponds to the mean value of the solution u over the square subdomain intersecting the outer interior layer whereas J_B and J_D integrate the convective flux through appropriate parts of boundary. By a strong hp-refinement, we obtained the reference values of these functionals as

$$J_V(u) = 0.20314158 \pm 10^{-8}, \tag{8.41}$$

$$J_B(u) = 0.07408122 \pm 10^{-8},$$

$$J_D(u) = 3.9670304 \pm 10^{-7}.$$

We note that these values slightly differ from those in [27]. Figure 8.19 also shows the solution of the adjoint problems corresponding to (8.38)–(8.40) with the functionals $J_V(u)$, $J_B(u)$ and $J_D(u)$.

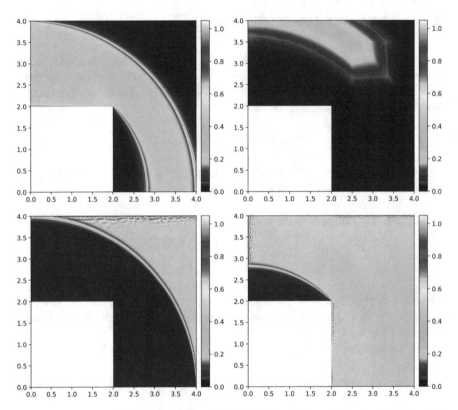

Fig. 8.19 Convection-dominated problem (8.38)–(8.40), the solutions of primal problem (top left) and the adjoint problems with J_V (top right), J_B (bottom left), and J_D (bottom right)

Figure 8.20 shows the convergence of the error estimator $|\eta^I|$ with respect to $\mathrm{DoF}^{1/3}$ and the computational time, respectively, for all tested adaptive techniques. Obviously, the hp-AMA method is superior to the h- and isotropic variants due to the presence of anisotropic features of the primal as well as adjoint problems.

Moreover, Fig. 8.21, left column, shows the comparison of the error estimators $|\eta^I|$, η^{II} with the error $|J(u - u_h)|$ for the hp-AMA method. We observe an exponential rate of the convergence and also a very good approximation of the error by both estimators. Figure 8.21, right column, presents the histogram of the distribution of the element error estimator η^I_K, $K \in \mathscr{T}_h^k$ on the last adaptation level. Distributions close to a Gaussian one are observed.

Moreover, Fig. 8.22 shows the isolines of the solution of the primal problem together with the final hp-meshes for the case with the functional J_V. We observe a strong mesh refinement along the part of the outer circular-shaped interior layer entering to the domain of interest $E := (2.5, 3.5) \times (2.5, 3.5)$. However, we observe also a refinement behind the square E which can be surprising since the adjoint solution is (almost) constant in front of E in the opposite direction of the convection

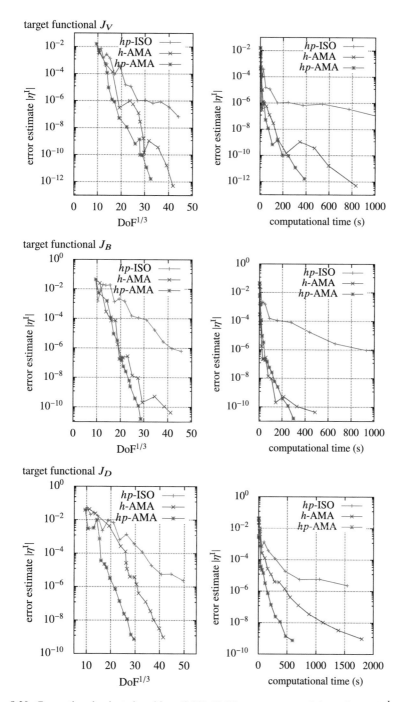

Fig. 8.20 Convection-dominated problem (8.38)–(8.40), convergence of the estimator $|\eta^1|$ with respect to DoF (left) and the computational time (right) for all tested adaptive strategies, target functionals J_V (first row), J_B (second row), and J_D (third row)

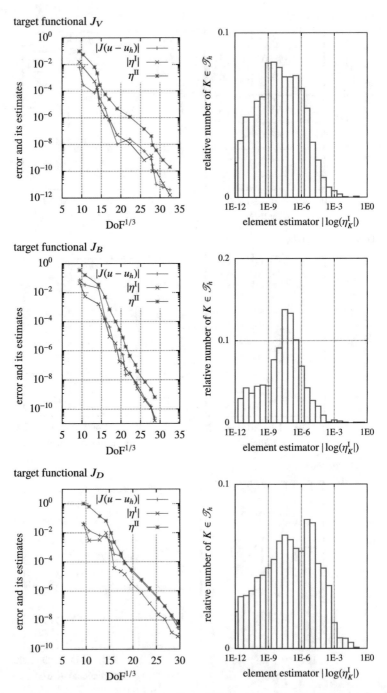

Fig. 8.21 Convection-dominated problem (8.38)–(8.40), convergence of estimators $|\eta^{\mathrm{I}}|$, η^{II} and error $|J(u - u_h)|$ for the hp-AMA method (left) and the histograms of the distribution of η^{I}_K, $K \in \mathscr{T}_h^k$ on the last adaptation level with the target functionals J_V (first row), J_B (second row), and J_D (third row)

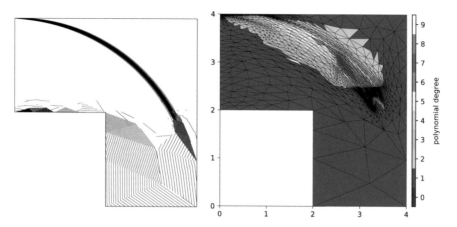

Fig. 8.22 Convection-dominated problem (8.38)–(8.40) with functional J_V, the isolines of the primal solution (left) and the final hp-mesh (right)

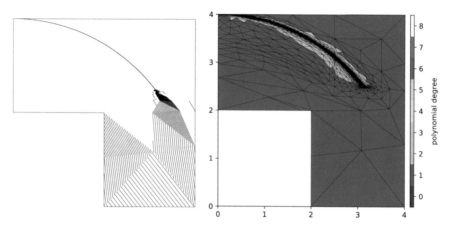

Fig. 8.23 Convection-dominated problem (8.38)–(8.40) with $\varepsilon = 10^{-6}$ and functional J_V, the isolines of the primal solution (left), the final hp-mesh (right)

field, cf. Fig. 8.19. Therefore, the adjoint residuals as well as the adjoint weights should be negligible. However, the considered diffusion $\varepsilon = 10^{-3}$ plays also a role which leads to the aforementioned mesh refinement. This argument is supported by Fig. 8.23, which shows the results of the same problem (with functional J_V) but with smaller diffusion $\varepsilon = 10^{-6}$. The refinement is just inside of E, the outer circular-shaped interior layer is thinner and lower polynomial approximation degrees are generated.

Furthermore, Fig. 8.24 shows the isolines of the solution and the final hp-meshes for functional J_B. We observe a strong anisotropic refinement along the entire outer circular-shaped interior layer since it attaches to the domain of region G_B, cf. (8.40).

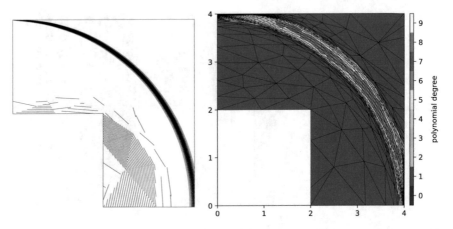

Fig. 8.24 Convection-dominated problem (8.38)–(8.40) with functional J_B, the isolines of the primal solution (left) and the final hp-mesh (right)

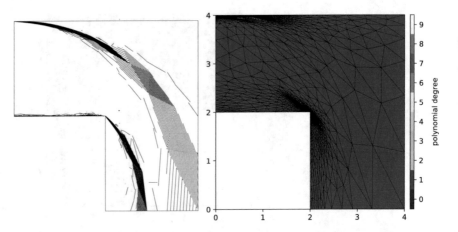

Fig. 8.25 Convection-dominated problem (8.38)–(8.40) with functional J_D, the isolines of the primal solution (left) and the final hp-mesh (right)

On the other hand, there is no refinement along the inner circular-shaped interior since it does not influence the value of the solution on G_B.

Figure 8.25 shows the results for the case with functional J_D. In contrary to the case with J_B, both interior layers act on G_D. However, we do not observe refinement along both interior layers since J_D is the mean value of u over G_D and, therefore, the smeared layers lead to (almost) the same value. We observe only strong refinements in regions where both interior layers begin, see zooms in Fig. 8.26.

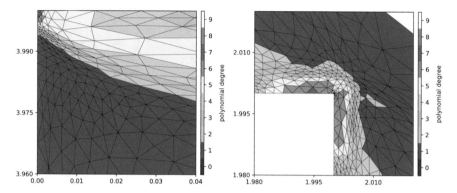

Fig. 8.26 Convection-dominated problem (8.38)–(8.40) with functional J_D, details of the final hp-mesh in regions where both interior layers begin

Chapter 9
Implementation Aspects

We present some implementation aspects of the anisotropic hp-mesh adaptation algorithms. In particular, in Sect. 9.1 we present higher-order reconstruction techniques which are required by our algorithms for estimating of the interpolation error. Moreover, in Sect. 9.2, we introduce a heuristic modification of the algorithm for the numerical solution of time-dependent problems.

9.1 Higher-Order Reconstruction Techniques

In this section, we describe two higher-order reconstruction techniques which are employed in previous chapters, namely in relations (5.130), (6.30), and (7.21). In particular, we consider a reconstruction $\mathscr{R} : S_{hp} \rightarrow S_{hp+1}$, cf. (7.50). We also discuss a possible reconstruction $\mathscr{R} : S_{hp} \rightarrow S_{h,p+2}$, which is employed in the hp-variant of the method by (6.30). These techniques allow us to obtain the approximations $u_h^+ \in S_{hp+1}$ (and possibly $u_h^+ \in S_{h,p+2}$) from the available approximate solution u_h. These reconstructions are employed, on the one hand, for the evaluation of the approximation of the derivatives of degree $p_K + 1$ (or $p_K + 2$) used in the interpolation error estimates on $K \in \mathscr{T}_h$, and, on the other hand, for the approximation of the primal and adjoint weak solutions in the framework of goal-oriented error estimates.

9.1.1 Weighted Least-Square Reconstruction

The first technique evaluates the higher-order reconstruction from the available piecewise polynomial approximation without a direct relation to the differential equation considered. This approach is well established for the one-dimensional case

© The Author(s), under exclusive license to Springer Nature Switzerland AG 2022
V. Dolejší, G. May, *Anisotropic* hp-*Mesh Adaptation Methods*, Nečas Center Series,
https://doi.org/10.1007/978-3-031-04279-9_9

due to the properties of orthogonal polynomials and it can be extended for structured rectangular grids. For simplicial grids the reconstructions are more delicate, we refer to works [118, 119]. However, the majority of works consider fixed polynomial degree over the whole computational domain which is not the case for hp-methods.

For each $K \in \mathscr{T}_h$, we define the patch

$$D_K = \{K' \in \mathscr{T}_h; \ K' \cap K \neq \emptyset\}, \qquad K \in \mathscr{T}_h, \tag{9.1}$$

which consists of the elements sharing at least a vertex with K, see Fig. 9.1 for the two-dimensional case. We note that $K \in D_K$. For each $K' \in D_K$, we define a weight $\omega_{K,K'} \in [0, 1]$, e.g., by

$$\omega_{K,K'} = \begin{cases} 1 & \text{for } K' = K, \\ \epsilon_1 & \text{for } K' \text{ sharing a face } (d = 3) \,/\, \text{edge } (d = 2) \text{ with } K, \\ \epsilon_2 & \text{otherwise}, \end{cases} \tag{9.2}$$

where we can set $\epsilon_1 = 1$ and $\epsilon_2 \leq 1$. If $\epsilon_2 = 1$, then all elements from D_K are taken with the same weights, if $\epsilon_2 = 0$, then only a small patch consisting of elements sharing a face $(d = 3)$ or an edge $(d = 2)$ with K is considered.

Moreover, let $s \in [0, 1]$, we define the norm $\|\cdot\|_K$ on K by

$$\|v\|_K := \left(\|v\|_{L^2(K)}^2 + s|v|_{H^1(K)}^2 \right)^{1/2}, \tag{9.3}$$

as a combination of the L^2-norm and H^1-seminorm.

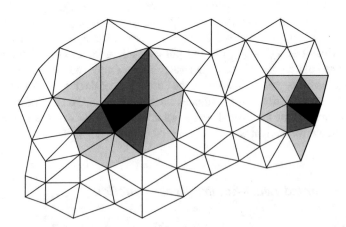

Fig. 9.1 Two-dimensional examples of patches D_K corresponding to interior (dark left) and boundary (dark right), elements having $\omega_{K,K'} = \epsilon_1$ are filled with the medium intensity and elements having $\omega_{K,K'} = \epsilon_2$ are filled with the light intensity

Let $u_h \in S_{hp}$ be the given function, typically the approximate solution obtained by DGM. Let $K \in \mathcal{T}_h$ and $p_K \in \mathbb{N}$ be the corresponding polynomial approximation degree. We denote by $P^{p_K+1}(D_K)$ the space of polynomials of degree $p_K + 1$ over the patch D_K.

We define the function $u_K^+ \in P^{p_K+1}(D_K)$ which minimizes the distance from $u_h|_{D_K}$ in the weighted norm of (9.3) by

$$u_K^+ = \underset{w_h \in P^{p_K+1}(D_K)}{\arg\min} \sum_{K' \in D_K} \omega_{K,K'} \|w_h - u_h\|_{K'}^2. \tag{9.4}$$

Finally, we assemble the higher-order reconstruction u_h^+ as an elementwise composition of $u_K^+|_K$ by

$$u_h^+ := \mathcal{R}u_h = \sum_{K \in \mathcal{T}_h} \chi_K u_K^+, \tag{9.5}$$

where χ_K is the characteristic function of $K \in \mathcal{T}_h$.

We studied the reconstruction operator (9.5) in [53] with setting $\epsilon_1 = 1$ and varying ϵ_2. Smaller value of ϵ_2 gives more accurate approximation in the sense $\|u - u_h\| \approx \|u_h^+ - u_h\|$, where u is the exact solution of the given problem. However, a small value of ϵ_2 is less stable for varying polynomial approximation degree and for highly anisotropic elements. Therefore, we employ values of ϵ_2 away from 0, typically 0.1 or 0.5. Concerning the choice of parameter s in (9.3), we did not observe a significant difference between the limit cases $s = 0$ and $s = 1$.

Remark 9.1 The procedure described above can be easily modified also for the reconstruction from S_{hp} to $S_{h,p+2} := \{v \in L^2(\Omega); \ v|_K \in P^{p_K+2}(K) \forall K \in \mathcal{T}_h\}$. In this case we replace (9.4) by

$$u_K^+ = \arg\min_{w_h \in P^{p_K+2}(D_K)} \sum_{K' \in D_K} \omega_{K,K'} \|w_h - u_h\|_{K'}^2. \tag{9.6}$$

The minimization task (9.6) is well defined if the input function $u_h|_{D_K}$ has more degrees of freedom than the dimension of D_K. This is usually true even for the case $\epsilon_2 = 0$ when D_K has only few nontrivial elements ($d + 2$ for elements without edge on the boundary of the computational domain). Naturally, this approximation by two polynomial degrees higher is less accurate. However, it can provide sufficient information for hp-mesh adaptation.

The reconstruction technique (9.4) and (9.5) is independent of the solved problem (i.e., the considered differential equation) which is not the case of the approach in the next section.

9.1.2   Reconstruction Based on the Solution of Local Problems

The idea of this reconstruction technique comes, e.g., from [9, 15], where the higher-order reconstructions come from the solution of local problems related to the original problem (cf. (7.54))

$$\text{find } u_h \in S_{hp} \text{ such that} \qquad a_h(u_h, \varphi_h) = \ell_h(\varphi_h) \quad \forall \varphi_h \in S_{hp}. \tag{9.7}$$

The local problems are defined on patches of elements Ω_i, $i = 1, \dots, M$ by a restriction of (9.7) on Ω_i with the boundary conditions on $\partial\Omega_i$ given by the approximate solution outside of Ω_i. For simplicity, we restrict ourselves to linear problems only. For a possible generalization to nonlinear problems, see Remark 9.4 below.

We employ this idea for DGM, where each patch Ω_i consists only of one $K \in \mathcal{T}_h$. Let $u_h \in S_{hp}$ be the solution of (9.7). For each $K \in \mathcal{T}_h$, we define function $u_K^+ \in S_{hp+1}$ such that

(i) $u_K^+|_{K'} := u_h|_{K'}$ for all $K' \neq K$,

(ii) $u_K^+|_K \in P^{p_K+1}(K)$,   (9.8)

(iii) $a_h(u_K^+, \varphi_h) = \ell_h(\varphi_h) \quad \forall \varphi_h \in V_K := \{v_h \in S_{hp+1}; \ v_h|_{\Omega \setminus K} = 0\}$,

where a_h is the form from (9.7). It means that u_K^+ is identical with u_h outside of K, it is a polynomial of degree $p_K + 1$ on K and fulfills (9.7) for test functions from space V_K. Finally, in the same spirit as in (9.5), we set

$$u_h^+ := \mathcal{R}u_h = \sum_{K \in \mathcal{T}_h} \chi_K u_K^+. \tag{9.9}$$

In comparison to (9.4) and (9.5), the approach (9.8) and (9.9) is free of user-dependent constants and it is directly related to the original problem. The implementation of (9.8) and (9.9) does not require an explicit setting of (9.8) since it is possible to employ the discretization of problem (9.7).

Let m_K denote the number of degrees of freedom corresponding to $K \in \mathcal{T}_h$,

$$m_K = \begin{cases} \frac{1}{2}(p_K + 1)(p_K + 2) & \text{for } d = 2, \\ \frac{1}{6}(p_K + 1)(p_K + 2)(p_K + 3) & \text{for } d = 3. \end{cases}$$

Moreover, let $\{\varphi_{h,K}^i, \ i = 1, \dots, m_K\}$ be the basis of the space $P^{p_K}(K)$, $K \in \mathcal{T}_h$ and

$$\boldsymbol{\varphi}_K := \left(\varphi_{h,K}^1, \dots, \varphi_{h,K}^{m_K}\right)^{\mathsf{T}} \in \mathbb{R}^{m_K}.$$

Extending $\varphi^i_{h,K}$, $i = 1, \dots m_K$ by zero outside of K, we construct the basis of S_{hp} as

$$\{\varphi^i_{h,K}; \ i = 1, \dots m_K, \ K \in \mathcal{T}_h\}.$$

Let $u_h \in S_{hp}$ be the solution of (9.7). We can write u_h as

$$u_h(x) = \sum_{K \in \mathcal{T}_h} \boldsymbol{u}_K \cdot \boldsymbol{\varphi}_K(x), \qquad (9.10)$$

where $\boldsymbol{u}_K \in \mathbb{R}^{m_K}$ consists of the basis coefficients of u_h in the basis corresponding to element $K \in \mathcal{T}_h$. Denoting $\boldsymbol{f}_K := \{\ell_h(\varphi^i_{h,K})\}_{i=1}^{m_K} \in \mathbb{R}^{m_K}$, $K \in \mathcal{T}_h$, we rewrite problem (9.7) in the block-matrix form (one block-row for each $K \in \mathcal{T}_h$)

$$\mathbb{A}_{K,K} \boldsymbol{u}_K + \sum_{K' \in N(K)} \mathbb{A}_{K,K'} \boldsymbol{u}_{K'} = \boldsymbol{f}_K \quad \forall K \in \mathcal{T}_h, \qquad (9.11)$$

where $\mathbb{A}_{K,K}$ are diagonal blocks (corresponding to a_h) of size $m_K \times m_K$, $\mathbb{A}_{K,K'}$ are the off-diagonal blocks of size $m_K \times m_{K'}$ and $N(K)$ is the set of elements sharing an edge ($d = 2$) or a face ($d = 3$) with $K \in \mathcal{T}_h$.

For each $K \in \mathcal{T}_h$, we write $u_K^+ = u_h + \tilde{u}_K$, where u_h is the approximate solution of (9.7) and $\tilde{u}_K \in V_K$, cf. (9.8), condition (iii). The function $\tilde{u}_K$ can be considered as a local higher-order correction to u_h.

Let $\varphi_{h,K} \in V_K$, using the linearity of a_h, identity (9.7) and condition (iii) in (9.8), we have

$$a_h(\tilde{u}_K, \varphi_{h,K}) = a_h(u_K^+, \varphi_{h,K}) - a_h(u_h, \varphi_{h,K}) = \ell_h(\varphi_{h,K}) - a_h(u_h, \varphi_{h,K})$$

$$=: r_h(u_h)(\varphi_{h,K}). \qquad (9.12)$$

Hence, for each $K \in \mathcal{T}_h$, we have the following problem

$$a_h(\tilde{u}_K, \varphi_{h,K}) = r_h(u_h)(\varphi_{h,K}) \quad \forall \varphi_{h,K} \in V_K. \qquad (9.13)$$

We denote

$$m_K^+ = \dim P^{p_K+1}(K) = \begin{cases} \frac{1}{2}(p_K + 2)(p_K + 3) & \text{for } d = 2, \\ \frac{1}{6}(p_K + 2)(p_K + 3)(p_K + 4) & \text{for } d = 3 \end{cases}$$

and choose a basis $\{\varphi^1_{h,K}, \dots, \varphi^{m_K}_{h,K}, \dots, \varphi^{m_K^+}_{h,K}\}$ of P^{p_K+1} as hierarchical extension of the basis $\boldsymbol{\varphi}_K$. Then (9.13) can be written in similar form to (9.11), where the off-diagonal terms are vanishing since $\tilde{u}_K = 0$ on all $K' \neq K$, namely

$$\mathbb{A}_{K,K}^+ \tilde{\boldsymbol{u}}_K = \boldsymbol{r}, \qquad (9.14)$$

where $\mathbb{A}_{K,K}^+ \in \mathbb{R}^{m_K^+ \times m_K^+}$ is the matrix $\mathbb{A}_{K,K}$ enlarged by $m_K^+ - m_K$ rows and columns, $\boldsymbol{r} \in \mathbb{R}^{m_K^+}$ is the vector with components $\boldsymbol{r}_i = r_h(u_h)(\varphi_{h,K}^i), i = 1, \ldots, m_K^+$ and $\tilde{\boldsymbol{u}}_K$ is the vector of basis coefficients defining the function $\tilde{u}_K$ on K. We note that first m_K components of $\boldsymbol{r}_i$ are vanishing up to the algebraic errors.

Therefore, in order to find the reconstruction (9.8) for each $K \in \mathscr{T}_h$, we have to assemble the block-diagonal block $\mathbb{A}_{K,K}^+$, evaluate the residual $r_h(u_h)(\cdot)$ in (9.12) for all basis functions of V_K, and solve the linear algebraic system (9.14). Then, we set

$$u_h^+ = u_h + \sum_{K \in \mathscr{T}_h} \tilde{u}_K. \tag{9.15}$$

Remark 9.2 In the framework of the goal-oriented error estimates in Chaps. 7 and 8, similar techniques can be used for the reconstruction of the adjoint solution, namely we replace (9.8) by the problem: find $z_K^+ \in S_{hp+1}$ such that

$$
\begin{array}{ll}
\text{(i)} & z_K^+|_{K'} := z_h|_{K'} \text{ for all } K' \neq K, \\
\text{(ii)} & z_K^+|_K \in P^{p_K+1}(K), \\
\text{(iii)} & a_h(w_h, z_K^+) = J_h(w_h) \quad \forall w_h \in V_K.
\end{array}
\tag{9.16}
$$

Again it is sufficient to solve a problem similar to (9.14) with the transposed matrix and we obtain updates $\tilde{z}_K$ for each $K \in \mathscr{T}_h$ and set $z_h^+ = z_h + \sum_{K \in \mathscr{T}_h} \tilde{z}_K$.

Remark 9.3 We note that reconstruction (9.8) and (9.9) satisfies the error equivalence (7.19) (in exact arithmetic only) which is a favorable property. Namely, due to the linearity of residuals, the Galerkin orthogonalities, (9.12) and (9.15) we have

$$
\begin{aligned}
r_h(u_h)(z_h^+) &= \sum_{K \in \mathscr{T}_h} r_h(u_h)(z_h^+|_K) = \sum_{K \in \mathscr{T}_h} r_h(u_h)(\tilde{z}_K) = \sum_{K \in \mathscr{T}_h} a_h(\tilde{u}_K, \tilde{z}_K) \\
&= \sum_{K \in \mathscr{T}_h} r_h^*(z_h)(\tilde{u}_K) = \sum_{K \in \mathscr{T}_h} r_h^*(z_h)(u_h^+|_K) = r_h^*(z_h)(u_h^+).
\end{aligned}
$$

Moreover, the equivalence also holds locally.

Remark 9.4 This method can be used even for nonlinear problems. Its definition is quite analogues to (9.8). However, in this case, relation (9.13) is not valid and the previous technique cannot be applied. Then, it is necessary to solve the local problems iteratively using a similar method as for the global problem. Finally, we note that for a nonlinear primal problem, the corresponding adjoint problem is always linear and then the approach based on (9.13) can be employed.

9.2   Anisotropic Mesh Adaptation for Time-Dependent Problems

Many physical processes are described by time-dependent partial differential equations where the sought solution depends on the space variable $x \in \Omega \subset \mathbb{R}^d$, $d = 2, 3$, and time variable $t \in (0, T)$, $T > 0$. For simplicity, we assume that Ω is fixed for $t \in (0, T)$, otherwise the following considerations have to be modified. The time-dependent problems are usually solved "sequentially," i.e., the time interval $(0, T)$ is split into $0 = t_0 < t_1 < \cdots < t_r = T$ and, when using a one-step method, employing the approximate solution on the time level t_{m-1}, we obtain an approximation on the time level t_m, $m = 1, \ldots, r$.

Some numerical techniques admit the use of different space discretizations (meshes) on different time levels, which is advantageous when specific transient phenomena are simulated, e.g., traveling waves or discontinuities. Anisotropic mesh adaptation for time-dependent problems has been developed in many research endeavors, e.g., [18, 38, 60, 70], to name just a few. The use of different meshes on different time levels requires some kind of recomputation of the approximate solution from one time level to the next. This recomputation is the cause of another source of error in the context of anisotropic mesh adaptation, namely when the meshes are not nested. See Fig. 9.2 illustrating meshes $\mathcal{T}_{h,m-1}$ and $\mathcal{T}_{h,m}$ at time levels t_{m-1} and t_m, respectively. An important aspect is the geometric conservation law, cf. [72, 91].

9.2.1   Space-Time Discontinuous Galerkin Method

In our computations, we employ the *space-time discontinuous Galerkin* (STDG) method which admits varying meshes in a natural way, see [45, Chapter 6]. Let

Fig. 9.2 Illustration of two meshes $\mathcal{T}_{h,m-1}$ (red) and $\mathcal{T}_{h,m}$ (blue) at time levels t_{m-1} and t_m, respectively, generated by an anisotropic mesh adaptation code

$\Omega \subset \mathbb{R}^d$, $d = 2, 3$ be the computational domain and $(0, T)$, $T > 0$ the time interval of our interest. We consider the following problem: Find $u : \Omega \times (0, T) \to \mathbb{R}$ such that

$$\partial_t u + \mathscr{L}u = f, \qquad (9.17)$$

where ∂_t denotes the partial derivative with respect to time, $\mathscr{L}$ is a differential operator with respect to the space variables, and f is a given function. The operator $\mathscr{L}$ can be possibly nonlinear. Problem (9.17) has to be accompanied by suitable initial and boundary conditions. In Sect. 10.3, we consider a more general problem with a nonlinear time derivative term.

Let $\mathscr{T}_{h,m}$ be a simplicial grid for the time level t_m and $I_m := (t_{m-1}, t_m)$, $m = 0, 1, \ldots, r$. We consider the space of space-time discontinuous functions

$$S_{h,p}^{\tau,q} := \left\{ \varphi : \Omega \times (0, T) \to \mathbb{R}; \ \varphi|_{K \times I_m} \in P_{p_K}^q (K \times I_m), \ K \in \mathscr{T}_{h,m}, \ m = 1, \ldots, r \right\}, \qquad (9.18)$$

where $P_{p_K}^q (K \times I_m)$ denotes the space of all polynomial functions on the space-time element $K \times I_m$ of degree p_K with respect to space and degree q with respect to time.

Moreover, let $a_{h,m} : S_{h,p}^{\tau,q} \times S_{h,p}^{\tau,q} \to \mathbb{R}$ formally denote the discretization of the operator $\mathscr{L}$ on the mesh $\mathscr{T}_{h,m}$, the right-hand side f and the boundary conditions. Then the STDG discretization of (9.17) reads: find $u_{h\tau} \in S_{h,p}^{\tau,q}$ such that

$$\int_{I_m} \left((\partial_t u_{h\tau}, v_h) + a_{h,m}(u_{h\tau}, v_h) \right) \, dt + (\{u_{h\tau}\}_{m-1}, v_h) = 0 \qquad (9.19)$$

$$\forall v_h \in S_{h,p}^{\tau,q}, \quad m = 1, \ldots, r,$$

where $(\cdot, \cdot)$ denotes the L^2-scalar product over Ω and $\{u_{h\tau}\}_{m-1}$ denotes the "time-jump" of $u_{h\tau}$ at t_{m-1}, $m = 1, \ldots, r$, i.e.,

$$\{u_{h\tau}(x)\}_{m-1} := u_{h\tau}(x)|_{m-1}^+ - u_{h\tau}(x)|_{m-1}^-, \qquad x \in \Omega, \qquad (9.20)$$

where

$$u_{h\tau}(x)|_{m-1}^{\pm} := \lim_{t \to t_{m-1}^{\pm}} u_{h\tau}(x, t), \qquad x \in \Omega. \qquad (9.21)$$

The approximations on two following time intervals are joined together in a weak form through the term

$$(\{u_{h\tau}\}_{m-1}, v_h) = \left(u_{h\tau}|_{m-1}^+, v_h \right) - \left(u_{h\tau}|_{m-1}^-, v_h \right). \qquad (9.22)$$

However, the second term on the right-hand side of (9.22) requires the integration of the product of $u_{h\tau}|_{m-1}^-$ (piecewise polynomial function on $\mathscr{T}_{h,m-1}$) and v_h (piecewise polynomial function on $\mathscr{T}_{h,m}$) which is technically difficult for non-matching grids, cf. Fig. 9.2. We can either evaluate the intersections of $K \in \mathscr{T}_{h,m}$ with all possible $K' \in \mathscr{T}_{h,m-1}$ and integrate over all arising segments or use a composite high-order numerical quadrature for each $K \in \mathscr{T}_{h,m}$. The latter approach is much easier to implement, but we have to be aware that $u_{h\tau}|_{m-1}^-$ is discontinuous on $K \in \mathscr{T}_{h,m}$ and the rate of convergence of the numerical quadrature is low even for high quadrature order.

9.2.2 Interpolation Error Estimates for Time-Dependent Problems

In principle, it would be possible to introduce the space-time projection of sufficiently regular functions to space $S_{h,p}^{\tau,q}$ and generalize the approach from Chaps. 3 and 4 also for time-dependent problems. However, we develop here a simpler technique where only the interpolation error with respect to the space coordinate is considered. We formulate the problem of the seeking optimal, or at least nearly optimal, approximation spaces with respect to interpolation operator π_h.

Let π_h be the (space) projection given by (6.10). Let $u \in C^\infty(\Omega \times (0, T)$ be the exact solution $u = u(x, t)$ of (9.17). For arbitrary but fixed $m = 1, \ldots, r$ and $t \in I_m$, we define the function $u(t) : \Omega \to \mathbb{R}$ by $u(t)(x) := u(x, t)$. Then $\pi_h u(t)$ is a piecewise polynomial function corresponding to the hp-mesh $\mathscr{T}_{h\boldsymbol{p},m}$. Our aim is to control the interpolation error

$$\|u(t) - \pi_h u(t)\|_W, \tag{9.23}$$

where $\|\cdot\|_W$ denotes a norm or seminorm of our interest, e.g., the $L^q(\Omega)$-norm, $q \in [1, \infty]$, or the broken $H^1(\mathscr{T}_h)$ seminorm.

Therefore, in virtue of Chaps. 5 and 6, we formulate the following task.

Problem 9.5 Let $u \in C^\infty(\Omega \times (0, T))$ and $0 = t_0 < t_1 < \cdots < t_r = T$ be a partition of $(0, T)$. For $\omega > 0$, find a sequence of hp-meshes $\mathscr{T}_{h\boldsymbol{p},m}$, $m = 1, \ldots, r$ such that

$$\|u(t) - \pi_h u(t)\|_W \leq \omega \qquad \forall t \in I_m, \; m = 1, \ldots, r \tag{9.24}$$

and the sum of the number of DoF over all hp-meshes $\mathscr{T}_{h,m}$, $m = 1, \ldots, r$ is minimal.

Remark 9.6 In Problem 9.5 the time partition t_m, $m = 0, \ldots, r$ is given a priori. However, the size of the time steps is chosen adaptively during the computational process. We discuss this aspect later.

Problem 9.5 requires the bound of the interpolation error for each $t \in I_m$, $m = 1, \ldots, r$, which is impossible in practical computations. Hence, we require (9.24) only at selected time instants. The simplest possibility is to require (9.24) only at the levels t_m, $m = 1, \ldots, r$. Moreover, we have to take into account that the function u is unknown and is thus approximated by a higher-order reconstruction $\mathcal{R}$ of the available approximate solution $u_{h\tau} \in S_{h,p}^{\tau,q}$, cf. (9.5). However, since we use the discontinuous approximation with respect to time, we consider the interpolation error from the left as well as from the right at each time instant, i.e., $u_{h\tau}|_{m-1}^{\pm}$, cf. (9.21). Then, instead of Problem 9.5, we have the following task.

Problem 9.7 Let $0 = t_0 < t_1 < \cdots < t_r = T$ be a partition of $(0, T)$. For $\omega > 0$, find a sequence of hp-meshes $\mathcal{T}_{hp,m}$, $m = 1, \ldots, r$ such that the corresponding approximate solution $u_{h\tau} \in S_{h,p}^{\tau,q}$ given by (9.19) satisfies

$$\eta_m^L := \left\| \mathcal{R}(u_{h\tau}|_{m-1}^+) - \pi_h(\mathcal{R}(u_{h\tau}|_{m-1}^+)) \right\|_W \leq \omega \qquad m = 1, \ldots, r, \qquad (9.25)$$

$$\eta_m^R := \left\| \mathcal{R}(u_{h\tau}|_m^-) - \pi_h(\mathcal{R}(u_{h\tau}|_m^-)) \right\|_W \quad \leq \omega \qquad m = 1, \ldots, r,$$

and the sum of the number of DoF over all hp-meshes $\mathcal{T}_{h,m}$, $m = 1, \ldots, r$ is minimal.

The practical solution of Problem 9.7 is the following. Assuming that both conditions in (9.25) are valid for time steps $m' = 1, \ldots, m - 1$. We initiate $\mathcal{T}_{h,m} := \mathcal{T}_{h,m-1}$ and perform the m-th time step given by solving (9.19). If both conditions in (9.25) are valid for m, then we go to the next time step. Otherwise, we propose a new hp-mesh by Algorithm 6.1, where functions $u_{h\tau}|_m^-$ and $u_{h\tau}|_{m-1}^+$ are taken as the input and we repeat the m-th step. This means that we define two metric fields, M_m^- using $u_{h\tau}|_m^-$ and M_m^+ using $u_{h\tau}|_{m-1}^+$, and the new metric for the m-th step is obtained by the metric intersection

$$M_m := M_m^+ \cap M_m^-. \qquad (9.26)$$

The intersection (9.26) of M_m^+ and M_m^- is defined in the following way: Let $x \in \Omega$ and $\mathbb{M}^{\pm} := M_m^{\pm}(x) \in Sym$. Further, let $\mathcal{E}_{\mathbb{M}^{\pm}}$ be the ellipses corresponding to $\mathbb{M}^{\pm}$, given by (2.15). Then we define an ellipse $\mathcal{E}^*$ such that $\mathcal{E}^* \subset \mathcal{E}_{\mathbb{M}^+} \cap \mathcal{E}_{\mathbb{M}^-}$ and $\mathcal{E}^*$ has the maximal possible volume. Finally, we set $M_m(x) = M^*$ where $M^* \in Sym$ is the matrix corresponding to $\mathcal{E}^*$.

9.2.3 Setting of the Time Step

Another important aspect of the numerical solution of (9.17) is the choice of the size of time step $\tau_m := t_m - t_{m-1}$, $m = 1, \ldots, r$. It should balance the discretization errors arising from the space and time discretizations. We refer to monograph [113] and the references cited therein. In this book we use the residual-based technique developed in [52] for the simulation of compressible flows which is able to identify the space and time discretization errors separately and also to propose the length of the next time step in order to balance both components of discretization error.

If condition (9.25) is violated for some m, it makes sense to repeat the computation on the same hp-mesh with shorter time step since the remeshing has higher computational costs in comparison to the reduction of the size of time step. However, if the estimate of the time error is significantly larger in comparison to the space error, then the remeshing is necessary.

9.2.4 Adaptive Solution of Time-Dependent Problem

A full adaptive algorithm for the numerical solution of time-dependent problems (9.17) is shown in Algorithm 9.1.

Algorithm 9.1: Adaptive solution of time-dependent problem

1: set tolerance $\omega > 0$
2: propose initial mesh hp-mesh $\mathcal{T}_{hp,0}$
3: propose initial time step τ_0
4: **for** $m = 0, 1, \ldots$ **do**
5:     solve problem (9.19) on time interval I_m and hp-mesh $\mathcal{T}_{hp,m}$
6:     evaluate $\eta_{I_m}^L$ and η_m^R by (9.25)
7:     **if** $\min(\eta_m^L, \eta_m^R) > \omega$ **then**
8:         **if** time error is too large **then**
9:             reduce τ_m and go to step 5
10:         **else**
11:             adapt mesh and go to step 5
12:         **end if**
13:     **else**
14:         time step m is accepted, $t_m := t_{m-1} + \tau_m$
15:         **if** $t_m \geq T$ **then**
16:             STOP the computation
17:         **end if**
18:         propose new time step τ_{m+1}
19:         set $\mathcal{T}_{hp,m+1} := \mathcal{T}_{hp,m}$
20:     **end if**
21: **end for**

Chapter 10
Applications

We present several additional applications of the anisotropic hp-mesh adaptation methods to more practical problems. In particular, we deal with compressible flow acting on an isolated profile, time-dependent viscous shock-vortex interaction, and porous media flows.

10.1 Steady-State Inviscid Compressible Flow

We consider the classical benchmark represented by a steady-state inviscid compressible flow around an isolated NACA0012 profile which is given by the parametrization

$$\Gamma_W := \left\{ \left[t, \pm \frac{0.2969\sqrt{t} - 0.126t - 0.3516t^2 + 0.2843t^3 - 0.1015t^4}{5} \right], \; 0 \le t \le 1 \right\}.$$

The domain occupied by a fluid is denoted by $\Omega \in \mathbb{R}^2$, its interior boundary is Γ_W, and the outer part of the boundary is an artificial inflow/outflow boundary denoted by Γ_{IO}, see Fig. 10.1.

The steady-state flow is described by the vector-valued function $\boldsymbol{u} : \Omega \to \mathbb{R}^4$ having components representing the density r, components of momentum $r\boldsymbol{v} = (rv_1, rv_2)$ and the total energy e of the fluid. The conservation laws lead to the *Euler equations* for the state vector $\boldsymbol{u}$ in the form (cf. also (7.120))

$$\sum_{s=1}^{2} \frac{\partial \boldsymbol{f}_s(\boldsymbol{u})}{\partial x_s} = 0, \tag{10.1}$$

© The Author(s), under exclusive license to Springer Nature Switzerland AG 2022
V. Dolejší, G. May, *Anisotropic* hp-*Mesh Adaptation Methods*, Nečas Center Series,
https://doi.org/10.1007/978-3-031-04279-9_10

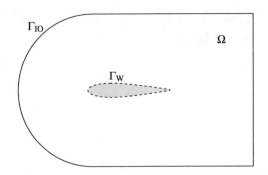

Fig. 10.1 The computational domain Ω with the profile Γ_W and the inflow/outflow boundary Γ_{IO}

where $f_s : \mathbb{R}^4 \to \mathbb{R}^4$, $s = 1, 2$ are the physical (inviscid) fluxes, for details we refer to [57, 58, 115]. The system is accompanied by the state equation for perfect gas and the definition of the total energy as the sum of internal and kinetic energy which give rise to the relation

$$e = \frac{p}{\kappa - 1} + \tfrac{1}{2}r|v|^2 \qquad \Longleftrightarrow \qquad p = (\kappa - 1)(e - \tfrac{1}{2}r|v|^2), \qquad (10.2)$$

where p is the pressure and $\kappa = 1.4$ is the constant.

The incoming flow is given by the far-field state vector $u_\infty = (r_\infty, r_\infty v_\infty, e_\infty)$ which is defined by the dimensionless density $r_\infty = 1$, the velocity $|v_\infty| = 1$, the angle of attack α (= angle between vector v_∞ and the axis x_1), and the inlet Mach number

$$M_\infty = |v_\infty| \left(\frac{r_\infty}{\kappa \, p_\infty} \right)^{1/2},$$

where p_∞ is the far-field pressure given by (10.2). Since problem (10.1) is hyperbolic, we have to prescribe the far-field boundary condition on Γ_{IO} with respect to the incoming/outcoming characteristics, see the references above, for example. On Γ_W, we prescribe the impermeability boundary condition $v \cdot n = 0$ where v is the velocity vector and n is the outer normal to Γ_W.

We are interested in the *drag* and *lift* coefficients c_D and c_L, respectively, representing the components of the force exerted by the fluid on the profile. These aerodynamic coefficients are given by the boundary integrals

$$c_D = \frac{1}{\tfrac{1}{2}r_\infty|v_\infty|^2} \int_{\Gamma_W} pn \cdot t \, dS, \qquad c_L = \frac{1}{\tfrac{1}{2}r_\infty|v_\infty|^2} \int_{\Gamma_W} pn \cdot t' \, dS, \qquad (10.3)$$

where $t = (\cos\alpha, \sin\alpha)^\mathsf{T}$ and $t' = (-\sin\alpha, \cos\alpha)^\mathsf{T}$ are vectors parallel and perpendicular to v_∞, respectively.

In the following we consider two flow regimes and apply the goal-oriented hp-mesh adaptation method from Chap. 8, namely Algorithm 8.2 with the target functionals given by the drag and lift coefficients, i.e., we put

$$J_D(u) := c_D, \qquad J_L(u) := c_L. \tag{10.4}$$

The formulation of the adjoint problem is outlined in Sect. 7.5. For the complete analysis we refer to [51].

10.1.1   Subsonic Flow

We consider the flow around NACA0012 profile with far-field Mach number $M_\infty = 0.5$ and the angle of attack $\alpha = 1.25°$. Since the resulting flow is subsonic (i.e., Mach number is strictly less than one in the whole domain Ω) and the flow is inviscid, the exact value of $J_D(u) = 0$ since there is no drag of the profile. On the other hand, the exact value of the lift coefficient has to be computed experimentally. We use the reference value of $c_L^{ref} = 1.757 \cdot 10^{-1} \pm 10^{-4}$ achieved by the hp-adaptive algorithm.

Figure 10.2 shows the decay of the error and its estimates with respect to DoF for of the drag and lift coefficients. For the case with J_D (the left figure), we observe that whereas both estimates converge exponentially ($\sim c \exp(-b \, \mathrm{DoF}^{1/3})$), i.e., the graphs are straight lines in the log-linear scale, the error $|J_D(u) - J_D(u_h)| = |J_D(u_h)|$ stagnates at the level slightly below 10^{-4}. It means that the drag coefficient

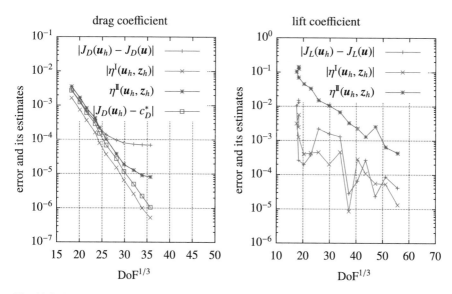

Fig. 10.2 Subsonic inviscid flow around the NACA 0012 profile ($M_\infty = 0.5$, $\alpha = 1.25°$), decay of estimates $|\eta^I|$, η^{II} and error $|J(u) - J(u_h)|$ with respect to the cube root of DoF for J equal to the drag (left) and lift (right) coefficients

does not converge to the exact value $J_D(u) = 0$ but to a positive value c_D^*. This effect was investigated in [110] using several codes with a strong global refinement. Each of the tested codes gave a small positive limit value c_D^* (obtained by the Richardson extrapolation). Therefore, in Fig. 10.2, left, we present also the quantity $|J_D(u_h) - c_D^*|$ with $c_D^* = 6.8 \cdot 10^{-5}$ which converges as expected.

Concerning the case with the lift coefficient c_L (Fig. 10.2, right), we see that the convergence is slower and it is not monotone, $|\eta^I|$ underestimates the error and, quite the other way, η^{II} overestimates it almost ten times. This may be caused by the weaker regularity of the adjoint solution for the lift coefficient where the reconstruction technique from Sect. 9.1 is not sufficiently accurate for low-regular solution. Nevertheless, although the convergence curves are not monotone, they support the exponential convergence of the hp-method.

Furthermore, Fig. 10.3, shows the distribution of the Mach number, the first component of the adjoint solutions z_h and the corresponding hp-meshes for the drag and lift coefficients on the final hp-mesh. These results indicate that the adjoint problem for the drag coefficient is quite smooth and the p_K are high for K in the surrounding of the profile Γ_W. On the other hand, the adjoint problem corresponding to the lift coefficient has less regularity due to "boundary layers" along the profile and consequently, strong h-refinement with low polynomial degree p_K appears for K close to Γ_W. This comparison confirms our conjecture concerning the regularity of the adjoint solutions for the lift coefficient.

10.1.2   Transonic Flow

We consider a faster flow with $M_\infty = 0.8$ and $\alpha = 1.25°$. This flow regime is transonic since the Mach number is greater than one in some subdomains of Ω and there appear solution discontinuities, called shock waves. In order to obtain a stable solution, we apply the shock-capturing technique based on the artificial viscosity whose amount is given by a jump indicator, see [45, Section 8.5] or [59]. Again, we consider both drag and lift coefficients as the quantity of interest. The reference values $J_D(u) = c_D^{ref} = 2.135 \times 10^{-2}$ and $J_L(u) = c_L^{ref} = 3.33 \times 10^{-1}$ were computed by the hp-anisotropic adaptation method.

Figure 10.4 shows the decay of the error and its estimates with respect to DoF for of the drag and lift coefficients. For both target functionals, $|\eta^I|$ underestimates and η^{II} overestimates the true error by a factor at most 10. We suppose that such overestimation is caused by the high-order reconstruction from Sect. 9.1, which is not sufficiently accurate for problems having a discontinuous solution.

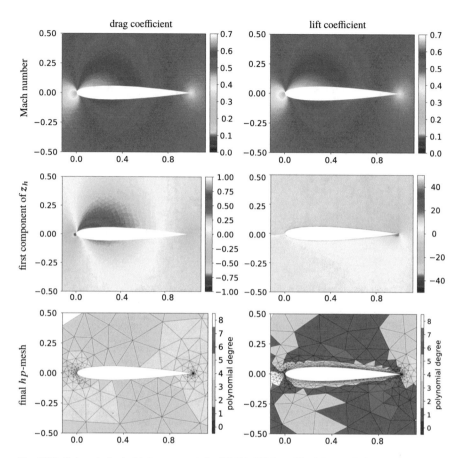

Fig. 10.3 Subsonic inviscid flow around the NACA 0012 profile ($M_\infty = 0.5$, $\alpha = 1.25°$), the Mach number distribution (first row), the distribution of the first component of the discrete adjoint solution (second row) on the final hp-mesh (third row) for target functionals J_D (left) and J_L (right)

Moreover, Fig. 10.5 shows the distribution of the Mach number, the first component of the adjoint solution and the corresponding final hp-grids. A strong h-refinement with low polynomial approximation degrees along both shock waves is observed. A sharp capturing of both waves is easy to see.

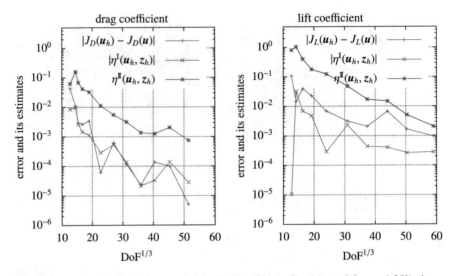

Fig. 10.4 Transonic inviscid flow around the NACA 0012 profile ($M_\infty = 0.8$, $\alpha = 1.25°$), decay of estimates $|\eta^I|$, η^{II} and error $|J(u) - J(u_h)|$ with respect to the cube root of DoF for J equal to the drag (left) and lift (right) coefficients

10.2   Viscous Shock-Vortex Interaction

This example represents an interaction between a plane weak shock wave with a single isentropic vortex, cf. [34, 65, 109]. The problem is described by the time-dependent compressible Navier–Stokes equations, which can be written in the form

$$\frac{\partial u}{\partial t} + \sum_{s=1}^{2} \frac{\partial f_s(u)}{\partial x_s} = \sum_{s=1}^{2} \frac{\partial R_s(u, \nabla u)}{\partial x_s} \qquad \text{in } \Omega \times (0, T), \tag{10.5}$$

where $u = (r, rv_1, rv_2, e)^T : \Omega \times (0, T) \to \mathbb{R}^4$ is the sought state vector and $f_s : \mathbb{R}^4 \to \mathbb{R}^4$, $s = 1, 2$ are the physical inviscid fluxes introduced in Sect. 10.1. Moreover, $R_s : \mathbb{R}^4 \to \mathbb{R}^4$, $s = 1, 2$ are the viscous fluxes. Problem (10.5) is accompanied by the equation of state (10.2) and the set of initial and boundary conditions, for details we refer again to [57, 58, 115].

The computational domain is a square $\Omega = (0, 2) \times (0, 2)$ with the periodic extension in the x_2 direction. A stationary plane shock wave is located at $x_1 = 1$. The prescribed pressure jump through the shock is $p_R - p_L = 0.4$, where p_L and p_R are the pressure values from the left and right of the shock wave, respectively, corresponding to the inlet (left) Mach number $M_L = 1.1588$. The reference density

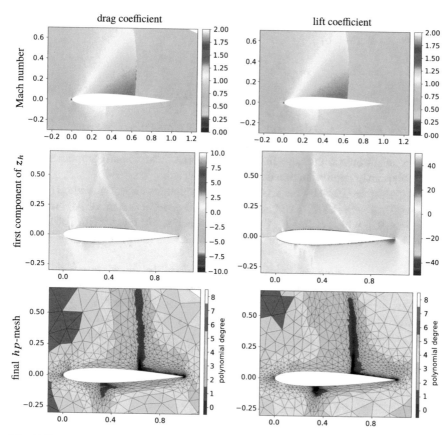

Fig. 10.5 Transonic inviscid flow around the NACA 0012 profile ($M_\infty = 0.8$, $\alpha = 1.25°$), the Mach number distribution (first row), the distribution of the first component of the discrete adjoint solution (second row) on the final hp-mesh (third row) for target functionals J_D (left) and J_L (right)

and velocity are those of the free uniform flow at infinity. In particular, we define the initial density, x_1-component of velocity and pressure by

$$\rho_L = 1, \ u_L = M_L \kappa^{1/2}, \ p_L = 1, \quad \rho_R = \rho_L K_1, \ u_R = u_L K_1^{-1}, \ p_R = p_1 K_2, \tag{10.6}$$

where

$$K_1 = \frac{\kappa+1}{2}\frac{M_L^2}{1+\frac{\kappa-1}{2}M_L^2}, \quad K_2 = \frac{2}{\kappa+1}\left(\kappa M_L^2 - \frac{\kappa-1}{2}\right). \tag{10.7}$$

Here, the subscripts $_L$ and $_R$ denote the quantities at $x < 1$ and $x > 1$, respectively, $\kappa = 1.4$ is the Poisson constant. The Reynolds number is 2000. An isolated

1Ɉ0▪

isentropic vortex centered at (0.5, 1) is added to the basic flow. The angular velocity in the vortex is given by

$$v_\theta = c_1 r \exp(-c_2 r^2), \quad c_1 = u_c/r_c, \quad c_2 = r_c^{-2}/2, \quad (10.8)$$

$$r = ((x_1 - 0.5)^2 - (x_2 - 1)^2)^{1/2},$$

where we set $r_c = 0.075$ and $u_c = 0.5$. Computations are stopped at the dimensionless time $T = 0.7$. Whereas the capturing of the shock wave requires a strong anisotropic refinement, the isentropic vortex is sufficiently smooth and then high-polynomial approximation degrees are expected.

We discretize (10.5) by the space-time discontinuous Galerkin method (cf. [45, Chapter 9], also Sect. 9.2.1) and apply the *time-dependent* variant of Algorithm 9.1 from Sect. 9.2, which optimizes the interpolation error in the $L^\infty((0, T); L^2(\Omega))$-norm. Figure 10.6 shows the pressure distribution and the hp-meshes for selected time instants.

We observe a strong anisotropic refinement along the shock wave including its curving when it interacts with the vortex. We observe that the generated polynomial degree is five which is caused by the smoothness of the wave due to the physical viscosity. Moreover, the vortex is captured by polynomial degree four or five using larger elements since it is smooth. Finally, when the solution is close to a constant state vector (e.g., in front of the shock wave), the lowest polynomial degree is generated.

10.3 Transient Flow Through a Nonhomogeneous Landfill Dam

Another application comes from the simulation of a flow though porous media. Considering a flow of air-water and assuming a constant pressure of the air, this problem can be described by the *Richards equation* [22, 102]

$$\frac{\partial \vartheta(\Psi - \gamma)}{\partial t} - \nabla \cdot (\mathbf{K}(\Psi - \gamma)\nabla\Psi) = 0 \quad \text{in } \Omega \times (0, T), \quad (10.9)$$

where $\Omega \subset \mathbb{R}^2$ is the domain occupied by the medium, $T > 0$ is the final physical time of interest, $\Psi : \Omega \times (0, T) \to \mathbb{R}$ is the sought function called *hydrostatic pressure*, γ is the vertical coordinate and the difference $\Psi - \gamma =: \psi$ is the so-called *pressure head*. Whereas ϑ and $\mathbf{K}$ are the functions of the pressure head ψ, the flow is driven by the gradient of the hydrostatic pressure $\nabla\Psi = \nabla\psi + (0, 1)^\mathsf{T}$ since it includes the gravity force in the vertical direction.

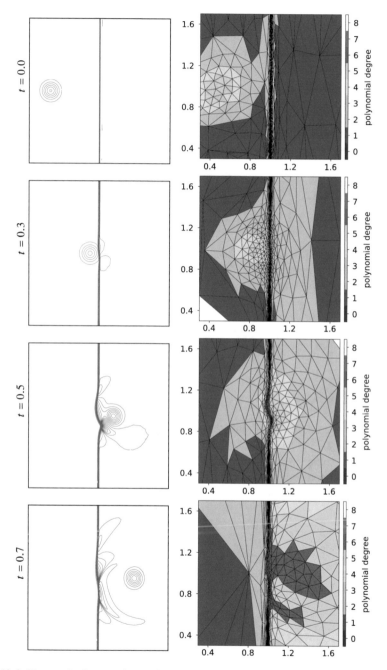

Fig. 10.6 Viscous shock-vortex interaction, pressure distribution (left) and the hp-mesh (right) at selected time levels

Moreover, ϑ and $\mathbf{K}$ are called the *active pore volume* and the *conductivity tensor*, respectively, they are nonlinear functions of the pressure head. These functions are given by constitutive relations, particularly,

$$\vartheta(\psi) := \theta(\psi) + \frac{S_s}{\theta_s} \int_{-\infty}^{\psi} \theta(s) \, ds, \tag{10.10}$$

where $S_s \geq 0$ is the specific aquifer storage, $\theta_s > 0$ is the saturated water content, and θ is the *water content* given by the van Genuchten's function [66]

$$\theta(\psi) = \begin{cases} \dfrac{\theta_s - \theta_r}{(1 + (-\alpha\psi)^n)^m} + \theta_r, & \text{for } \psi < 0, \\ \theta_s & \text{for } \psi \geq 0, \end{cases} \tag{10.11}$$

where θ_r is the residual water content, m and n are pore size distribution parameters, and α is usually expressed as the inverse of air entry value. The unsaturated hydraulic conductivity is given by the Mualem function [96] which in combination with (10.11) reads

$$K_r(\psi) = \begin{cases} \dfrac{\left(1 - (-\alpha\psi)^{mn}(1 + (-\alpha\psi)^n)^{-m}\right)^2}{(1 + (-\alpha\psi)^n)^{m/2}} & \text{for } \psi < 0, \\ 1.0 & \text{for } \psi \geq 0, \end{cases} \tag{10.12}$$

$$\mathbf{K}(\psi) = \mathbf{K}_S K_r(\psi)$$

where $\mathbf{K}_S$ is the saturated hydraulic conductivity, and parameters m, n, and α were already explained in (10.11).

This problem has to be accompanied by an initial condition for Ψ and boundary conditions. We consider the Dirichlet and Neumann boundary conditions together with the so-called outflow boundary condition

$$\psi := \Psi - \gamma \leq 0, \quad -\mathbf{K}(\Psi - \gamma)\nabla\Psi \cdot \boldsymbol{n} \geq 0, \quad \psi(\nabla\Psi \cdot \boldsymbol{n}) = 0, \tag{10.13}$$

which means that the pressure head ψ cannot be positive, the fluid cannot enter into the medium since outside there is no fluid and the fluid can exit the medium only if the pressure head $\psi = 0$. The Richards equation (10.9) belongs among the degenerate parabolic problems since ϑ and $\mathbf{K}$ can either vanish or blow up.

We solve this equation in [47] by the space-time discontinuous Galerkin method. We refer to this work for the discretization and some implementation aspects concerning, e.g., the realization of the outflow boundary condition (10.13), the adaptive choice of the time steps, and adaptive stopping criteria for algebraic solvers. We employ again the time-dependent anisotropic hp-mesh adaptation Algorithm 9.1.

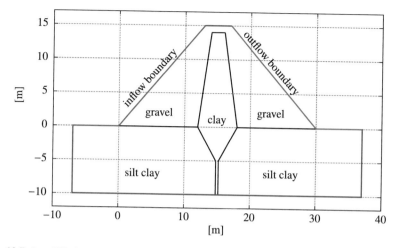

Fig. 10.7 Landfill dam, the geometry of the problem and the type of prescribed boundary conditions

Table 10.1 Parameters for landfill dam problem

	Gravel	Clay	Silt clay
α [m^{-1}]	250.0	0.8	2.0
n [–]	1.41	1.2	1.41
m [–]	0.291	0.167	0.291
K_s [m·s^{-1}]	4.63×10^{-7}	5.55×10^{-9}	1.24×10^{-8}
θ_s [–]	0.6	0.38	0.45
θ_r [–]	0.0	0.06	0.067
S_s [m^{-1}]	0.01	0.01	0.01

We present a simulation through a nonhomogeneous landfill dam which consists three materials (coarse sand, clay, and silt clay). Figure 10.7 shows the corresponding geometry, the units are meters. This figure also shows the considered type of boundary conditions: we prescribe the Dirichlet boundary condition $\psi = 15$ m on the inflow boundary, the outflow boundary condition (10.13) on the outflow part of the boundary, and the homogeneous Neumann condition on the rest of boundary. The material parameters appearing in the van Genuchten's function (10.11) and the Mualem function (10.12) are given in Table 10.1. At time $t = 0$ we prescribe the initial condition $\Psi = -0.2 + y$ and perform the computation till the final physical time $T = 5$ days. The inconsistency of the initial and boundary conditions leads to the propagation of a wave, which is fast for small time t and then its speed is decreasing. We apply the anisotropic hp-mesh adaptation algorithm and expect a refinement in the front of the wave and also on the interfaces of different materials.

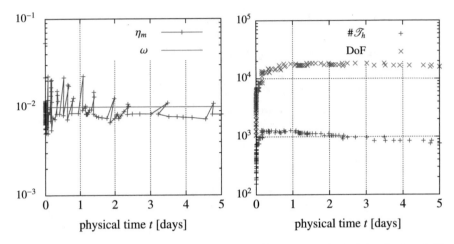

Fig. 10.8 The dependence of the interpolation error with respect to physical time in days (left) and the number of elements and DoF during the mesh adaptation process (right)

Figure 10.8, left, shows the tolerance $\omega = 10^{-2}$ and the interpolation error

$$\eta_m := \max(\eta_m^L, \eta_m^R), \quad m = 0, 1 \ldots, r, \tag{10.14}$$

where η_m^L and η_m^R are given by (9.25). We observe that $\eta_m > \omega$ for several time level m and then the hp-mesh is adapted and the size of the time step is typically decreased. Furthermore, Fig. 10.8, right, shows the number of the mesh elements $\#\mathcal{T}_h$ and the number of degrees of freedom DoF for each time level. We observe that after an initial increase of both quantities (since we start with a coarse mesh), DoF is almost constant whereas $\#\mathcal{T}_h$ is slowly decreasing. It is caused by the fact that the solution is smoother for larger time t and hence higher polynomial degrees are employed.

Finally, Fig. 10.9 shows the distributions of the pressure head ψ and the corresponding anisotropic hp-meshes at selected time levels. A strong hp-refinement with low polynomial approximation degrees in the vicinity of the interior interfaces and the propagating waves is obvious.

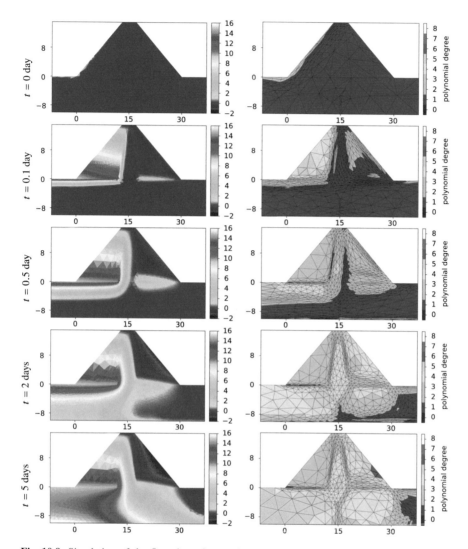

Fig. 10.9 Simulation of the flow through a nonhomogeneous landfill dam, distributions of the pressure head ψ (left) and the corresponding hp-meshes (right) at selected time levels

Conclusion

In this book, we presented the concept of the metric-based anisotropic hp-mesh adaptation method for the numerical solution of partial differential equations. Anisotropic hp-adaptation offers a great deal of flexibility in the adaptive construction of meshes and corresponding finite element spaces. This approach is based on interpolation error estimates including the geometry of mesh elements and their shape and size optimization with respect to the interpolation error. Whereas the optimization of the shape of elements is done locally, the size optimization is global employing the continuous mesh and error model with the aid of variational calculus.

Anisotropic hp-adaptation was also incorporated in the framework of the goal-oriented error estimates and mesh adaptation for linear as well as nonlinear problems. The presented numerical experiments demonstrate the potential of this technique, namely its superiority to the isotropic (hp-)adaptation and h-variant of the anisotropic adaptation.

Although the technical aspects were shown, and the numerical experiments were carried out, by discontinuous Galerkin method, the proposed anisotropic hp-mesh adaptation can be employed (with some modifications) also with different discretization schemes, like conforming finite element methods, residual distribution schemes, etc.

On the other hand, we should mention the weak points and open questions of this approach. The mesh optimization is defined in order to minimize the number of degrees of freedom (DoF). However, although the arising (nonlinear as well as linear) algebraic systems are smaller, their numerical solution is more complicated since the strong refinement and hp-adaptation typically increase the conditioning of the systems and cause slower convergence by iterative solvers. It is an open problem how much the reduction of DoF also reduces the computational costs.

Furthermore, we are aware that not all considerations are based on rigorous analysis, but some heuristics are developed. However, we believe that the presented methods and strategies can be interesting for both further research and possible practical applications.

© The Author(s), under exclusive license to Springer Nature Switzerland AG 2022   243
V. Dolejší, G. May, *Anisotropic* hp-*Mesh Adaptation Methods*, Nečas Center Series,
https://doi.org/10.1007/978-3-031-04279-9

References

1. Adams, R.A., Fournier, J.J.F.: Sobolev Spaces. Academic Press, New York (2003)
2. Aguilar, J.C., Goodman, J.B.: Anisotropic mesh refinement for finite element methods based on error reduction. J. Comput. Appl. Math. **193**(2), 497–515 (2006)
3. Ainsworth, M., Rankin, R.: Guaranteed computable bounds on quantities of interest in finite element computations. Int. J. Numer. Methods Eng. **89**(13), 1605–1634 (2012)
4. Ait-Ali-Yahia, D., Baruzzi, G., Habashi, W.G., Fortin, M., Dompierre, J., Vallet, M.: Anisotropic mesh adaptation: towards user-independent, mesh-independent and solver-independent CFD. II. Structured grids. Int. J. Numer. Methods Fluids **39**, 657–673 (2002)
5. Apel, T.: Anisotropic Finite Elements: Local Estimates and Applications. Teubner, Stuttgart, Leipzig (1999)
6. Arioli, M., Liesen, J., Miedlar, A., Strakoš, Z.: Interplay between discretization and algebraic computation in adaptive numerical solution of elliptic PDE problems. GAMM-Mitt. **36**(1), 102–129 (2013)
7. Arnold, D.N., Brezzi, F., Cockburn, B., Marini, L.D.: Unified analysis of discontinuous Galerkin methods for elliptic problems. SIAM J. Numer. Anal. **39**(5), 1749–1779 (2002)
8. Aubry, R., Löhner, R.: Generation of viscous grids at ridges and corners. Int. J. Numer. Methods Eng. **77**(9) (2009)
9. Babuska, I., Rheinboldt, W.C.: Error estimates for adaptive finite element computations. SIAM J. Numer. Anal. **15**(4), 736–754 (1978)
10. Babuška, I., Suri, M.: The p- and hp- versions of the finite element method. An overview. Comput. Methods Appl. Mech. Eng. **80**, 5–26 (1990)
11. Babuška, I., Suri, M.: The p- and hp-FEM a survey. SIAM Rev. **36**, 578–632 (1994)
12. Babuška, I., Aziz, A.K.: On the angle condition in the finite element method. SIAM J. Numer. Anal. **13**(2), 214–226 (1976)
13. Balan, A., Woopen, M., May, G.: Adjoint-based hp-adaptivity on anisotropic meshes for high-order compressible flow simulations. Comput. Fluids **139**, 47–67 (2016)
14. Bangerth, W., Rannacher, R.: Adaptive finite element methods for differential equations. Lectures in Mathematics. ETH. Birkhäuser Verlag, Zürich (2003)
15. Bank, R.E., Weiser, A.: Some a posteriori error estimators for elliptic partial differential equations. Math. Comput. **44**(170), 283–301 (1985). http://www.jstor.org/stable/2007953
16. Bartoš, O., Dolejší, V., May, G., Rangarajan, A., Roskovec, F.: Goal-oriented anisotropic hp-mesh optimization technique for linear convection-diffusion-reaction problem. Comput. Math. Appl. **78**(9), 2973–2993 (2019)
17. Becker, R., Rannacher, R.: An optimal control approach to a-posteriori error estimation in finite element methods. Acta Numer. **10**, 1–102 (2001)

V. Dolejší, G. May, *Anisotropic* hp-*Mesh Adaptation Methods*, Nečas Center Series,
https://doi.org/10.1007/978-3-031-04279-9

18. Belme, A., Dervieux, A., Alauzet, F.: Time accurate anisotropic goal-oriented mesh adaptation for unsteady flows. J. Comp. Phys. **231**(19), 6323–6348 (2012)
19. Braack, M., Ern, A.: Coupling multimodelling with local mesh refinement for the numerical computation of laminar flames. Combust. Theory Model. **8**(4), 771–788 (2004)
20. Brandts, J., Korotov, S., Křížek, M.: Simplicial Partitions with Applications to the Finite Element Method. Springer International Publishing, Cham (2020)
21. Brezzi, F., Fortin, M.: Mixed and Hybrid Finite Element Methods. Springer Series in Computational Mathematics, vol. 15. Springer, New York (1991)
22. Buckingham, E.: Studies on the movement of soil moisture. USDA Bureau Soils – Bull. **38** (1907)
23. Cao, W.: On the error of linear interpolation and the orientation, aspect ratio, and internal angles of a triangle. SIAM J. Numer. Anal. **43**(1), 19–40 (2005)
24. Cao, W.: Anisotropic measures of third order derivatives and the quadratic interpolation error on triangular elements. SIAM J. Sci. Comput. **29**(2), 756–781 (2007)
25. Cao, W.: Sn interpolation error estimate on anisotropic meshes in $\mathcal{R}^n$ and optimal metrics for mesh refinement. SIAM J. Numer. Anal. **45**(6), 2368–2391 (2007)
26. Cao, W.: An interpolation error estimate in R^2 based on the anisotropic measures of higher order derivatives. Math. Comp. **77**(261), 265–286 (2008)
27. Carpio, J., Prieto, J., Bermejo, R.: Anisotropic "goal-oriented" mesh adaptivity for elliptic problems. SIAM J. Sci. Comput. **35**(2), A861–A885 (2013)
28. Carson, H., Darmofal, D., Galbraith, M., Allmaras, S.: Analysis of output-based error estimation for finite element methods. Appl. Numer. Math. **118**, 182–202 (2017)
29. Ceze, M., Fidkowski, K.J.: Anisotropic hp-adaptation framework for functional prediction. AIAA J. **51**(2), 492–509 (2012)
30. Chen, L., Sun, P., Xu, J.: Optimal anisotropic meshes for minimizing interpolation errors in L^p-norm. Math. Comp. **76**, 179–204 (2007)
31. Ciarlet, P.G.: The Finite Elements Method for Elliptic Problems. North-Holland, Amsterdam, New York, Oxford (1979)
32. Clavero, C., Gracia, J.L., Jorge, J.C.: A uniformly convergent alternating direction (HODIE) finite difference scheme for 2D time-dependent convection-diffusion problems. IMA J. Numer. Anal. **26**, 155–172 (2006)
33. Coulaud, O., Loseille, A.: Very high order anisotropic metric-based mesh adaptation in 3D. Proc. Eng. **163**, 353–365 (2016)
34. Daru, V., Tenaud, C.: High order one-step monotonicity-preserving schemes for unsteady compressible flow calculations. J. Comput. Phys. **193**(2), 563–594 (2004)
35. Demkowicz, L., Rachowicz, W., Devloo, P.: A fully automatic hp-adaptivity. J. Sci. Comput. **17**(1–4), 117–142 (2002)
36. Dervieux, A., Loseille, A., Alauzet, F.: High-order adaptive method applied to high-speed flows. In: West-East High Speed Flow Field Conference (2007)
37. Di Pietro, D., Ern, A.: Mathematical Aspects of Discontinuous Galerkin Methods. Mathematiques et Applications, vol. 69 (Springer, Berlin, Heidelberg, 2012)
38. Dobrzynski, C., Frey, P.: Anisotropic Delaunay mesh adaptation for unsteady simulations. In: Garimella, R.V. (ed.) Proceedings of the 17th International Meshing Roundtable. Springer, Berlin, Heidelberg (2008)
39. Dolejší, V.: Anisotropic mesh adaptation for finite volume and finite element methods on triangular meshes. Comput. Vis. Sci. **1**(3), 165–178 (1998)
40. Dolejší, V.: ANGENER – Anisotropic Mesh Generator, Examples. Charles University, Prague, Faculty of Mathematics and Physics (2000). https://msekce.karlin.mff.cuni.cz/~dolejsi/AMA/ama.htm
41. Dolejší, V.: ANGENER – Anisotropic mesh generator, in-house code. Charles University, Prague, Faculty of Mathematics and Physics (2000). https://msekce.karlin.mff.cuni.cz/~dolejsi/angen/
42. Dolejší, V.: Anisotropic mesh adaptation technique for viscous flow simulation. East-West J. Numer. Math. **9**(1), 1–24 (2001)

43. Dolejší, V., Bartoš, O., Roskovec, F.: Goal-oriented mesh adaptation method for nonlinear problems including algebraic errors. Comput. Math. Appl. **93**, 178–198 (2021)
44. Dolejší, V., Ern, A., Vohralík, M.: hp-adaptation driven by polynomial-degree-robust a posteriori error estimates for elliptic problems. SIAM J. Sci. Comput. **38**(5), A3220–A3246 (2016)
45. Dolejší, V., Feistauer, M.: Discontinuous Galerkin Method – Analysis and Applications to Compressible Flow. Springer Series in Computational Mathematics, vol. 48. Springer, Cham (2015)
46. Dolejší, V., Felcman, J.: Anisotropic mesh adaptation and its application for scalar diffusion equations. Numer. Methods Part. Differ. Equ. **20**, 576–608 (2004)
47. Dolejší, V., Kuráž, M., Solin, P.: Adaptive higher-order space-time discontinuous Galerkin method for the computer simulation of variably-saturated porous media flows. Appl. Math. Modell. **72**, 276–305 (2019)
48. Dolejší, V., May, G., Rangarajan, A., Roskovec, F.: A goal-oriented high-order anisotropic mesh adaptation using discontinuous Galerkin method for linear convection-diffusion-reaction problems. SIAM J. Sci. Comput. **41**(3), A1899–A1922 (2019)
49. Dolejší, V., Roos, H.G.: BDF-FEM for parabolic singularly perturbed problems with exponential layers on layer-adapted meshes in space. Neural Parallel Sci. Comput. **18**(2), 221–235 (2010)
50. Dolejší, V., Roskovec, F.: Goal-oriented error estimates including algebraic errors in discontinuous Galerkin discretizations of linear boundary value problems. Appl. Math. **62**(6), 579–605 (2017)
51. Dolejší, V., Roskovec, F.: Goal-oriented anisotropic hp-adaptive discontinuous Galerkin method for the Euler equations. Commun. Appl. Math. Comput. **4**, 143–179 (2021)
52. Dolejší, V., Roskovec, F., Vlasák, M.: Residual based error estimates for the space-time discontinuous Galerkin method applied to the compressible flows. Comput. Fluids **117**, 304–324 (2015)
53. Dolejší, V., Solin, P.: hp-discontinuous Galerkin method based on local higher order reconstruction. Appl. Math. Comput. **279**, 219–235 (2016)
54. Dolejší, V., Tichý, P.: On efficient numerical solution of linear algebraic systems arising in goal-oriented error estimates. J. Sci. Comput. **83**(5) (2020)
55. Dompierre, J., Vallet, M.G., Bourgault, Y., Fortin, M., Habashi, W.G.: Anisotropic mesh adaptation: towards user-independent, mesh-independent and solver-independent CFD. Part III. unstructured meshes. Int. J. Numer. Methods Fluids **39**(8), 675–702 (2002)
56. Ern, A., Guermond, J.L.: Theory and Practice of Finite Elements. Springer, New York (2004)
57. Feistauer, M.: Mathematical Methods in Fluid Dynamics. Longman Scientific & Technical, Harlow (1993)
58. Feistauer, M., Felcman, J., Straškraba, I.: Mathematical and Computational Methods for Compressible Flow. Clarendon Press, Oxford (2003)
59. Feistauer, M., Kučera, V.: On a robust discontinuous Galerkin technique for the solution of compressible flow. J. Comput. Phys. **224**, 208–221 (2007)
60. Fidkowski, K.J., Luo, Y.: Output-based space-time mesh adaptation for the compressible Navier-Stokes equations. J. Comput. Phys. **230**(14), 5753–5773 (2011)
61. Formaggia, L., Micheletti, S., Perotto, S.: Anisotropic mesh adaption in computational fluid dynamics: application to the advection-diffusion-reaction and the Stokes problems. Appl. Numer. Math. **51**(4), 511–533 (2004)
62. Formaggia, L., Perotto, S.: New anisotropic a priori error estimates. Numer. Math. **89**(4), 641–667 (2001)
63. Fortin, M., Vallet, M.G., Dompierre, J., Bourgault, Y., Habashi, W.G.: Anisotropic mesh adaptation: Theory, validation and applications. In: Désidéri, J.A., Hirsch, C., Le Tallec, P., Pandolfi, M., Périaux, J. (eds.) Computational Fluid Dynamics '96, pp. 174–180. Wiley, Chichester, Paris (1996)
64. Frey, P.J., Alauzet, F.: Anisotropic mesh adaptation for CFD computations. Comput. Methods Appl. Mech. Eng. **194**, 5068–5082 (2005)

65. Fürst, J.: Modélisation numérique d'écoulements transsoniques avec des schémas TVD et ENO. Ph.D. thesis, Université Mediterranée, Marseille and Czech Technical University Prague (2001)

66. van Genuchten, M.T.: Closed-form equation for predicting the hydraulic conductivity of unsaturated soils. Soil Sci. Soc. Am. J. **44**(5), 892–898 (1980)

67. George, P., Hecht, F., Vallet, M.: Creation of internal points in Voronoi's type method. control adaptation. Adv. Eng. Softw. Workstat. **13**(5–6), 303–312 (1991)

68. Georgoulis, E.H., Hall, E., Houston, P.: Discontinuous Galerkin methods on hp-anisotropic meshes I: a priori error analysis. Int. J. Comput. Sci. Math. **1**(2-3), 221–244 (2007)

69. Giles, M., Süli, E.: Adjoint methods for PDEs: a posteriori error analysis and postprocessing by duality. Acta Numer. **11**, 145–236 (2002)

70. Guégan, D., Allain, O., Dervieux, A., Alauzet, F.: An L^∞-L^p mesh-adaptive method for computing unsteady bi-fluid flows. Int. J. Numer. Methods Eng. **84**(11), 1376–1406 (2010)

71. Gui, W., Babuška, I.: The h, p and h-p versions of the finite element method in 1 dimension. III. The adaptive h-p version. Numer. Math. **49**(6), 659–683 (1986)

72. Guillard, H., Farhat, C.: On the significance of the geometric conservation law for flow computations on moving meshes. Comput. Methods Appl. Mech. Eng. **190**(11), 1467–1482 (2000)

73. Habashi, W.G., Dompierre, J., Bourgault, Y., Ait-Ali-Yahia, D., Fortin, M., Vallet, M.G.: Anisotropic mesh adaptation: towards user-independent, mesh-independent and solver-independent CFD. Part I: general principles. Int. J. Numer. Methods Fluids **32**(6), 725–744 (2000)

74. Hadamard, J.: Le problème de Cauchy et les équations aux derivées partielles linéaires hyperboliques. Herman, Paris (1932)

75. Harriman, K., Gavaghan, D.J., Süli, E.: The importance of adjoint consistency in the approximation of linear functionals using the discontinuous Galerkin finite element method. Tech. rep., Oxford University Computing Laboratory (2004)

76. Harriman, K., Houston, P., Schwab, C., Süli, E.: hp-version discontinuous Galerkin methods with interior penalty for partial differential equations with nonnegative characteristic form. Rec. Adv. Sci. Comput. Part. Differ. Equ. **330**, 89–119 (2003)

77. Hartmann, R.: Adjoint consistency analysis of discontinuous Galerkin discretizations. SIAM J. Numer. Anal. **45**(6), 2671–2696 (2007)

78. Hartmann, R., Houston, P.: Adaptive discontinuous Galerkin finite element methods for nonlinear hyperbolic conservation laws. SIAM J. Sci. Comp. **24**, 979–1004 (2002)

79. Hecht, F.: BAMG: bidimensional anisotropic mesh generator.   Tech. rep., INRIA-Rocquencourt, France (1998)

80. Houston, P., Schwab, C., Süli, E.: Discontinuous hp-finite element methods for advection-diffusion problems. Tech. Rep. 2000-07, SAM, ETH Zürich (2000). https://www.sam.math.ethz.ch/sam_reports/reports_final/reports2000/2000-07_fp.pdf

81. Houston, P., Schwab, C., Süli, E.: Discontinuous hp-finite element methods for advection-diffusion-reaction problems. SIAM J. Numer. Anal. **39**(6), 2133–2163 (2002)

82. Jech, T.J.: The Axiom of Choice. Dover Books on Mathematics. Courier Corporation, North Chelmsford (2008)

83. Jourlin, M., Fillere, I., Becker, J.M., Labourne, M.J.: Shape and metrices. In: Chetverikov, D., Kropatsch, W. (eds.) Computer Analysis of Images and Patterns: 5th International Conference, CAIP '93 Budapest, Hungary, September 13–15, 1993. Lecture Notes in Computer Science, pp. 254–258. Springer, New York (1993)

84. Kuzmin, D., Korotov, S.: Goal-oriented a posteriori error estimates for transport problems. Math. Comput. Simul. **80**(8, SI), 1674–1683 (2010)

85. Laug, P., Borouchaki, H.: The BL2D Mesh Generator: Beginner's Guide, User's and Programmer's Manual. Research Report RT-0194, INRIA (1996). https://hal.inria.fr/inria-00069977

86. Laug, P., Borouchaki, H.: BL2D-V2: isotropic or anisotropic 2D mesher. INRIA (2002). https://www.rocq.inria.fr/gamma/Patrick.Laug/logiciels/bl2d-v2/INDEX.html

87. Loseille, A., Alauzet, F.: Continuous mesh framework part I: well-posed continuous interpolation error. SIAM J. Numer. Anal. **49**(1), 38–60 (2011)

88. Loseille, A., Alauzet, F.: Continuous mesh framework part II: validations and applications. SIAM J. Numer. Anal. **49**(1), 61–86 (2011)

89. Loseille, A., Dervieux, A., Alauzet, F.: Fully anisotropic goal-oriented mesh adaptation for 3D steady Euler equations. J. Comput. Phys. **229**(8), 2866–2897 (2010)

90. Loseille, A., Löhner, R.: Boundary layer mesh generation and adaptivity. In: 49th AIAA Aerospace Sciences Meeting including the New Horizons Forum and Aerospace Exposition (2011)

91. Ma, R., Chang, X., Zhang, L., He, X., Li, M.: On the geometric conservation law for unsteady flow simulations on moving mesh. Proc. Eng. **126**, 639–644 (2015). Frontiers in Fluid Mechanics Research

92. Mallik, G., Vohralík, M., Yousef, S.: Goal-oriented a posteriori error estimation for conforming and nonconforming approximations with inexact solvers. J. Comput. Appl. Math. **366**, 112367 (2020)

93. Meidner, D., Rannacher, R., Vihharev, J.: Goal-oriented error control of the iterative solution of finite element equations. J. Numer. Math. **17**, 143–172 (2009)

94. Melenk, J.M.: hp-Finite Element Methods for Singular Perturbations. Lecture Notes in Mathematics, vol. 1796. Springer, Berlin (2002)

95. Mitchell, W.F.: A collection of 2D elliptic problems for testing adaptive grid refinement algorithms. Appl. Math. Comput. **220**, 350–364 (2013)

96. Mualem, Y.: A new model for predicting the hydraulic conductivity of unsaturated porous media. Water Resources Res. **12**(3), 513–522 (1976)

97. Nečas, J.: Les Méthodes Directes en Théorie des Equations Elliptiques. Academia, Prague (1967)

98. Nochetto, R., Veeser, A., Verani, M.: A safeguarded dual weighted residual method. IMA J. Numer. Anal. **29**(1), 126–140 (2009)

99. Quarteroni, A., Valli, A.: Numerical Approximation of Partial Differential Equations. Springer Series in Computational Mathematics, vol. 23. Springer, Berlin (1994)

100. Rangarajan, A., May, G., Dolejší, V.: Adjoint-based anisotropic hp-adaptation for discontinuous Galerkin methods using a continuous mesh model. J. Comput. Phys. **409**, 109321 (2020)

101. Rannacher, R., Vihharev, J.: Adaptive finite element analysis of nonlinear problems: Balancing of discretization and iteration errors. J. Numer. Math. **21**(1), 23–61 (2013)

102. Richards, L.A.: Capillary conduction of liquids through porous mediums. J. Appl. Phys. **1**(5), 318–333 (1931)

103. Richter, T., Wick, T.: Variational localizations of the dual weighted residual estimator. J. Comput. Appl. Math. **279**, 192–208 (2015)

104. Rivière, B.: Discontinuous Galerkin Methods for Solving Elliptic and Parabolic Equations: Theory and Implementation. Frontiers in Applied Mathematics. SIAM, New York (2008)

105. Roskovec, F.: Goal-oriented a posteriori error estimates and adaptivity for the numerical solution of partial differential equations. Ph.D. thesis, Charles University, Prague (2019). https://dspace.cuni.cz/bitstream/handle/20.500.11956/111302/140079106.pdf

106. Schwab, C.: p- and hp-Finite Element Methods. Clarendon Press, Oxford (1998)

107. Simpson, R.B.: Anisotropic mesh transformations and optimal error control. Appl. Numer. Math. **14**, 183–198 (1994)

108. Šolín, P., Demkowicz, L.: Goal-oriented hp-adaptivity for elliptic problems. Comput. Methods Appl. Mech. Eng. **193**, 449–468 (2004)

109. Tenaud, C., Garnier, E., Sagaut, P.: Evaluation of some high-order shock capturing schemes for direct numerical simulation of unsteady two-dimensional free flows. Int. J. Numer. Methods Fluids **126**, 202–228 (2000)

110. Vassberg, J.C., Jameson, A.: In pursuit of grid convergence for two-dimensional Euler solutions. J. Aircraft **47**(4), 1152–1166 (2010)

111. Venditti, D., Darmofal, D.: Grid adaptation for functional outputs: application to two-dimensional inviscid flows. J. Comput. Phys. **176**(1), 40–69 (2002)

112. Venditti, D., Darmofal, D.: Anisotropic grid adaptation for functional outputs: Application to two-dimensional viscous flows. J. Comput. Phys. **187**(1), 22–46 (2003)
113. Verfürth, R.: A Posteriori Error Estimation Techniques for Finite Element Methods. Numerical Mathematics and Scientific Computation. Oxford University Press, Oxford (2013)
114. Watkins, D.S.: Fundamentals of Matrix Computations. Pure and Applied Mathematics, Wiley-Interscience Series of Texts, Monographs, and Tracts. Wiley, New York (2002)
115. Wesseling, P.: Principles of Computational Fluid Dynamics. Springer, Berlin (2001)
116. Yano, M., Darmofal, D.L.: An optimization-based framework for anisotropic simplex mesh adaptation. J. Comput. Phys. **231**(22), 7626–7649 (2012)
117. Zienkiewicz, O.C., Wu, J.: Automatic directional refinement in adaptive analysis of compressible flows. Int. J. Numer. Methods Engrg. **37**(13), 2189–2210 (1994)
118. Zienkiewicz, O.C., Zhu, J.Z.: The superconvergent patch recovery and a-posteriori error estimates. Part 1: The recovery technique. Int. J. Numer. Methods Eng. **33**, 1331–1364 (1992)
119. Zienkiewicz, O.C., Zhu, J.Z.: The superconvergent patch recovery and a-posteriori error estimates. Part 2: Error estimated and adaptivity. Int. J. Numer. Methods Engrg. **33**, 1365–1382 (1992)

Index

Symbols
hp-mesh, 133, 193

A
Adjoint consistency, 157, 162, 172
Adjoint problem, 157, 161, 171
    discrete, 181
Anisotropic bound of the interpolation error
    function, 62, 82, 98
Anisotropy (of tetrahedron), 35, 78
Anisotropy (of triangle), 28, 44
Anisotropy of function, 47
Approximate adjoint problem, 162, 164
Approximate solution, 6, 169

C
Conforming grid, 9
Consistency, 6, 156, 161, 169
Continuous interpolation error, 107, 134
Continuous-mesh error estimate, 108, 136

D
Discontinuous Galerkin method (DGM), 168
Discrete adjoint problem, 171, 172
Dual weighted residual, 176

E
Error identity
    adjoint, 158, 174
    primal, 158, 174

Euclidean length, 20
Euler equations, 184, 229

G
Galerkin orthogonality, 156, 170
    adjoint, 157, 174
Geometry (of ellipse), 28
Geometry (of tetrahedron), 35, 78
Geometry (of triangle), 28, 44
Goal-oriented error estimates, 164
    type I, 175, 183, 185
    type II, 177, 183, 186
    type III, 187, 192

H
Homogeneous function, 46

I
Interpolation error, 10
Interpolation error function, 45, 62, 79, 83, 90, 140
Interpolation operator, 45, 78
Isotropic mesh, 21

L
Level set function, 47
Local density DoF function, 134
Localization of error estimates, 176, 183

© The Author(s), under exclusive license to Springer Nature Switzerland AG 2022
V. Dolejší, G. May, *Anisotropic* hp-*Mesh Adaptation Methods*, Nečas Center Series,
https://doi.org/10.1007/978-3-031-04279-9

M
Mesh adaptive algorithm, 2
Mesh density distribution, 106, 134

N
Number of degrees of freedom, 1, 168
Numerical flux, 178

O
Orientation (of ellipse), 28
Orientation (of tetrahedron), 35, 78
Orientation (of triangle), 28, 44

P
Primal problem, 156, 161

Q
Quantity of interest, 155, 157, 161, 171, 180

R
Reference tetrahedron, 33
Reference triangle, 25
Residual equivalence, 158

S
Shape (of ellipse), 28
Shape (of tetrahedron), 35, 78
Shape (of triangle), 28, 44
Singular value decomposition, 26, 34, 43, 77
Size (of ellipse), 28
Size (of tetrahedron), 35, 78
Size (of triangle), 28, 44
Space-time discontinuous Galerkin (STDG), 223, 236
Steiner ellipse, 30
Symmetric interior penalty Galerkin (SIPG) method, 168, 178

T
Tetrahedrization, 9
Triangulation, 9

Printed in the United States
by Baker & Taylor Publisher Services